Lecture Notes in Bioengineering

Lecture Notes in Bioengineering (LNBE) publishes the latest developments in bioengineering. It covers a wide range of topics, including (but not limited to):

- Bio-inspired Technology & Biomimetics
- Biosensors
- Bionanomaterials
- Biomedical Instrumentation
- Biological Signal Processing
- Medical Robotics and Assistive Technology
- Computational Medicine, Computational Pharmacology and Computational Biology
- Personalized Medicine
- Data Analysis in Bioengineering
- Neuroengineering
- Bioengineering Ethics

Original research reported in proceedings and edited books are at the core of LNBE. Monographs presenting cutting-edge findings, new perspectives on classical fields or reviewing the state-of-the art in a certain subfield of bioengineering may exceptionally be considered for publication. Alternatively, they may be redirected to more specific book series. The series' target audience includes advanced level students, researchers, and industry professionals working at the forefront of their fields.

Indexed by SCOPUS, INSPEC, zbMATH, SCImago.

Ann-Malin Schmidt

Novel Methodologies for Digital Investigations of Clothing-Body-Interactions

Advancing Development of Tightly Fitted Clothing Using 4D Scan Data and Finite Element Analysis Based on Deformable Human Body Models

Springer

Ann-Malin Schmidt
Institute of Textile Machinery and High Performance Material Technology (ITM)
Dresden University of Technology (TU Dresden)
Dresden, Germany

ISSN 2195-271X ISSN 2195-2728 (electronic)
Lecture Notes in Bioengineering
ISBN 978-3-031-91793-6 ISBN 978-3-031-91794-3 (eBook)
https://doi.org/10.1007/978-3-031-91794-3

This Springer imprint is published by the registered company Springer Nature Switzerland AG
The registered company address is: Gewerbestrasse 11, 6330 Cham, Switzerland

Ann-Malin Schmidt
Dresden, Deutschland

The present work was submitted as a dissertation on July 31, 2024, to the Faculty of Mechanical Science and Engineering at the Dresden University of Technology under the title "Novel Methodologies for Digital Investigations of Clothing-Body-Interactions—Advancing Development of Tightly Fitted Clothing using 4D Scan Data and Finite Element Analysis Based on Deformable Human Body Models" and was successfully defended in Dresden on December 16, 2024.

Reviewers:
Prof. Dr.-Ing. habil. Yordan Kyosev
Prof. Dr.-Ing. Christian Kaiser

Foreword

4D scanning (high-speed 3D body scanning) is an emerging technology that revolutionises research on the shape and movement of the human body and its interaction with clothing. Combining advanced hardware and software, it opens new possibilities for analysing changes in body geometry, textile displacements and deformations. However, due to the novelty and cost of this technology, there are currently very few software solutions available for studying these interactions, which are crucial for optimising clothing design and product development. Additionally, the vast amounts of data generated by 4D scanning hold great potential for validating finite element method (FEM) simulations of the human body and extracting material characteristics—yet research in this area remains limited to a few individual studies.

In this work, Ann-Malin Schmidt applies her expertise in FEM modelling, Python programming, mechanics, textiles and clothing to develop groundbreaking algorithms for evaluating body motion captured through 4D scanning. Her contributions include the extraction of technologically significant features, the analysis of human soft tissue behaviour and the integration of these findings into numerical simulations using FEM software. Her pioneering work in this field was recognised with the Innovation Award 2024 in the category of "Digitalization + AI" at the TexProcess International Trade Fair.

This book offers a comprehensive and meticulously detailed exploration of 4D body scanning and its applications. By bridging multiple disciplines, it lays a strong foundation for future research into clothing-body interaction, high-speed body scanning and numerical simulation—paving the way for further advancements in this exciting field.

Prof. Dr.-Ing. habil. Yordan Kyosev
Institute of Textile Machinery and High
Performance Material Technology (ITM)
Dresden University of Technology (TU Dresden)
Dresden, Germany

Acknowledgements "What area of research are you interested in?" people asked me as I finished my studies expressed my intention to pursue a Ph.D. "Preferably something related to textiles, sports/biomechanics, simulation, and high usability," I replied. Little did I know that a detour through the simulation of fibre-reinforced plastics would lead me precisely to the topic I had envisioned from the start.

First and foremost, I would like to thank my professors who provided me with insight into the world of research. Special thanks to Prof. Kyosev, who enabled me to work on a research topic I could pursue with full enthusiasm, and for the continuous support in both my professional and personal development. Thanks to Prof. Cherif, who facilitated my start at ITM, a renowned textile engineering institute. I also extend my gratitude to Prof. Kaiser, who served as a constructive second examiner and provided valuable advice to improve my work.

On my journey to completing my dissertation, I am deeply grateful to many people who accompanied me along the way. I can't name everyone here, but I am truly appreciative of the positive, uplifting and supportive environment around me: Thanks to Alon for all your support, especially your emotional support, motivating words and firm belief in my potential. Thank you for having my back, understanding the many hours of work, and enduring the sometimes very technical breakfast conversations. Thanks to my father, who was always available with good advice. Thank you to my mother and sister for your genuine interest in my dissertation and your support. Thanks to my friends: to those from my time in Aachen, who have always been there for me despite the distance and long periods apart, and to those from my time in Dresden, who supported me with open ears and many sporting activities. Thanks to my colleagues, who made my everyday research life more enjoyable.

The greatest enemy of knowledge is not ignorance,
it is the illusion of knowledge.
(attributed to Hawking, n.d.)

The author acknowledge the financial support by the Federal Ministry of Education and Research of Germany in the programme of "Souverän. Digital. Vernetzt.". Joint project 6G-life, project identification number: 16KISK001K and by the IGF research projects IGF 22435 BR of the "Forschungsvereinigungen Forschungskuratorium Textil e. V.", which was funded through the AiF within the programme for supporting the "Industrielle Gemeinschaftsforschung (IGF)" from funds of the Federal Ministry for Economic Affairs and Energy (BMWi) by a resolution of the German Bundestag.

Competing Interests The author has no competing interests to declare that are relevant to the content of this manuscript.

Abstract

Tightly fitted clothing items interact with the body. Body-clothing interactions introduce pressures in the area of contact, which lead to deformations of clothing items and the body. The functionality of clothing is dependent on individual body characteristics of the wearer in combination with the properties of the clothing. Knowledge of body-clothing interactions on an individual level is necessary to improve functional clothing and enhance the clothing development process as a whole.

This research work investigates how the development process of tighly fitted, functional clothing can be improved through the use of 4D scan data and individualised, deformable human body finite element models (ID-HB-FEM). Novel methods and tools to analyse 4D scan data effectively are developed to improve the development process of tightly fitted clothing. The novel 4D scan data methods enable the experimental analyses of clothing-body interactions. The (semi-)automated 4D scan data analysis framework for the analysis of dynamic clothing-body interactions is developed, called BodyTex4D. BodyTex4D is implemented in Python. BodyTex4D is flexibly applicable to track motions, measure length changes and deformations of any body region visible by a 4D scanner. To analyse the clothing-body interaction digitally ID-HB-FEM and clothing Finite Element Models (CLOTH-FEM) based on original cut pattern are devised. Generalised methods for the ID-HB-CLOTH-FEM set up are presented and evaluated. The IndentoPRO for static analyses and 4D scan data for dynamic analyses are integrated as new material calibration methods for the ID-HB-FEMs. A novel dressing method to combine ID-HB-FEMs and CLOTH-FEMs by thermal shrinkage is derived. The focus of the Finite Element (FE) model development methodology is on usability in clothing development. 4D scan data is integrated into the modelling steps of an ID-HB-CLOTH-FEM. Applied are both developed experimental and digital clothing-body interaction methods to the example of breast-bra interactions in static T-poses and dynamic running or jumping motions.

The developed 4D scan data analysis framework, BodyTex4D, proves to be highly usable in the development process of tightly fitted, functional clothing. In the static breast-bra interaction analysis, the comparability of T-poses is improved by picking comparable frames based on similar breathing states of the breathing cylce. Effects of bra materials, cut pattern and sizes are made rigorously and efficiently analysable by the improvement of comparability. In the dynamic breast-bra interaction analysis, BodyTex4D enabled a breast deformation analysis when wearing different bras in comparable states of movement. A new metric for classifying the bra support in respect to the application scenario and the wearers body characteristics is developed and implemented in BodyTex4D. The new metric (amplification factor) proves its usability by classifying bras with different cut pattern, materials and sizes in their support for running on a treadmill at different velocities.

The developed ID-HB-CLOTH-FEM is evaluated and proves its usability for each modelling step. The IndentoPRO and 4D scan data based material calibration methods lead to an indivialisation of deformable body parts in ID-HB-FEMs. The novel dressing method by thermal shrinkage is proven to be stable and fast. Transferability to different development and research scenarios, such as breast bra analysis, proves the usability of the presented methodology. Four different bra models are remodelled and merged with an ID-HB-FEM in static T-poses. The FE models are validated to be accurate by using 4D scan data, BodyTex4D and pressure analysis. 4D scan data and BodyTex4D improves the model validation by selection comparable breathing states of 4D scanned T-pose in the same bra as modelled. In the dynamic breast-bra FE models, 4D scan data shows advantageous in the breast calibration and model validation by comparing the breast deformation behaviour of the FE models with 4D scanned breast deformations. Analyses of pressure distributions in FE models show clear advantages over experimental analyses.

In summary, the developed methodologies for the 4D scan data analysis and development of ID-HB-CLOTH-FEMs show high usability for the clothing item development process. The combination of 4D scanning technology with FEM has disruptive potential for the development of tightly fitted clothing items. The methods enable efficient analyses and comparisons of different bra models and their interactions with bodies of different individuals in static and dynamic states. The findings can be used for (individualised) bra development.

Kurzfassung

In dieser Forschungsarbeit wird untersucht, wie der Entwicklungsprozess von eng anliegender, funktioneller Bekleidung durch die Nutzung von 4D-Scandaten und individualisierten verformbaren Finite Element Menschmodellen verbessert werden kann. Eng anliegende Bekleidung interagiert mit dem Körper. Die Bekleidung und der Körper werden verformt und Drücke entstehen auf dem Körper. Resultat dieser Interaktionswirkung sind Funktionalitäten der Bekleidung wie z.B. die Unterstützung der Venentätigkeit bei Kompressionsstrümpfen oder das Stützen der Brust in Bewegung. Dabei ist die Bekleidungsfunktionalität stark von den individuellen Körpereigenschaften der TrägerInnen und der Bekleidungseigenschaften abhängig. Wissen über die individuellen Körper-Bekleidungsinteraktionen ist notwendig, um die funktionelle Bekleidung und den Entwicklungsprozess von Bekleidung im Allgemeinen zu verbessern.

In dieser Forschungsarbeit wird ein (semi-)automatisiertes 4D-Scandatenanalyse-Framework für die Analyse von dynamischen Körper-Bekleidungs-Interaktionen entwickelt und generalisierte Methodiken vergestellt, um individualisierte, verformbare Mensch-Bekleidungs-Finite-Elemente-Modelle (Mensch-Bekleidungs-FEM) aufzubauen. Das 4D-Scandatenanalyse-Framework (BodyTex4D) ist in Python programmiert und flexibel auf verschiedene Körperregionen anwendbar. Sowohl generalisierte Methoden für die Analyse der Körper-Textil-Interaktion in Bewegung als auch anwendungsspezifische Methoden wurden entwickelt und implementiert. Ein Fokus der anwendungsspezifischen Methoden ist die Verbesserung der Vergleichbarkeit verschiedener Textil-Körperinteraktionen. Die numerische, digital Analyse der Mensch-Bekleidungs-Interaktion umfasst die Entwicklung von generalisierten Methoden für den Aufbau von Mensch-Bekleidungs-FE-Modellen. Das IndentoPRO wird für die Materialkalibrierung der individualisierten, verformbaren Menschmodelle für statische Analysen und 4D Scandatan für dynamische Analysen eingesetzt. Eine neue digitale Anziehmethode wird über einen thermischen Schrumpf eines Bekleidungsstücks realisiert. Methoden zur Erhöhung der Modellgenauigkeit durch die Nutzung von 4D-Scandaten und BodyTex4D werden präsentiert. Der Fokus in der Modellentwicklung liegt auf der Nutzbarkeit und Verbesserung der industriellen Bekleidungsentwicklung. Sowohl die Methoden des BodyTex4D als auch die zur

Erzeugung des Mensch-Bekleidungs-FEM werden auf das Beispiel der statischen und dynamischen Analyse der Brust-BH-Interaktion angewandt.

Das entwickelte 4D-Scandatenanalyse-Framework (BodyTex4D) zeigte eine hohe Nutzbarkeit in dem Entwicklungsprozess von eng anliegender funktioneller Bekleidung. In der statischen Brust-BH-Analyse kann die Vergleichbarkeit der T-Pose durch die automatisierte Analyse des Atmungszustandes erzielt werden. In der dynamischen Analyse kann die Brustdeformation beim Tragen verschiedener BHs in vergleichbaren Bewegungszuständen analysiert werden. Es wird eine neue anwendungsspezifische Methode zur Klassifizierung des BH-Supports auf Basis einer Bewegungsanalyse der Brustentwicklung und in BodyTex4D implementiert. BodyTex4D beweist sich als hoch-effizientes Werkzeug für die Analyse von verschiedenen BH-Modellen in Interaktion mit verschiedenen Körpern in statischen Posen und dynamischen Bewegungen. Erkenntnisse für die (individualisierte) BH-Entwicklung können abgeleitet werden. Die entwickelten individualisierten Brust-BH-FE Modelle zeigen eine hohe Genauigkeit in der Darstellung verschiedener Einflüsse von BH-Schnittmustern und -Materialien. Die Nutzung von 4D-Scandaten verbessert die Geometrieentwicklung der Menschmodelle, die Materialkalibrierung der dynamischen FEMs und die Validierung der statischen und dynamischen FEMs. Analysen von Druckverteilungen in Finite Element Modellen zeigen klare Vorteile gegenüber experimentellen Analysen. Zusammenfassend weisen die entwickelten Methodiken eine hohe Brauchbarkeit für den Bekleidungsentwicklungsprozess auf. Beide Methoden sind ausgelegt für eine Übertragung auf unterschiedliche Personen, Kleidungsstücke und Anwendungsszenarien. Das Kombinieren von 4D-Scantechnologie mit FEM hat großes Potential für die Entwicklung von körpernaher Bekleidung.

Contents

Abbreviations

AI	Artificial Intelligence
ANN	Artificial Neuronal Network
API	Advanced programming interface
bpm	Beats per minute
C2M distance	Cloud-to-Mesh distance
circum	Circumference
CLOTH-FEM	Clothing Finite Element Model
CT	Computer Tomography
EU size	European size
FEM	Finite Element Method
FFT	Fast Fourier Transformation
fps	Frames per second
FT	Fourier Transformation
ICP	Iterative Closest Point
ID-HB-CLOTH-FEM	Individualised, Deformable Human Body—Clothing Finite Element Model
ID-HB-FEM	Individualised, Deformable Human Body Finite Element Model
IGES	Initial Graphics Exchange Specification
IR	Infrared
JPG	Joint Photographic experts Group
L	Litre
m	metre
ML	Machine Learning
MRI	Magnetic Resonance Imaging
MTL	Material Template Library format
OBJ	Object (File format from Wavefront)
Pa	Pascal
PLY	Polygon File Format
PMM	Pectoralis major muscle
PNG	Portable Network Graphics

RGB	Red Green Blue
RGBA	Red Green Blue Alpha
ROI	Regions of interest
sec	Second
STEP	Standard for the Exchange of Product Model Data
$\Delta circum\_breath$	Difference of the maximal underbust circumference while breathing without a bra
F	External forces
K	Stiffness
f	Face
kmh^{-1}	Kilometre per hour
v	Vertex
vt	Vertex texture
x	Displacement
ρ	Density
υ	Poission's ratio

List of Figures

List of Tables

List of Listings

Part I
Introduction, Objectives and Literature Review

Chapter 1
Introduction and Objectives

Abstract The introductory chapter provides an overview of the necessity of developing new methods to improve close-fitting clothing. The structure of the research work is presented, which is divided into three parts. The first part includes a literature review on scanning technology, finite element analysis and their applications in the analysis of breast-bra interactions. The second part introduces methods for evaluating 4D scan data and applying them to breast-bra analysis. The third part presents methods for developing finite element models to digitally analyse clothing-body interactions.

Keywords 4D scanning · FEM · Soft avatar · Digital clothing development · Breast-bra

This research work investigates how novel measurement technologies and digitisation can improve the development of tightly fitted, functional clothing. Tightly fitted, functional clothing interacts with the wearer's body. The interaction can cause body deformations, pressures or restricts soft tissue from moving uncomfortably. Examples are bras restricting the breast's movement to prevent pain (Fig. 1.1a), compression socks, which apply pressure onto the legs to increase blood circulation and decrease the severity of varicose veins or swellings (Fig. 1.1b) or shape-wear altering the form of the bottom and belly to increase comfort and look. Badly fitted clothing leads to loss of functionality and discomfort.

For the development of tightly fitted, functional clothing, it is essential to adapt the development to the individual's body shape and tissue characteristics. As the clothing has to fulfil its functionality in movement and still be comfortable, it is mandatory to understand how the clothing behaves and interacts with a body in motion. Therefore, information about body deformation due to muscle activation and soft tissue redistribution is essential in the clothing development process.

To acquire data on clothing-body interactions, 3D scanners are the state of the art if any measurement tool is used at all. Most clothing developments are experience-based. A technology to capture the body and clothing-body interaction in motion is missing in clothing development. **The 4D scanner** is a new technology that captures object's surfaces in motion. Therefore, 4D scanners are a promising tool for developing highly functional and individual clothing items. However, this technol-

A.-M. Schmidt, *Novel Methodologies for Digital Investigations of Clothing-Body-Interactions*, Lecture Notes in Bioengineering,
https://doi.org/10.1007/978-3-031-91794-3_1

Fig. 1.1 Tightly fitted clothing require individualised and improved clothing development processes

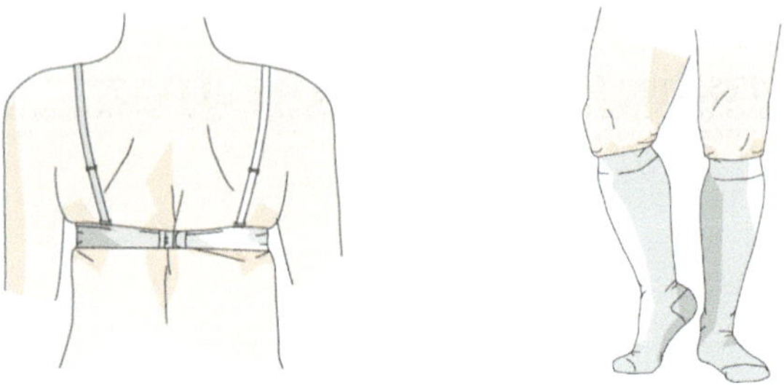

(a) Uncomfortable, bad-fitted bra **(b)** Tightly fitted compressive socks

ogy has not yet been implemented due to the lack of data utilization. 4D scanners produce huge amounts of data. Until now, no fundamental research has been done on 4D data processing methods to automated processing of 4D scan data for clothing development purposes.

Currently, manufacturers mainly execute clothing development iteratively and physically. Manufacturers produce prototypes, test prototypes on test subjects and reiterate until the desired functionality is achieved. No rigorous scientific method is applied. Physical, iterative methods are time-consuming, cost-intensive and wasteful development processes without much room for individualisation. Individualisation processes are linked to a high manual effort. This manual work is not easily reproducible. The market offers clothing software that tries to reduce manual prototyping through digital clothing development.

However, current clothing software products are limited in representing material properties, cut pattern constructions (seams, seam allowance) and the representation of individualised, deformable human body models. Non-deformable avatars are till now generally used in the digital clothing development, but they are unsuitable in developing tightly fitted, functional clothing as they do not represent the real bio-mechanics of human soft tissue properties.

Deformable human body models, which deform close to a real body and have individualised material properties, are required to develop tightly fitted, functional clothing. The **Finite Element Method (FEM)** is a widespread technology to calculate accurately deformations, pressure, stresses and strains. FEM is a promising technology to use in the tight clothing-body interaction.

This research work presents fundamental research in the development of

1. generalised methods for 4D scan data analyses,
2. generalised methods for digital clothing development using individualised, deformable human body finite element models and real clothing cut patterns,
3. deformable body-clothing FE models utilising 4D scan data
4. an automatable process from scan measurements to an individualised FE model of a human body model wearing tightly fitted clothing.

To exemplarily apply generic methods listed above to analyse 4D scan data and generate an individualised, deformable human body model, breast-bra interactions are analysed experimentally and digitally. Most women are wearing a bra every day. In a study by McGhee and Steele McGhee and Steele (2010), it has been determined that 85 % of woman had problems with the fit of the bra. Not having a well-fitted bra leads to discomfort. As bra discomfort, insufficient support, pain caused by to tight bra-bands, sliding of the bra strap and pressure peaks in the area of bra straps are identified (Pei et al. 2021). Discomfort can lead to back pain, shoulder pain and neck pain (Greenbaum et al. 2003). The importance of a well-fitted bra increases with activity. In a study by Lorentzen and Lawson (1987), half of the participants experienced breast pain and discomfort during exercise. The dominant reason for the discomfort during activities is intense breast movement (Lorentzen and Lawson 1987).

Breast-bra interactions are complex due to the high individuality of the female breast, the non-linear tissue properties of the breast, the non-linear material behaviour of textiles and the complexity of the clothing construction process. Understanding breast behaviour in motion, breast-bra interactions in static poses and dynamic motions is the groundwork for bra development. Current methods to perform groundwork are limited to locally placed acceleration sensors or static 3D scans. The digital bra development is barely possible due to the lack of deformable, individualised human bodies.

Consequently, the research field of breast-bra interaction is an important, under-explored field of research.

This work is structured along the following parts (Fig. 1.2).

Part I consists of a literature review which covers the theoretical background of 3D and 4D scan technologies followed by current applications in the breast-bra interaction analysis. The theoretical background of numerical models is described, and application of FEM in the breast-bra analysis is discussed.

Part II experimentally investigates body deformations and the interaction of clothing with a deformable human body. 4D scan datasets of bodies in motion to capture deformations with and without clothing are acquired and analysed. A 4D scan data analysis framework (BodyTex4D) is developed and described (Chap. 4). General methods to analyse movements, length or shape changes and application-specific methods, for example, to classify the bra support are developed within this framework, which can be used for an automated 4D scan data analysis. Putting the framework to use, in the next two chapters, the breast-bra interactions are analysed in a static pose (Chap. 5) and in motion (Chap. 6). The framework includes novel comparability criteria to find comparable poses and comparable body movement states. Chapters 5 and 6 investigate differences in breast deformation due to various bras, and prove that 4D scan data analysis improves the investigation.

Part III focuses on a finite element model modelling the interaction of tightly fitted clothing with a deformable human body. In Chap. 9, the development of individualised, deformable human body models (ID-HB-FEM), the development of digital clothing items (CLOTH-FEM) and the development of methods to dress ID-HB-FEM in tightly fitted clothing are described. Various methods for each modelling

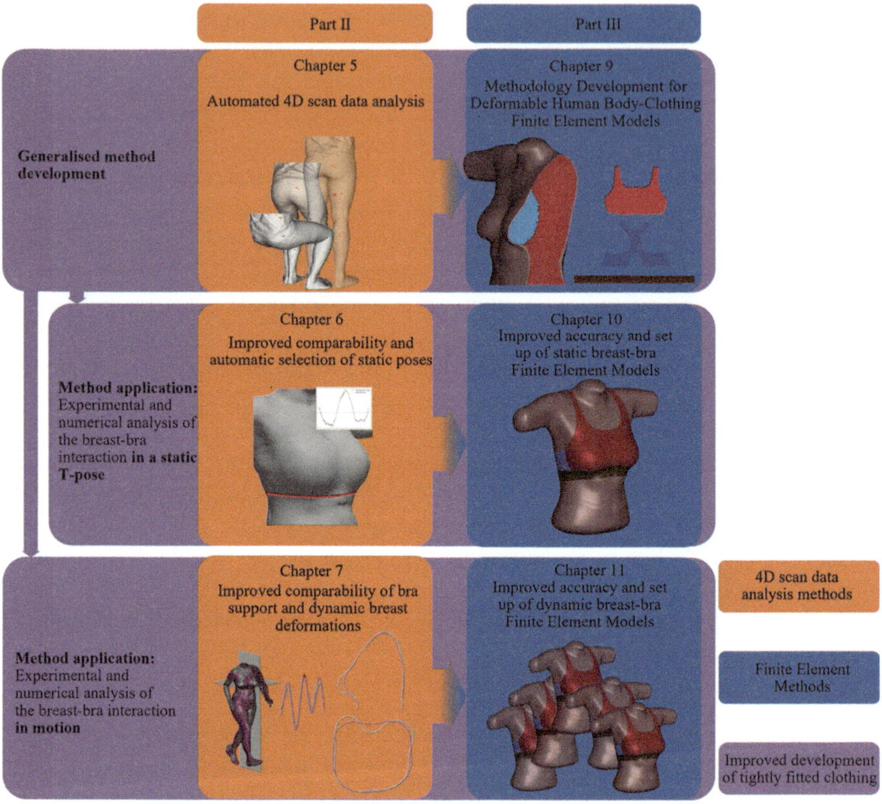

Fig. 1.2 Overview of the research work: Connecting chapters and sections

step are presented, tested and evaluated for usability in clothing development, leading to fundamental research results in digital clothing development. New approaches to develop FE models based on 4D scan data are presented. Using the generalised FE modelling methods, in Chap. 10, bra fitting, breast deformation and pressure distribution over the surface of breasts for four bras are analysed digital and evaluated with experimental results. The digital model is based on static 4D scans. The deformation and fitting validation is based on 4D scan data. Using the developed BodyTex4D, comparable static poses are selected, and the clothing-ID-HBM is improved due to the increased accuracy of the validation process. Chapter 11 presents a breast-bra FE model in motion. The breast material calibration and model validation is performed using 4D scan data, which improves the model accuracy. This part firstly combines FEM and 4D scan data.

References

Greenbaum AR et al (2003) An investigation of the suitability of bra fit in women referred for reduction mammaplasty. Br J Plastic Surg 56(3):230–236. ISSN: 0007-1226. https://doi.org/10.1016/s0007-1226(03)00122-x

Lorentzen D, Lawson L (1987) Selected sports bras: a biomechanical analysis of breast motion while jogging. Phys Sports Med 15(5):128–139. ISSN: 0091-3847. https://doi.org/10.1080/00913847.1987.11709355

McGhee DE, Steele JR (2010) Optimising breast support in female patients through correct bra fit. A cross-sectional study. J Sci Med Sport 13(6):568–572. https://doi.org/10.1016/j.jsams.2010.03.003

Pei J et al (2021) The detection of the upper boundary of breasts using 4D scanning technology. Int J Fashion Des Technol Educ 14(1):1–11. ISSN: 1754-3266. https://doi.org/10.1080/17543266.2020.1829097

Chapter 2
Literature Review

Abstract This chapter provides a comprehensive literature review on 3D and 4D scanning technologies as well as finite element analysis (FEA). It explains the terminology and data formats associated with 3D and 4D scanning technologies. As an example of how 4D scanning technology can be used in clothing development, the literature is further explored for applications in breast-bra interaction. For finite element analysis, fundamental modelling steps are outlined, and current research gaps in the study of breast-bra interaction are identified. The chapter concludes with a summary of key research gaps and potential advancements in 4D scanning technology and finite element analysis in the field of clothing development.

Keywords 4D scanning · FEM · Soft avatar · Digital clothing development · Breast-bra

The literature review presents the theoretical background and current state of research in scan technologies, Finite Element Methods (FEM), and analysis of breast-bra interactions (Fig. 2.1). Literature published until December 2023 is reviewed.

Section 2.1 outlines the theory of 3D and 4D scan technologies. Scanning terms, the functionality of scanners, and the structure of the export data formats are explained. This description improves the knowledge base for understanding the developed method for analysing 4D scan data later. Section 2.2 presents the current state of research of 4D scan technology to analyse the breast-bra interaction. Section 2.3 describes general steps to generate a FE model, which improves the understanding of the later developed clothing-body FE model. Based on these described modelling steps, the literature is reviewed for models investigating the breast-bra interaction in Sect. 2.4. Finally, in Sect. 2.5, the main finding of the literature review is concluded, and the research gaps are determined.

2.1 Theoretical Background of 3D and 4D Scan Technology

Scanning is the process of creating a virtual image of a physical object (Haleem et al. 2022). Scanning technology is used, for example, in classical mechanical engineering to check the quality of components, for medical (Javaid et al. 2021) or

A.-M. Schmidt, *Novel Methodologies for Digital Investigations of Clothing-Body-Interactions*, Lecture Notes in Bioengineering,
https://doi.org/10.1007/978-3-031-91794-3_2

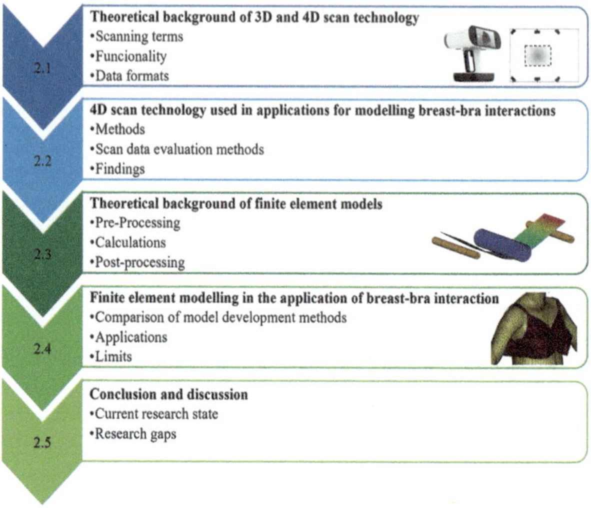

Fig. 2.1 Overview of sections in the literature review

textile applications (Špelic 2020). In this subchapter, firstly the general scan process is described. The definitions further used in this research work are defined and explained in Sect. 2.1.1. Secondly, the 3D scanning technology is described in Sect. 2.1.2. Finally, the 4D scanning technology is outlined in Sect. 2.1.3.

2.1.1 Scan Process and Scan Terms

The overall sequence of the scan process is shown in Fig. 2.2. An object is captured in a static pose or dynamic movement during the scan process. The acquisition of dynamic scan data is only possible with 4D scanning. A 4D scanner has a fourth dimension: time. Several 3D recordings are captured over time. After scanning, 3D objects are processed as meshes in a data processing step. The raw or processed data

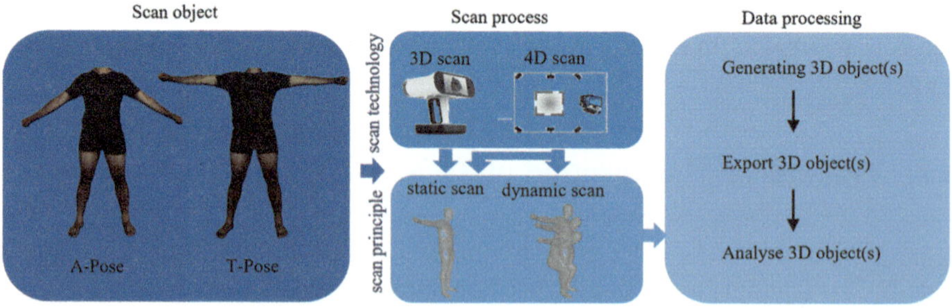

Fig. 2.2 Overview of the workflow of a scan process

can be saved in different file formats such as *.obj or *.ply. A large number of software such as MeshLab (Cignoni et al. 2008) exist to visualise the data. Analysing the data is done manually, with the, for example, scanner softwares or by self-developed, individual methods (Schmidt et al. 2023; Kyosev et al. 2022).

Following brief descriptions of scanning terms are given. In particular, the terms relevant to this thesis will be defined.

Scan subject or scan object: The subject/object being scanned is called a scan subject/object. Scan objects can represent human and non-human objects. Human scans can be full-body scans or scans of specific body parts. Non-human objects can be manufactured parts or archaeological finds.

Scan pose: The term scan pose is used when scanning humans. The person's posture during the scanning process is called the scan pose. Classical scan poses are the A-pose and T-pose, as shown in Fig. 2.2.

Scan process: The scan process refers to virtually capturing the object to be scanned in its assumed scan pose. Different scanning techniques can perform this process. In the following, the scan process of 3D and 4D scanning techniques will be discussed in particular.

Static scans: Static scans are performed without movements of the scanned object. After the scan process is completed, the scan pose can be changed, and a new scan process can be started. 3D scanners are only able to perform this type of scan process.

Dynamic movement scans: Dynamic movement scans capture the movement of the scan subject over time. 4D scanners can perform this type of scan process.

Scan series: A collection of dynamic movement scans is called a scan series. For example, if three dynamic movement scans are made of three different movements of the same scan subject, these dynamic scans can be combined into one scan series.

Post-processing: After scanning, the acquired data is processed into 3D objects by the scanner manufacturer's software or by custom algorithms. The 3D objects are exported. The file export formats can vary depending on the application.

2.1.2 3D Scan Technology

3D scanning techniques are contact-based (tactile) and non-contact-based (optical) (Haleem et al. 2022). Contact-based measurements involve the tactile collection of points on the scanned object using a mechanical device, such as a coordinate measuring machine (CMM) (Hammad Mian and Al-Ahmari 2014). Non-contact scanning techniques can be divided into passive and active 3D scanning techniques. Passive scanning systems work without additional media, such as light. They process mostly pure image data. In stereoscopy, 3D objects are created from the different perspectives of two cameras. Active scanning techniques use structured light. Examples are time-of-flight, laser or stripe light scanners. Light in the infrared range, invisible to the human eye, is often used as an in-depth sensor. Triangulation is used to calculate the 3D coordinates (Lukas 2020; Lanman and Taubin 2009). Laser scanning and triangulation are explained in Sect. 2.1.2.1.

As an example of a 3D scanner, the Artec Leo 3D scanner by Artec3D (Artec 3D 2023) is further described (Sect. 2.1.2.2). The Artec Leo is a triangulation-based structured light scanner.

2.1.2.1 Laser Scanning using Point Pattern Projection and Triangulation

Laser scanners use structured light. This light can be a line or a point pattern. If a point pattern is used, a projector projects points in a certain pattern with known point spacing onto the scan object (Fig. 2.3a). The projection light is mostly in the infrared range. Due to the three-dimensional scanned object, the point pattern on the scanned object is distorted. An infrared camera picks up this distorted point pattern. The depth coordinates of the object can then be calculated from the known initial point distance of the point pattern, the distance between the projector and the infrared camera, and the recorded distorted point pattern. The calculation method used for this is triangulation, as seen in Fig. 2.3b (Lukas 2020). Besides the 3D surface of an object, a 3D scanner can also capture the texture of an object.

2.1.2.2 Technical Setup of the Artec Leo 3D Scanner

The Artec Leo 3D scanner from Artec3D (Fig. 2.4) (Artec 3D 2023) is a triangulation-based structured light 3D scanner. The 3D scanner is handheld and portable (Fig. 2.4). The 3D scanner has a projector, which projects the structured light in point patterns onto the object to be scanned. The projector is the VCSL laser emitter. The camera (Fig. 2.4 left) receives the projected point pattern. It calculates the size and shape of the scanned object using the triangulation method (Sect. 2.1.2.1). The Artec Leo 3D scanner has a port for MicroSD cards, charging, and Ethernet. The scanned objects

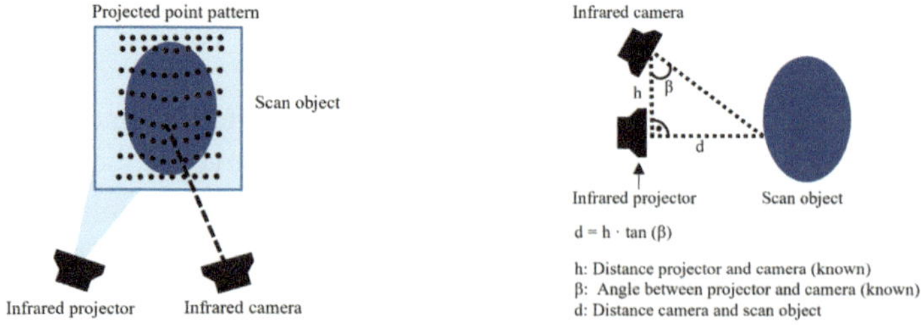

(a) Point pattern projection (figure modified from Lukas 2020) (b) Triangulation (figure modified from Lukas 2020)

Fig. 2.3 Principles of a laser scanner using point pattern projection

Fig. 2.4 Artec Leo 3D scanner

Table 2.1 Data export formats of the Artec Leo 3D scanner (Artec 3D 2023)

Export formats	Data format
3D meshes	OBJ, PLY, WRL, STL, AOP, ASC, PTX, E57, XYZRGB
CAD formats	STEP, IGES
Measurements	CSV, DXF, XML

can be viewed directly on a display (Manual of Artec Leo 2023). The export formats are described in Sect. 2.1.2.3. The Artec Leo has a resolution of up to 0.2 mm and a 3D point accuracy of up to 0.1 mm. Its volume capture zone is 160 000 mm^2 (Artec 3D 2023).

2.1.2.3 Data Export Formats of the Artec Leo 3D Scanner

The Artec Leo's data export formats are divided into 3D polygon meshes, 3D point clouds, CAD formats, and measurement formats. In Table 2.1 an overview of the data export formats is given. In this work, the export format OBJ is used, and described further in Sect. 2.1.4.

2.1.3 4D Scan Technology

Unlike the 3D scan, the 4D scan can capture dynamic movements besides static poses. Multiple 3D scans are taken over time, incorporating the fourth dimension, time. In the following, the technical setup (Sect. 2.1.3.1) of a 4D scanner is explained using the example of the Move4D scanner by IBV (Parrilla et al. 2019).

2.1.3.1 Technical Setup

The Move4D scanner consists of several camera modules. The camera modules are placed around the scan area/scan volume (Fig. 2.5a). Using 12 camera modules, a scan volume of 2 m × 2 m × 2.8 m can be covered. In the standard configuration, two cameras are placed on top of each other to cover the lower and upper scan volume accurately. The remaining camera modules are placed as evenly as possible around the scan volume if scanner space allows. An example of a camera set up can be seen in Fig. 2.5a and from the camera views of the six cameras on the backside can be seen in Fig. 2.5c. The rate of capturing can be set to up to 178 frames per second (fps). Each module comprises three video cameras and an infrared (IR) projector. The three video cameras are divided into two infrared (IR) cameras and one red-green-blue (RGB) camera (Fig. 2.5b).

The Move4D scanner is a photogrammetry-based 3D/4D capture system (Parrilla et al. 2019). The camera modules create a point cloud, each from their point of view from the scan object simultaneously. The image processing unit superimposes and combines these point clouds to produce a single point cloud of the scanned object. The camera module's good coverage of the entire scan volume is essential for an accurate point cloud that covers the full surface of scanned object. The RGB camera

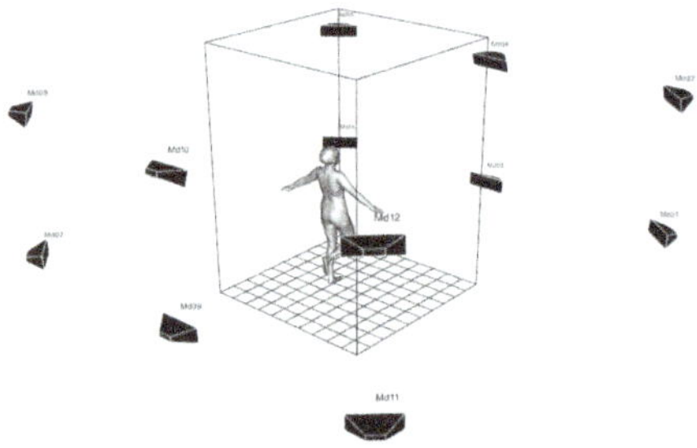

(a) 4D Scanner: set up

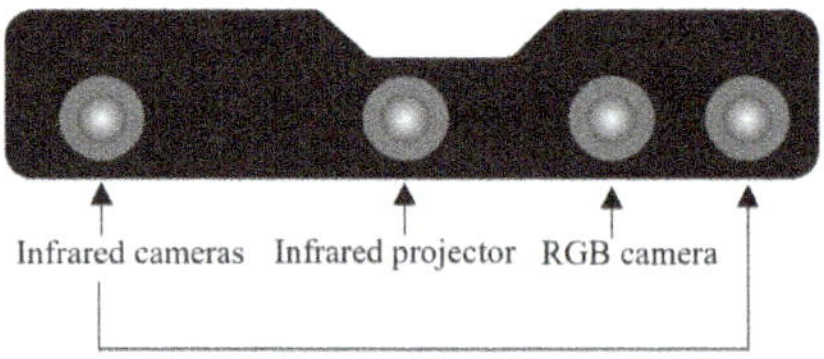

(b) Camera module

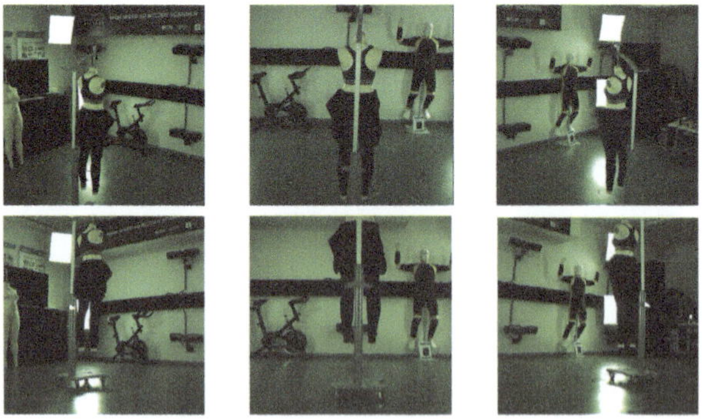

(c) Camera vision of six cameras at one side

Fig. 2.5 Overview of a **a** 4D scanner setup, **b** camera module and **c** camera visions captures the scanned object's colour values and generates the texture. By repeatedly capturing images (frames) over time the motion of 3D objects can be captured as one point cloud is created per frame (Fig. 2.6). The resolution is 1 mm.

2.1.3.2 Data Processing and Export Formats

The data generated by the Move4D scanner can be exported and used in different data types. The three data types are

- point cloud,
- homologous mesh and
- non-homologous mesh.

The **point cloud (raw data)** contains the raw data. The point clouds of each camera module are unfiltered and exported. The point cloud is exported using Polygon File Format (*.ply file). The colour values are stored within this *.ply file.

The **homologous mesh** can be created in the Move4D software based on the raw data and is a watertight (Fig. 2.8a), over several frames' consistence mesh. This data processing method requires an A-pose (Fig. 2.2) and can only be used for human scan subjects. An internal algorithm is used to identify precise marker points, further called key-vertex-IDs (Fig. 2.7) and semantic lines in the captured point cloud. A rig is placed (fitted) in the point cloud and a mesh of approximately 50.000 nodes is fitted to the raw data of the scanned body surface. This mesh and skeleton remain the same for an entire scan series. For each captured frame, the mesh and skeleton is newly

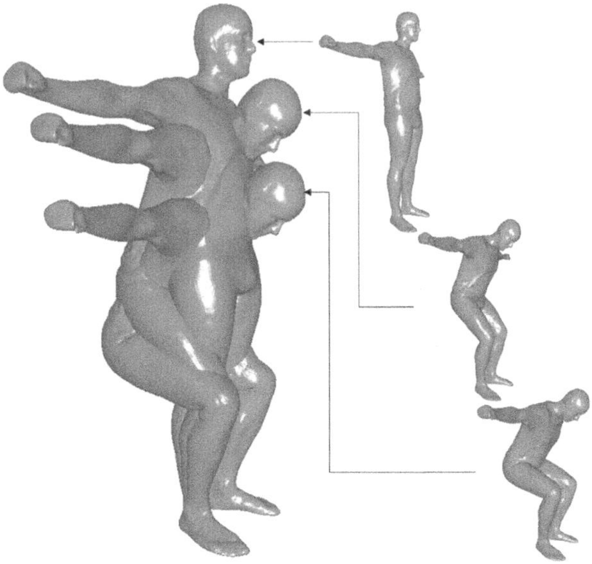

Fig. 2.6 Three 3D objects exported from a dynamic movement scan

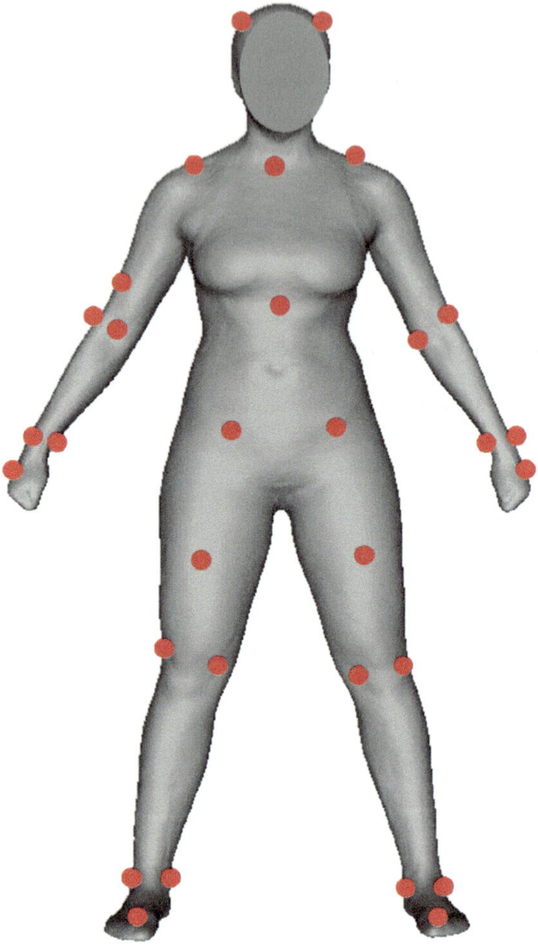

Fig. 2.7 Key-vertex-IDs of an avatar's front side of the Move4D scanner (Adapted from Move 4D User guide 2023)

fitted to the raw data of the scanned body surface. Due to the homology of the mesh, the mesh vertices remain the same, describing always the same point on the body's surface. Move4D released in 2022 a new software version using the homologous mesh to obtain dynamic measurements. At pre-set only lengths on the body can be exported from the dynamic movement (Uriel et al. 2022). The measurement curves are based on ISO 8559. The pre-defined key-vertex-IDs can be read in the User Manual (Move 4D User guide 2023). Individualisation of this measurement area or analysis method is till now not possible.

The **non-homologous mesh** processes the raw point cloud to a surface mesh. For each frame, a new mesh is computed. This mesh is not watertight and does not stay consistent. Areas without captured data are causing defects resulting in an incomplete mesh. Unlike the homologous mesh, the non-homologous mesh is not generated

(a) Homologous, watertight mesh **(b)** Non-homologous mesh

Fig. 2.8 Homologous and non-homologous mesh of the Move4D scanner

based on markers, so the vertex-IDs of the same body locations change frame by frame (Fig. 2.8b).

2.1.4 Data Export File Formats

Both 3D scanners and 4D scanners use *.obj and *.ply file formats to export data. As only the *.obj-file format is used for this work, its structure is described in Sect. 2.1.4.2. Before, a concise definition of a mesh is given in Sect. 2.1.4.1.

2.1.4.1 Mesh Definition

Meshes in the context of scan data represent the surfaces of scan objects. A mesh consists of points, also called nodes or vertices. Multiple vertices can be connected, forming a face and a surface element. Connecting several faces leads to a mesh. Figure 2.9 shows an example of a unit of a mesh. The example shown in Fig. 2.9 consists of three vertices (purple dots) connected by three edges (blue lines). Each vertex has its specific vertex-ID. Thus, the face is described by these three vertices and defined by its specific face ID.
The colour values, called texture, can be assigned to a mesh. Various methods exist to assign the texture. Colour values can be assigned to vertices or faces. If the colour is assigned by face, the face has a homogeneous colour (Fig. 2.10a). Is the colour defined per vertex, a face can have multiple (Fig. 2.10b). The definition of the texture

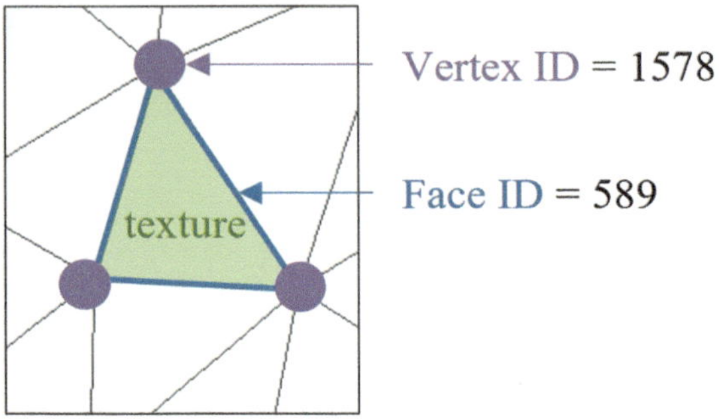

Fig. 2.9 Vertices and faces of a mesh

(a) Face texture visualised by one colour value per face

(b) Face texture visualised by one colour value per vertex

Fig. 2.10 Differences in texture definitions for describing texture of faces

value depends on the export file format and the software that visualises the scanned object.

2.1.4.2 *.obj-Files with Texture Data

Objects scanned in 3D or 4D, including texture data, can be saved using Wavefront Technologies' open-source, human-readable OBJECT (*.obj) file format. The structure of an *.obj file can be seen in Fig. 2.11. Scan data with texture information comes with corresponding *.obj, *.mtl and *.png files. The *obj file contains information of the computed mesh of the scanned object. Within the *.obj, the material information is referred to and saved in a **m**aterial **t**emplate **l**ibrary format (*.mtl). The *.mtl file stores information about the texture stored as a **p**ortable **n**etwork **g**raphic (*.png) file or a **j**oint **p**hotographic **e**xperts Group (.jpg). Subsequently, *.obj-files provide all information of a 3D scan object.

The nodes or vertices of the mesh are defined in a *.obj file by the indicator v. After the v, three numbers are describing the specific vertex's x-, y- and z- coordinates. Each vertex has a unique ID. In the *.obj file, each vertex has its own line. The

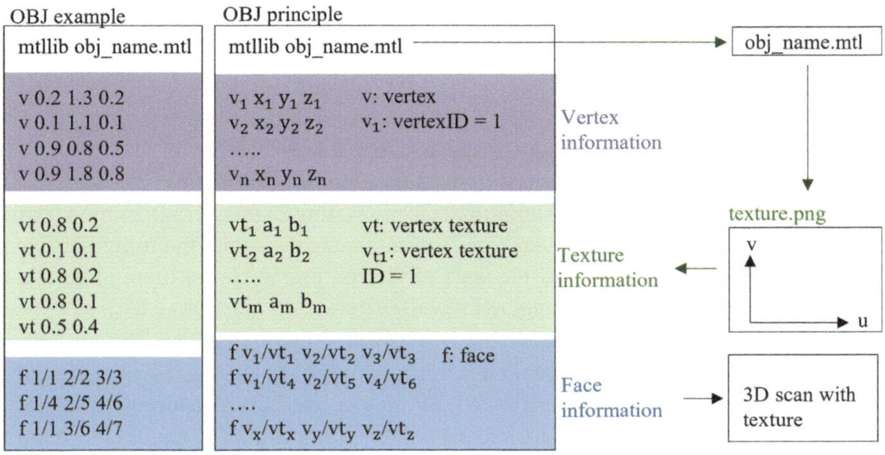

Fig. 2.11 Structure of a *.obj-file format

vertex-ID is determined by the line numbering, whereby the first vertex coordinate that appears has a vertex-ID of one.

Texture information is stored in the next section of the *.obj file. A `vt` marks the texture information for vertex texture. Two numbers u and v (Fig. 2.11) define a colour value after the `vt`. These coordinates refer to a pixel within the linked *.jpg or *.png file. As the vertex-IDs, the vertex textures also have individual IDs. The vertex-texture-IDs are numbered based on the appearance order line by line.

The face information is stored in the next section of the *.obj file (blue section in Fig. 2.11). The face information is indicated by a `f` at the line start. Three vertices define a face if a triangular mesh is stored in the *.obj file (Fig. 2.9). One face is defined by three vertex-IDs and three vertex-texture-IDs for a triangular mesh. The vertex texture ID is behind the vertex-ID separated by a backslash ($f v_n/vt_m$). If no texture data is available, just three numbers indicate that the three vertex-IDs that define the face are stored. With this face information and the link to the texture values, the 3D scan can be visualised.

2.2 4D Scan Technology in the Application of Breast-Bra Interaction

3D scanning to measure bodies exists for several years now. 3D scanning is the state of the art in the clothing development process. 3D body scanning technology is used in the garment industry for size charts, analysing body shapes and fitting clothing (Nguyen and Tran 2021). As analysis based on 3D scanners is state of the art, literature discussing purely 3D scan technologies is not reviewed by this work. 4D scanning is a new technology, and till now, only very few industrial applications are known. Concerning analyses and modelling breast-bra interactions, to the authors knowledge, only Pei et al. used a 4D scanner with focus on breasts only (Pei et al. 2021a, b).

Pei et al. (2021a, b, c) have the same test setup and the same test subjects. Test subjects were running on the spot while wearing a bra. The frame rate is set to 10 fps. The females wore a bra consisting of a moulded contour, cups with underwire, unpadded or their own bras. A running cycle consisting of an average of nine frames is selected after the test subject has familiarised themselves with the running pattern. Three keyframes are picked. At the start of the running cycle, in the middle and at the end. The alignment of the selected frames is based on markers, which have been placed on the test subjects' bodies and is performed manually. The bust, underbust and overbust measurements are taken according to ISO 8559-1:2017. How the measurements are exactly made is not stated. Figures of overbust measurement suggest a non-horizontal measurement which leads to the assumption of a manual localisation of the measurement area.

Pei et al. (2021a) determined that no significant differences between the dynamic and static scans could be seen between the overbust and underbust measurements. The paper did not analyse breast deformation. The authors state limitations in the running speed varsity of the test subjects and the manual evaluation process, which was time-consuming. The authors state that automating the process of evaluation is necessary.

Reusing the same scan data, Pei et al. investigated in Pei et al. (2021c) the upper breast boundaries. The scan data is processed in MATLAB to have the same number of points for each scan. The breast is 100 times vertically and horizontally sliced. It is assumed that the authors did not have the option to export the scan data with a homologous mesh. The breast boundaries are determined due to a comparison between the slices of different frames.

Pei et al. (2019b, 2020) and Pei et al. (2019a) used the same 3D dataset to analyse the breast-bra interaction in a static pose. Pei et al. (2019a) sliced the scan data into horizontal and vertical slices, which created a net structure to identify asymmetries of the breast. Only scans of nude test subjects are used. The same approach is used in Pei et al. (2020). Measurements are performed based on the sliced scan data, and an algorithm estimates a fitting bra size. Nevertheless, the algorithm does not take the bra's cut pattern or material into account. Pei et al. (2019b) scanned 46 females wearing two differently designed bras. Data processing is done analogues to Pei et al. (2021c). The two bras tested have different material properties. One bra is made out of a high-elastic, non-supportive bra, and the other bra is a cup bra. Differences in the breast deformation and asymmetry of the breasts could be detected. Limitations were stated in the comparability of body poses.

In conclusion, to the author's knowledge, Pei et al. are the only researchers using 4D scan data for experimental analyses of the breast-bra interaction until now (December 2023). The author's publication in this field (Schmidt et al. (2023)) is not included. The dynamic breast-bra interactions are not analysed in the reviewed literature. Only the breast movement behaviour when wearing the same bra. Additionally, the selection of comparable frames is not based on the state of motion the test subjects is in. In these studies, it is not assured that the running cycle of the frame selection reflects the same motion. Additionally, Pei et al. stated that the 4D scan data processing automation is missing. This is confirmed in further literature and software reviews. To the author's knowledge, the only 4D scan data processing option is by the Move4D scanner software. However, this software is licensed, misses interfaces to Python or MATLAB, and is limited in processing the software pre-set analysis tools. Body circumferences and lengths on the body are pre-defined on the body and cannot be changed in height or position. Subsequently, automating the scan data analysis process based on this software is not feasible. The breast-bra interaction exists in static analyses. Nevertheless, the studies state uniformly the non-comparability of the static scans as major limitation.

Summarising, the following research gaps within the literature review are determined:

- Automated 4D scan data analysis tools.
- Methods for dynamic breast-bra interaction based on 4D scan data to analyse deformations of breasts in motion.
- Methods to select comparable frames of dynamic 4D scan datasets.
- Methods to improve the comparability of static scan datasets.

2.3 Theoretical Background of Finite Element Models (FE Models)

The **F**inite **E**lement **M**ethod (FEM) is a computer-aided method. A finite number of elements is used to discretise a geometry (Okereke and Keates 2018). To create a Finite Element model (FE model), the workflow can be divided into three main steps: pre-processing, calculation and post-processing. Figure 2.12 shows the overall process. Each FEM step is explained below.

Klein (2015) describes the time effort of the operator in the pre-processing step as the highest, followed by the post-processing step. The calculation step takes the most computational time but no time from the operator. FEM can be used for dynamic or thermal analysis, biomedical studies, fracture mechanics or problems of discontinuous mechanics. Okereke and Keates (2018) describe the need for FEM as it "opens the possibility for problems that cannot always be solved analytically to be solved numerically".

2.3.1 Pre-processing

The pre-processing can be divided into the three main steps of generating the geometries, meshing the geometries, and modelling the material properties and boundary conditions as parameter setup (Fig. 2.13) (Okereke and Keates 2018). Pre-processing is the most important step of FEM since it sets up the model leading to results (Klein 2015). The three steps are explained in more detail in the following.

2.3.1.1 Modelling the Geometry

The first step of the pre-processing is to model the geometry of the object of interest. 3D geometries from CAD software such as SolidWorks (SolidWorks 2023) or Inventor (Autodesk Inventor 2024) can typically be imported to FEM software tools to support the pre-processing stage. Common import formats are **St**andard for the **E**xchange of **P**roduct Model Data (STEP) files or **I**nitial **G**raphics **E**xchange **S**pecification (IGES) files. Alternatively, the 3D geometry can be modelled man-

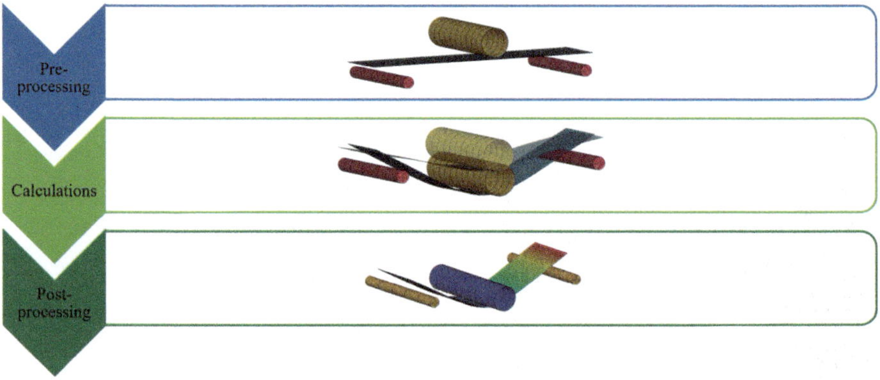

Fig. 2.12 Workflow to set up a finite element model

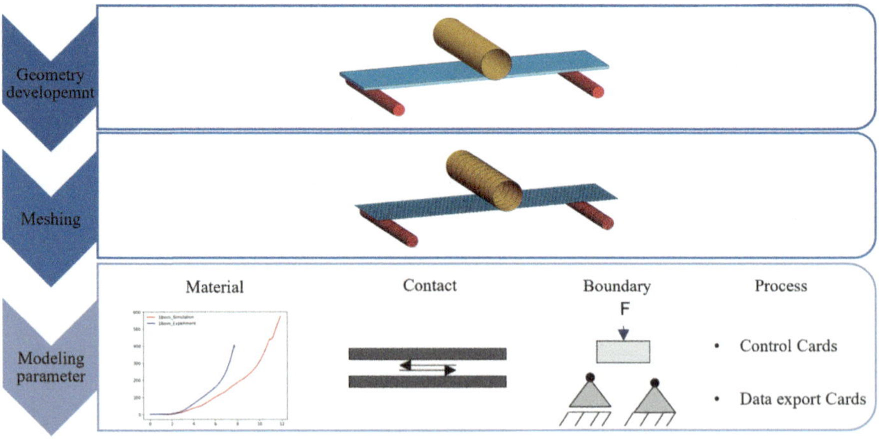

Fig. 2.13 Pre-processing modelling steps

(a) Experimental setup **(b)** Simplified geometrical model

Fig. 2.14 Geometry development of a 3-point bending test

ually within the FEM software during the pre-processor stage. Refinement of the geometry is performed to adapted regions of interest according to the target of the simulation at hand. Areas of high importance are represented in a higher level of detail. Objects less relevant, further away from the region of interest, are omitted and/or highly abstracted. Further simplifications can be made by exploiting symmetries and/or further boundary conditions. Exploiting symmetries leads to shorter computational times. The refinement step of the pre-processing stage has a great influence on the results of the computation stage. Subsequently, an operator with a great deal of experience is required (Klein 2015). Figure 2.14 shows an example of an abstracted model. A 3-point bending test is modelled. The experimental setup is shown in Fig. 2.14a. Figure 2.14b shows the simplified model.

2.3.1.2 Meshing: Discretisation of Geometries

The next pre-processing step discretises the created geometry into a finite number of elements. Each element is assembled by connecting nodes. Properties of meshes are element density, element dimension, -shape and shape functions. The mesh properties have a significant impact on the results.

Mesh density is the number of elements per unit area. The mesh density affects the convergence, accuracy, and computational time. A mesh with bigger elements has a lower mesh density. A lower mesh density can lead to inaccuracy or even wrong results. A higher mesh density with a finer mesh tends to lead to a higher convergence but also increases computation time (Okereke and Keates 2018).

Element dimension can differ in one-dimensional (1D) elements, two-dimensional (2D) elements, and three-dimensional (3D) elements. The 1D element is a line element. The 2D element is a surface element, and the 3D element is a volume element (Klein 2015; Okereke and Keates 2018; DIANA User Manual 2015; Wai et al. 2013).

Shape functions are the approach/interpolation functions for the displacements between the nodes (Zienkiewicz et al. 2005). Based on the displacements the stresses, strains and forces on the geometry are calculated (s. Sect. 2.3.2). The shape function depends on parameters like the element's number of nodes and the degree of freedom of a node. If a shape function uses linear interpolation, the element is called a first-order element. Shape functions using a quadratic interpolation function are called second-order elements. Due to the quadratic interpolation function, these elements can be curved. Specific element types or category categories can be defined depending on the chosen shape function.

Typical element shapes and types for each element dimension are described in the following. **Element shapes** describe the shape of the element, for example, how many edges they have.

2.3.1.3 Meshing 1D Elements: Element Shapes and Types

The element shape of the 1D element is the line. First-order line elements consist of two nodes connected by a straight line (Fig. 2.15a). Line elements in second-order consist of three nodes and are defined by a quadratic curve (Fig. 2.15b) (DIANA User Manual 2015; Wai et al. 2013).

Element types of 1D elements are trusses and beams. A criterion for an element to be declared as a truss or beam is that the cross section of the bar is relatively small compared to the bar axis. Trusses can have axial deformations (Fig. 2.15d). Beams can have axial deformation, shear deformation, curvature and torsion (Fig. 2.15e). Using beams, axial forces, shear forces and moment can be described (DIANA User Manual 2015). Truss nodes only have three translational degrees of freedom (DOF). Beam nodes also have rotational degrees of freedom and thus have six (three rotational plus three translational) DOFs per node (Wai et al. 2013; Link 2014).

2.3.1.4 Meshing 2D Elements: Element Shapes and Types

2D elements are surface elements. The 2D elements come in triangular and quadrilateral shapes (Wai et al. 2013). A first-order triangular 2D element has three nodes (Fig. 2.16a). A second-order triangular 2D element has six nodes (Fig. 2.16b). A first-order quadrilateral element has four nodes (Fig. 2.16d). A second-order quadrilateral element has eight nodes (Fig. 2.16e) (DIANA User Manual 2015).

The 2D elements are divided into the subtypes: plates, membranes and shells. For these element types, the thickness of the element is relatively small compared

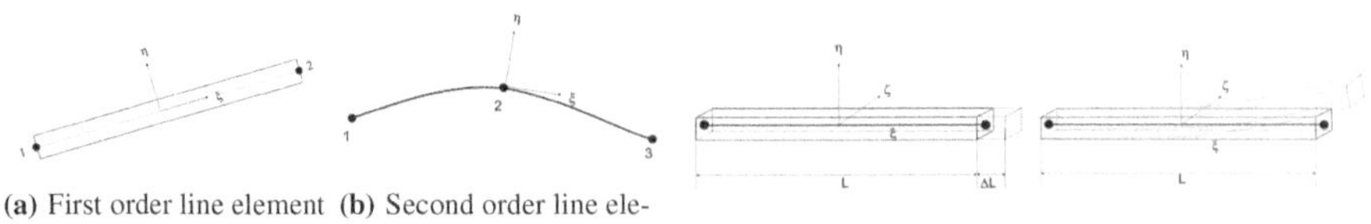

(a) First order line element (b) Second order line element

(c) 1D element shape and -formulation

(d) Truss element (e) Beam element

(f) 1D element types

Fig. 2.15 1D element mesh parameter

to the plane dimensions (Klein 2015). Plates can resist orthogonal loads to the plane through bending effects. In this context, both torque and moments are constrained to act exclusively within the plane of the surface, meaning their vectors lie parallel to the surface and do not have orthogonal components. The same is true for shear forces. Membranes cannot resist moments and thus cannot provide any bending stiffness. Axial forces parallel to the surface and shear forces can act on membranes. Shell elements combine the properties of membranes and plates. They exhibit both bending and shear deformation. Consequently, orthogonal and parallel loads, moments, and torques can act on shell elements (Fig. 2.17) (Vendl Craveiro et al. 2021; Klein 2015).

2.3.1.5 Meshing 3D Elements: Element Shapes and Types

3D elements are volumetric elements and also called solids. Shapes of 3D elements are hexahedrons, tetrahedrons and pentahedrons. The first-order hexahedron has eight nodes, also called an eight-node solid (brick) element. The first-order tetrahedron has four nodes and the first-order pentahedron has six nodes. The first-order elements are based on a linear interpolation. For the hexahedron and tetrahedron strains ε and stresses σ are constant for ε_{xx} and σ_{xx} in x-direction and vary linear in y- and z-directions. ε_{yy} and σ_{yy} are constants in y-direction and vary linearly in x- and z-directions. ε_{zz} and σ_{zz} are constants in z-direction and vary linearly in x- and y-directions (DIANA User Manual 2015). Higher order 3D elements consist of more nodes. The hexahedron second order has 20 nodes, the tetrahedron 13 nodes and the

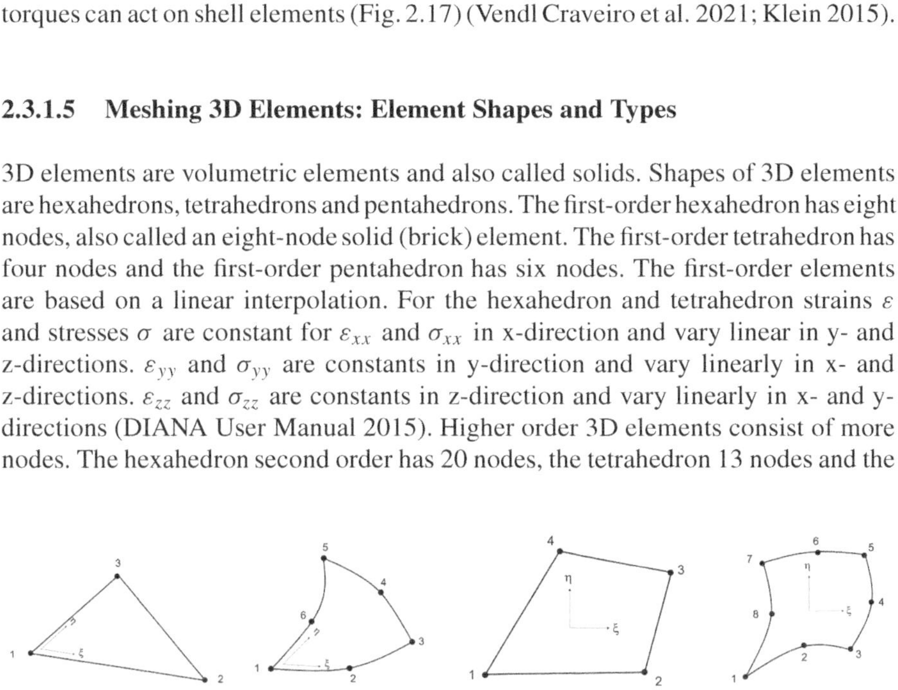

(a) First order **(b)** Second order **(d)** First order **(e)** Second order

(c) Triangular shape **(f)** Quadrilateral shape

Fig. 2.16 2D element: element shapes

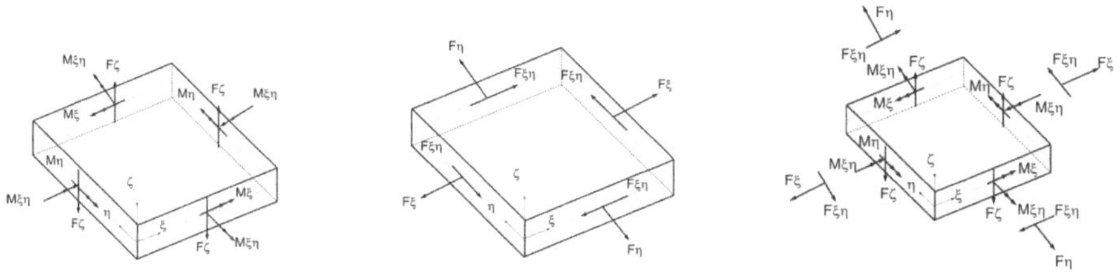

(a) Plate **(b)** Membran **(c)** Shell

Fig. 2.17 2D element: element types

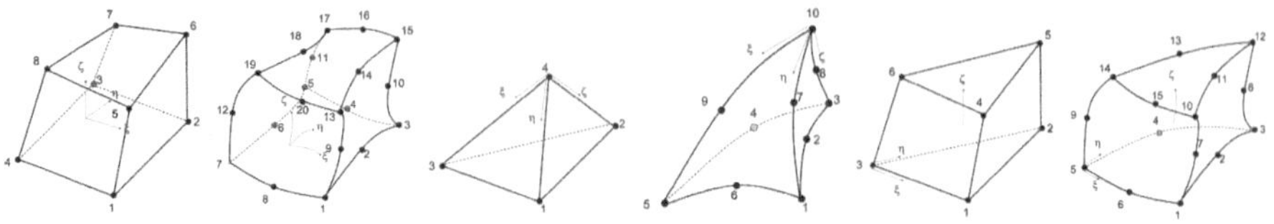

(a) First order (b) Second order (c) Hexahedron shape

(d) First order (e) Second order (f) Tetrahedron shape

(g) First order (h) Second order (i) Pentahedron shape

Fig. 2.18 3D element: element shapes

pentahedron 15 nodes. Stresses and strains are calculated for higher order elements based on a quadratic interpolation function. This quadratic interpolation function leads for the hexahedron and tetrahedron to element strains ε and stresses σ which vary linear for ε_{xx} and σ_{xx} in x-direction, for ε_{yy} and σ_{yy} in y-direction and ε_{zz} and σ_{zz} in z-direction. The strains and stresses vary quadratic for ε_{xx} and σ_{xx} in y- and z-directions, for ε_{yy} and σ_{yy} in x- and z-directions and for ε_{zz} and σ_{zz} in x- and y-directions. The strain and stress distribution for the 15-node pentahedron vary approximately linearly over the element volume. It can be seen that the order and the interpolation function influence the calculation results in terms of accuracy and computational times (DIANA User Manual 2015). The degrees of freedom of the solid elements are only the three translation directions (Wai et al. 2013; Okereke and Keates 2018) (Fig. 2.18).

2.3.1.6 Meshing: Element Applications and Summary

Regarding detail and computational time, 2D elements are placed between 1D and 3D elements and are often chosen for meshing planar and volumetric homogeneous objects (Wai et al. 2013). Using 3D elements usually increases the level of detail and computational time. Thickness adaption of 3D elements is not as easy as for 2D elements as all nodes have to be moved (Klein 2015).

Elements of a higher order are often used to model a curved contour or to increase the quality of calculation results. If lower order elements should be used for complex contours, the mesh can be generated finer to model a curved contour or increase calculation accuracy. However, this finer mesh is limited to the usability of the computational time and memory capacity. The computational time increases with the number of mesh elements. Using elements higher order, the mesh does not have to be as fine as with linear elements (Klein 2015).

2.3.1.7 Modelling Parameter Setup: Material, Boundary, Contact and Process Parameters

The FE model's behaviour also depends on the material parameter, boundary conditions, initial forces and velocities, contact and process parameters.

The material parameter indicates how the meshed geometry behaves. The displacements and thus stresses, strains and forces are calculated by involving the material. Material parameters like density, elastic modulus, Poisson ratio and many others can be set depending on the material's original behaviour. In LS-DYNA Keyword User's Manual (2020), a detailed description of how materials are included in the later calculation process can be seen. The material parameters should be as close as possible to the original ones. Material parameters are mostly determined due to experimental tests. The computational time will differ depending on the material model's complexity.

The boundary conditions define how an object is connected to its surroundings and can set initial states as velocities for dynamic calculations. Contacts are defined to prevent intersections of objects by setting up self-contact or contact with other objects. Process parameters are keeping the model stable by using control cards and defining the export data files.

2.3.2 Calculation

Common FE software to calculate FE models are ANSYS, LS-DYNA, NASTRAN, ABAQUS and NX. All named FE software tools are similar. For example, they are all capable of strength, stability, flow or acoustic analysis. The FE software tools differ in calculating magnetic fields, explicit dynamic analyses, or modelling viscoelastic materials (Klein 2015). Every FE software represents the simulation task as an equation system. The strength of the software lies in solving complex equation systems automatically. The equation system is solved after the displacement. This displacement information can reverse-calculate strain, stress and reaction forces. This leads to relations between stiffness, masses, forces and displacements (formula 2.1) (Klein 2015).

$$\begin{gathered}(Stiffness) \times (Displacement) = (Externalforces) \\ K \times x = F\end{gathered} \tag{2.1}$$

To solve the differential equations implicit and explicit methods are available. The implicit method is characterised by solving the system's behaviour through the current time step t and the future time step $t + \delta t$.

The explicit method is characterised by solving the system's future behaviour based on the current time step t. The explicit method is often used for dynamic behaviours and impact problems (Okereke and Keates 2018).

2.3.3 *Post-processing*

The last stage of FEM is the post-processing step. The results of the calculation are evaluated. The deformations of the geometric structure, stresses, strains and resulting forces acting on the model at different points in steps are analysed and evaluated. Typically, coloured surfaces indicate the evaluation parameter such as pressure or stress or graphs are used to visualise the results to the operator. The results have to be checked for their validity–for example, if the order of magnitude lays within realistic boundaries (Klein 2015). This can be done by using experimental data or plausibility tests.

2.4 Finite Element Modelling in the Application of Breast-Bra Interaction

The **F**inite **E**lement **M**ethod (FEM) presents itself as a powerful tool of investigating the interaction of tightly fitted clothing with the human body. FEM can be applied for compression socks or tight jeans interact with the soft tissue of the leg (Dubuis et al. 2012), shape-wear interacts with the soft tissue of the hip and belly area, or bras interact with breasts (Cai et al. 2016; Li et al. 2003; Liang et al. 2019). Numerical breast models were developed for a variety of fields of research. The main research fields are medicine, biomechanics and minor the textile industry. For the medicine sector, a numerical analysis is used to assist with breast surgery (Myung et al. 2019), determining material parameters of breast tissue (Rajagopal 2007), recreating breast compression in X-ray mammography (Chen et al. 2013) and predicting breast cancer (Rajagopal et al. 2008). In biomechanics and textile technology, breast movements and deformations are analysed mainly (Chen et al. 2013).

The following section presents a literature review using FEM to investigate the breast-bra interaction. The literature is sorted based on the general FE model development steps (Sect. 2.3). The reviewed literature discusses breast FE models with and without bra interaction. The literature sources of FE models developed to interact with bras are marked *cursive*.

2.4.1 *Anatomical Background of a Breast*

The female breast lies on the pectoralis major muscle (PMM), between the second and sixth ribs (Zhang et al. 2022; Patrick 2004). The breast is inhomogenous and consists of various types of soft tissue (Cai et al. 2016). Fat, glandular tissue and skin are parts of the breast. The mechanical properties are dependent on genetic factors, age, nutrition (Cai et al. 2016) and varieties of hormonal levels (Steele 2011). Hormones of a female human change over time. Changes occur during the menstrual cycle, while and after breastfeeding, during menopause, in stressful times, and when gaining or losing weight. While ageing, the reproductive hormone oestrogen decreases, causing the breast's skin and soft tissue to become less elastic, and the density reduces. Zhang et al. (2022) have investigated that fat tissue dominates the breast composition after the age of 55 years.

Breasts differ enormously in size and shape. McGhee and Steele (2011) measured the breast volume within 104 females a range of 100–1900 mL per breast. The main difference is the relative amount of contained fat tissue (Cai et al. 2016). Also, the breast appearance varies. The breast appearance can vary by breast separation, fullness, shape, nipple direction and many more. Using three to five subtypes per breast appearance category more than 960 breast types can be defined (HerRoom 2024). Considering this is already a simplified breast classification the complexity of analysing breasts emerges (Pei et al. 2021b).

2.4.2 *Development of a Human Body Finite Element Model with Deformable Breasts*

In the following based on Sect. 2.3 describing FEM, the literature of human body models with deformable breasts is reviewed. Geometry development, meshing methods and material calibration are discussed in detail. Literature related to the application of breast-bra interaction and breast modelling is reviewed. Literature related to bra development is marked red.

2.4.2.1 Geometry Development

The first step in building a numerical model is generating its geometrical structure. Two main methods are documented by the literature reviewed. Firstly, the breast is captured by medical devices based on Magnetic Resonance Imaging (MRI) or Computer Tomography (CT) technology. Secondly, the breast is captured by a 3D scanner.

The great advantage of capturing the breast by MRI or CT is to retrieve the exact internal composition of the breast, such as the fat tissue, fibro glandular tissue and skin components next to the pure shape of the surface. A high level of detail and highly individualised models can be built up (Rankin 2014). However, the availability of MRI equipment is low, and the costs are very high. Also, most CT scans are made while the patient lies down (MacArthur et al. 2011) causes a breast shape that is hard to compare with the breast shape and properties while standing or in motion.

3D scans can capture the body's surface statically. Compared to MRI or CT scans, it is impossible to capture any inner structure of the breast. Nevertheless, the availability of 3D scans is higher, and costs are lower. Studies using 3D scanners had the test subjects in a standing pose with the arms spread away from the body in a A-

Table 2.2 Comparison and evaluation of technologies to capture the body geometry (cursive text: clothing-related models)

Evaluation criteria	MRI, CT	3D scan
Level of detail	Very high	Low
Costs	Very high	Medium
Availability	Low	Medium-high
Sources	MacArthur et al. (2011), Han et al. (2012), Ruiter et al. (2002), Rajagopal et al. (2008), Rankin (2014), Eder et al. (2014)	*Liang et al.* (2019), *Bel-Brunon et al.* (2014), *Harrison et al.* (2018), *Sun et al.* (2021), *Hong et al.* (2019a; 2011), *Zhang et al.* (2020), *Perez Del Palomar et al.* (2008), *Ruixin et al.* (2020), Li et al. (2003), Yiqing (2016), Rajagopal (2007), Zain-Ul-Abdein et al. (2013), Chen et al. (2013), Sun et al. (2019b)

or T-pose (*Perez Del Palomar et al.* 2008; *Sun et al.* 2021; *Ruixin et al.* 2020; *Hong et al.* 2011; *Yiqing* 2016). *Sun et al.* (2021) added the description that the test subject should breathe normally. Bel-Brunon et al. (2014) and Liang et al. (2019) performed a standing scan underwater to generate a gravity-free initial breast geometry.

Table 2.2 compares MRT/CT with 3D scan methods regarding the level of detail of the breast, the costs and availability. Sources using one or the other method are listed. Sources marked cursive points to studies that developed a breast model that interacted with clothing. FE models based on 4D scans could not be seen in the literature.

Based on the *cursive* marked sources, it can be summarised that only the 3D scan approach is used in clothing-related breast modelling. This is due to the lower costs and higher availability of scan technology in non-medical applications.

2.4.2.2 Meshing

Reviewed literature showed many different part segmentation and meshing methods. The models differ in modelling the thorax, breast and skin parts. A difference in modelling the parts homogeneous or heterogeneous could be seen. Modelling the thorax heterogeneous includes bones and/or muscle tissue (Zhang et al. 2022). Modelling the breast heterogeneously includes the different kinds of breast tissue (MacArthur et al. 2011). An additional difference lies in adding a detailed skin model or not. Some source just modelled the breast itself (Han et al. 2012; Zhang et al. 2007).

Table 2.3 displays part sectioning variants and the mesh element types. Observed element types were volume or surface elements for thorax and breasts. Clothing-related numerical models are marked cursive.

Table 2.3 displays that a large variety of meshing the breast area exists. All bra-related models have a homogenised thorax/body and breast part. Zhang et al. (2022) modelled the effect of arm movement on breast deformation, hereby, including muscle and bones in the analysis leading to a heterogeneous thorax and breast part. Even though this model aims to improve the bra design, breast-bra interactions were not modelled. The other bra-related models differ in their skin modelling. *Liang et al.* (2019), *Hong et al.* (2011), *Bel-Brunon et al.* (2014), *Ruixin et al.* (2020), *Yiqing* (2016), *Zhang et al.* (2022) have modelled the skin layer using shell elements. *Perez Del Palomar et al.* (2008), *Sun et al.* (2021), *Sun et al.* (2019a), *Harrison et al.* (2018), *Li et al.* (2003) did not model a skin layer. A skin layer with volume elements could not be seen. An equal number of breast-bra interaction studies were modelled with and without skin layers. A higher number of heterogeneous breast models can be seen when looking at non-bra-related breast models. Models with and without modelling the thorax could be seen. These breast-only models are mainly used for medicine purposes. Summarising, the variety in breast and body modelling is high and there is no standard method in the application of bra development.

Apart from part sectioning and the element types, a difference in using volume elements could be determined. Liang et al. (2019), *MacArthur et al.* (2011), *Bel-Brunon et al.* (2014) are thickening the outer surface with volume elements as it is exemplary shown in Fig. 2.19b. *Sun et al.* (2019b), Hong et al. (2011) using a full solid model as shown in Fig. 2.19c.
As derived from Table 2.3, most bra-related breast models are meshed with volumetric elements. Mainly, tetrahedral or hexahedral elements were used. Most commonly used element types are summarised in Table 2.4. Tetrahedral elements are dominated as volume elements. All bra-related models are using tetrahedral elements. Both linear and quadratic element formulations are used in the literature (Rankin 2014).

The element sizes are varying between 5 and 10 mm. *Sun et al.* (2021) used a mesh size of 5 mm for its tetrahedral mesh, *Ruixin et al.* (2020) and *Liang et al.* (2019) used a mesh size of 6 mm, *Perez Del Palomar et al.* (2008) used a mesh size

(a) Horizontal cross-section **(b)** Thicked volume mesh **(c)** Full volume mesh

Fig. 2.19 Volume meshing methods determined in the literature for the breast-bra modelling

Table 2.3 Overview of meshing methods for the breast and body part (cursive text: clothing-related models)

Body part	Modelling type	Element type	Modelling variants							
			A	B	C	D	E	F	G	H
Thorax	Homogeneous	Solid	x	x				x		
		Shell					x			
	Heterogeneous	Solid			x					
		Shell								
Breast	Homogeneous	solid	x	x	x					
		shell				x	x			
	Heterogeneous	Solid						x	x	
		Shell								x
Skin	Modelled	Solid								x
		Shell	x		x			x		
	Not modelled			x		x	x		x	
Sources			Chen et al. (2013), Zain-Ul-Abdein et al. (2013), *Liang et al.* (2019), *Hong et al.* (2011), *Bel-Brunon et al.* (2014), *Ruixin et al.* (2020), *Yiqing* (2016)	*Perez Del Palomar et al.* (2008), *Sun et al.* (2021), *Sun et al.* (2019a), *Harrison et al.* (2018), *Li et al.* (2003), Sun et al. (2019b)	Zhang et al. (2022)	Eder et al. (2014), Ruiter et al. (2002), *Li et al.*(2003)	Rajagopal et al. (2008)	MacArthur et al. (2011)	Tanner et al. (2003), Rankin (2014)	Han et al. (2012), Tanner et al. (2006)

Table 2.4 Element type specifications of numerical breast (-bra) models (cursive text: clothing-related models)

	Volume elements		Surface elements	
	Tetrahedral	Hexahedral	Triangular	Quadrilateral
No specifications:	*Liang et al.* (2019), *Perez Del Palomar et al.* (2008), *Sun et al.* (2021), *Sun et al.* (2019a), *Hong et al.* (2011), *Yiqing* (2016), Rajagopal (2007), Eder et al. (2014), Zhang et al. (2007)	Rajagopal et al. (2008), Cai et al. (2016), Li et al. (2003), Azar et al. (2000), Ruiter et al. (2006), Samani et al. (2001)	MacArthur et al. (2011), Hong et al. (2011)	Membrane elements for skin: Samani et al. (2001)
Linear:	Rankin (2014)			
Quadratic:	*Ruixin et al.* (2020), Tanner et al. (2003), Tanner et al. (2006)			

of 8 mm and Rankin (2014) used a mesh size of 10 mm. To the author's knowledge, the other sources did not publish the mesh element types or mesh sizes used.

2.4.2.3 Material Calibration

The following part investigates breast soft tissue material models. When a skin has been modelled, also the skin parameters will be mentioned. Firstly, the experimental investigation methods are described, which are used for the material calibration. Secondly, the material models used in numerical models are presented.

Two material measurement methods were determined; one method is ex vivo and one in vivo, meaning using a non- or living human body. Ex vivo material calibration is based on tensile or compression tests of the breast tissue which was removed from a dead body. The exact parameters can be measured, but the material properties have changed while the tissue was taken from the body (Krouskop et al. 1998). Material parameters from ex vivo analysis are mainly used in medical applications for heterogeneous breast models.

In vivo material calibration is performed on a living test subject. *Bel-Brunon et al.* (2014) mention performing experimental and numerical indentation tests without naming the exact indentation test experiments executed. *Zhang and Yeung* (2003) used data taken from literature.

Another common breast material calibration method in bra-related applications is based on vibration. Firstly, marker points are attached to the breast. Secondly, the breast is lifted by hands. Thirdly, the breasts are released, and cameras capture the movement of the breast until it stands still. Material models and their damping coefficients can be derived based on this experiment. The calibration by vibration analysis is found in *Ruixin et al.* (2020), *Cai et al.* (2018) and *Yiqing* (2016). The calibration by vibration analysis method can be classified as a dynamic material calibration method.

To transfer the experimental material data into the numerical models, various material models are used to describe breast tissue deformation. The main material models found in the literature are:

- Linear-elastic material models (*Hong et al.* 2011; *Li et al.* 2003; Zhang et al. 2007).
- Hyper-elastic material models:
  - Neo-Hook (Chen et al. 2013; Rajagopal et al. 2008; MacArthur et al. 2011) and
  - Mooney-Rivlin (*Liang et al.* 2019; *Perez Del Palomar et al.* 2008; *Sun et al.* 2019b; 2021; 2019a; *Ruixin et al.* 2020).
- Exponential material models (Azar et al. 2002).

The linear-elastic material model is the simplest as the stress is modelled proportional to the strain. The material model setup is simple as just the density ρ, elastic modulus E and Poisson's ratio ν are needed. This material model tends to be accurate for small strains to model hyper-elastic material behaviour. It is used as a linearisation approach (Rankin 2014).

In Table 2.5, parameters of linear-elastic material models are shown. As the modelling method (part sectioning, element types) influences the model properties next to the material properties, the table also provides information about the model meshing approaches for comparability.

Hyper-elastic material models like the Neo-Hook and Moonley-Rivlin models can model non-linearities. Ruiter et al. (2004) and MacArthur et al. (2011) achieved good breast deformation behaviour using the Neo-Hook model. Both studies researched applications in the medical area. Eder et al. (2014) compared various material models and parameter variations with the same numerical model. The numerical model is a volumetric breast meshed with tetrahedral elements. The breast tissue and breast skin were modelled as a homogenous material. Eder et al. (2014) found out that most material models in the literature model the breast as too stiff. The parameter of the material model by Tanner et al. (2006) and the material model by Rajagoapal et al. (2008) lead to the best deformation results observed so far. But Tanner et al. underestimated the material model, and Rajagoapal et al. overestimated the deformation model. Eder et al. suggest using a better balanced material-deformation model. Tanner et al. and Eder et al. used a Neo-Hookean model. Table 2.6 shows the material parameter set of a Neo-Hook material model.

Moonley-Rivlin material models were used by *Liang et al.* (2019); *Perez Del Palomar et al.* (2008); *Sun et al.* (2019b; 2021; 2019a) and *Ruixin et al.* (2020). *Bel-Brunon*

Table 2.5 Linear, elastic material models used in the deformable breast modelling

Source	Body part	Element type	Density kg m^{-3}	Elastic modulus Pa	Poisson's ratio-
Li et al. (2003)	Rigid body	Surface	$1579 \cdot 13^2$	810431	0.3
	Deformable breast	Surface	2250	24.52	0.45
Hong et al. (2011)	Rigid body	Volume	150	4790	0.3
	Deformable breast	volume		24.5	0.45
	Deformable skin surface			0.46	
Zhang et al. (2007)	Deformable breast	Volume		15	0.495

Table 2.6 Neo-Hook material models used in the deformable breast modelling

Source	Body part	Element type	Density kg m^{-3}	Shear modulus kg m^{-3}	Poisson's ratio -
Chen et al. (2013)	Deformable breast	Volume	1000	0.34	0.495
	Deformable skin–Hypodermis	surface	970	0.17	0.495
	Deformable skin–Dermis	surface	1100	18.8	0.495
	Deformable skin–Epidermis	surface	1000	0.35	0.495
Rajagopal et al. (2008)	Deformable breast	volume		C1 = 0.08 kPa	
Rajagopal et al. (2008)	Deformable breast	volume		C1 = 0.13 kPa	

et al. (2014) modelled the skin using the Moonley-Rivlin material model. *Harrison et al.* (2018) used a generalised Rivlin-type polynomial model. Model parameters of the breast material information are not given. More material constants are needed to confirm the model. The hyper-elastic material model Mooney-Rivlin generally requires five different material contacts: C10, C01, C11, C02, and C20. Table 2.7 presents examples of material parameters used in the breast modelling by Moonley-Rivlin models. It should be noted that the sources (*Sun et al.* 2019b; 2021; 2019a) were written by the same authors. *Perez Del Palomar et al.* (2008) used the same breast values as *Sun et al.* (2019b) did.

Eder et al. (2014) and Tanner et al. (2006) show that hyper-elastic material models are more accurate than linear-elastic material models if the application exceeds the

Table 2.7 Moonley-Rivlin material models used in the deformable breast modelling

Source	Body part	Element type	Density kg m^{-3}	C10 kPa	C01 kPa	C11 kPa	C02 kPa	C20 kPa	D1 kPa
Liang et al. (2019), *Ruixin et al.* (2020)	Deformable breast	Thicked volume	/	0.094	0.108	0.82	1.18	0.84	
Perez Del Palomar et al. (2008)	Deformable breast	Thicked volume	1000	1.059	2.1	0.535	0.609	0.918	0.5
Sun et al. (2021), Sun et al. (2019a)	Deformable breast	Thicked volume	1000	0.05	0.052	0.375	0.63	0.78	
Sun et al. (2019b)	Deformable breast	Full volume	1000	0.3	0.31	2.25	3.8	4.72	

Hookean material behaviour in a stress-strain graph. However, both analyses were for the application in medical research and not bra-related.

The common parameter all material models rely on is breast density. Depending on the modelling approach, a homogenised breast and skin density or single densities for the different breast tissues are needed. Determining the real densities is difficult as ex vivo measurements change the material properties. Palomar et al. (2011) determined breast tissue densities for the different types of fat and fibroglandular tissue. Cai et al. (2016) reviewed breast densities of various researchers and determined a breast density range of 685–1017 kg m^{-3}. Further reviewed literature shown in Tables 2.5, 2.6 and 2.7 leads to the conclusion that the upper boundary for breast's density could be as high as 2250 kg m^{-3}. This wide range of the breast density illustrates once more the difficulty of modelling breasts. Also, the Young modulus shows a wide range of 0.69–56 kPa as Cai et al. (2016) discovered. Zain-Ul-Abdein et al. (2013) confirm the wide range of the elastic modulus of breast tissue.

Tanner et al. (2006) additionally discovered that breast models are more sensitive to Poisson's ratio than to elastic properties. Roose et al. (2005) confirms the elastic modulus to have little influence on breast deformation. Tanner et al. also discovered the boundary condition for a compression of less than 20% being more important than the breast tissue material parameter. Linear-elastic material models are found to be as accurate as non-linear material models for compressions lower than 20%. Additionally, a finer mesh of the breast did not change the accuracy of the breast deformation in the analysis of Tanner et al. (2006; 2003)

In conclusion, there is no standard breast material model as of today. The mechanical properties of the breast tissue vary in wide ranges. Rajagopal et al. (2008) used material models which differ depending on the meshing approach, mesh elements and test subjects. Only Poisson's ratio has been modelled similarly. The value was mainly modelled as quasi-incompressible between 0.45 and 0.5 (0.45: *Li et al.* 2003, 0.49: Zhang et al. 2022; Cai et al. 2016, 0.495: Chen et al. 2013; Tanner et al. 2003; Zhang et al. 2007, 0.5: MacArthur et al. 2011; Ruiter et al. 2002). Finally, boundary conditions and Poisson's ratio have a noticeable influence on breast deformation.

2.4.3 Development of a Bra Finite Element Model

The literature on bra modelling regarding geometry development, meshing and material calibration is reviewed. Although other tight clothing numerical models exist, only the bra-related modelling approach is focused on. As only bra-related literature is reviewed in the following, this type of literature is not marked with cursive anymore.

2.4.3.1 Geometry Development

3D scan is the main method to construct the shape of a worn bra as FE model (Sun et al. 2019a, 2021). Based on visual impression and no test subject descriptions, the same method is assumed for Li et al. (2003) and Yiqing (2016). Bel-Brunon et al. (2014) applied the same method to a manikin wearing a bra. After the 3D scan of the worn bra, areas not being the bra are deleted until only the bra is left. This bra is meshed and used as a bra model.

Liu et al. (2008) have a body-paint approach. A bra is drawn on the naked upper body, and the upper body shape is used as the bra shape. Liang et al. (2019) applied on the numerical body a negative gravity load to generate a gravity-free model. Based on this gravity-free model, the bra was constructed. The only source using realistic cut patterns of a real bra is the software VitalFit (VitalMechanics 2024; Harrison et al. 2018). Bra's 2D cut pattern-based can be constructed. The 3D geometry is assembled automatically by the software, but the algorithms are the intellectual property of VitalMechanics (VitalMechanics 2024).

2.4.3.2 Meshing

Most bra models have been meshed as thin shell elements (Bel-Brunon et al. 2014; Li et al. 2003; Yiqing 2016; Cai et al. 2016; Harrison et al. 2018; Sun et al. 2021). Some specified the element types as being quadrangular (Liang et al. 2019; Perez Del Palomar et al. 2008; Ruixin et al. 2020). In the study of Liu et al. (2008) a bra pad is added with hexahedral volumetric elements. Hong et al. (2011) modelled bras element-free-based as they calculated forces on the body. A load is applied normally to the body's surface. The element sizes could be seen in a range of 4–6 mm, when it was documented (Perez Del Palomar et al. 2008; Sun et al. 2021; Liang et al. 2019).

2.4.3.3 Material Calibration

The setup of the material model of the bras is divided in information about the experimental investigation and the transfer to a numerical model. Tensile tests are the most common form of experiments for material property investigations reported by researchers. How the tests were performed in detail and which data were used for the numerical model development were not documented most of the time.

Sun et al. (2021) performed tensile tests for the material identification. Zhang et al. (2020) modelled just bra straps and performed a tensile tests on the straps. Guo et al. (2021) described that tensile tests on cups have been performed. Ruixin et al. (2020) sewed each bra material part together to generate a loop and tested the loops in a cyclic tensile test. Liu et al. (2008) and Bel-Brunon et al. (2014) mentioned that they did "dynamic mechanical analysis" or analysed "material parameter by mechanical tests" but did not elaborate on it further.

For the transfer of the experimental results into the numerical mode, two different material models have been used to model the bra material properties. Liang et al. (2019) and Ruixin et al. (2020) are modelling the same bra and are using a two-coefficient Mooney-Rivlin material model. Parameters can be seen in Table 2.8. Perez Del Palomar et al. (2008) and Li et al. (2003) are using an isotropic, linear-elastic material model. Parameters of the material models can be seen in Table 2.8. Guo et al. (2021); Sun et al. (2021) and Ruixin et al. (2020) reported that they used an isotropic and linear-elastic material model. Other researchers or the software VitalFit (Harrison et al. 2018; VitalMechanics 2024) did not specify their material model.

Table 2.8 Material models used for bra finite element models identified in the literature

Sources	Bra part	Element type	Mooney-Rivlin			Isotropic, linear elastic	
			C10 kPa	C01 kPa	Densitykg m^{-3}	Elastic modulus kPa	Poisson's ratio ν
Liang et al. (2019), Ruixin et al. (2020)	Cup	Surface	200	–40			
	Back panel	Surface	197				
	Elastic band	Surface	200	–57			
	Strap	Surface	300	–87.5			
Li et al. (2003)	Bra	Surface			455	34.3	0.32
Perez Del Palomar et al. (2008)	Bra	Surface			1000	720	0.3
	Pads	Volume			45	330	0.32

2.4.4 *Combination of Human Body and Bra Finite Element Model*

The human body and bra models must be combined to investigate the breast-bra interaction numerically. A combination of the human body and bra models is done through a dressing process. The following methods for the dressing process are described (Sect. 2.4.4.1). After successfully applying the dressing process, the breast-bra model can be analysed statically and dynamically. Section 2.4.4.2 gives an overview of approaches for analysing the breast-bra interaction static and dynamic and how these analyses were performed. Section 2.4.4.3 describes validation methods of the designed FE models.

2.4.4.1 Dressing Process

A challenge while creating a breast-bra model is that the bra is geometrical smaller than the breast. Importing the bra model to the human body model in their initial states often leads to unavoidable collisions of the human body mesh and bra mesh (Liang et al. 2019).

Li et al. (2003) and Perez Del Palomar et al. (2008) created the bra model mirroring the 3D scanned body surface. Penetration is assumed not to have been a challenge since the bra geometry is the same as the body surface. This assumption is strengthened because no dressing method is described.

Sun et al. (2021, 2019a) described dressing bras with closures by closing those in the numerical model (Fig. 2.20a). A detailed description is not given. It is described that "many steps" are needed.

Another dressing approach is to reduce the body geometry geometrically and then expand it to its original size. The body is pushed against the bra during the expanding process, and a deformation occurs. This expansion can be performed by geometric scaling (Liang et al. 2019; Ruixin et al. 2020) or by thermal expansion (Yiqing 2016). The bras stay the same size during the process. After the dressing process, gravity is activated to model a more realistic breast-bra interaction.

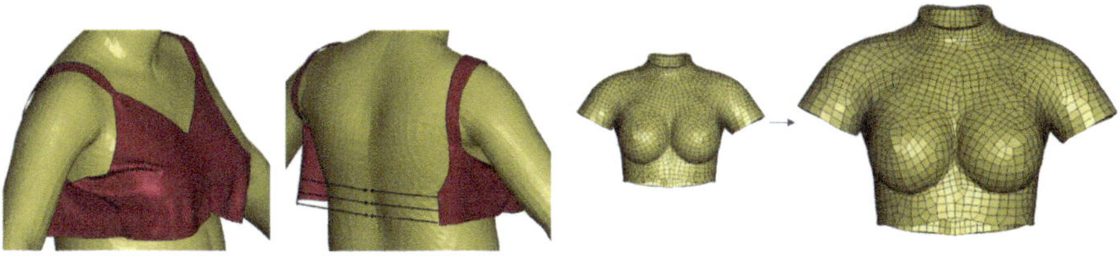

(a) Dressing method by using clamps **(b)** Dressing method by expansion

Fig. 2.20 Dressing methods for dressing a bra identified in literature

Bel-Brunon et al. (2014) names that "a lot of steps" are used to fit the bra to the body. A more specific fitting method is not named. Zhang et al. (2020) analysed only the bra straps and did not need a bra fitting process. VitalBody (Harrison et al. 2018) dressing process without access to the commercial software is not known.

In summary, only a small number of fitting methods have been used in FE modelling up to now.

2.4.4.2 Static and Dynamic Breast-Bra Interaction

In the breast-bra interaction models, two model types could be determined. The models analyse a static and the models analyse a dynamic breast-bra interaction. The models which keep the body from moving are considered static, and the models moving the body are considered as dynamic. The following sources are identified as static and dynamic models:

- Static models: Guo et al. (2021), Hong et al. (2011), Ruixin et al. (2020), Sun et al. (2019a), Sun et al. (2021).
- Dynamic models: Bel-Brunon et al. (2014), Harrison et al. (2018), Zhang et al. (2020), Perez Del Palomar et al. (2008), Li et al. (2003), Liang et al. (2019), Yiqing (2016).

In the static interaction analysis, validation and analysis methods are performed directly after dressing. Validation methods are described in Sect. 2.4.4.3.

Bel-Brunon et al. (2014) performed a dynamic breast-bra interaction. The human body model is moved by captured vertical displacement and rotational data. Tracking points were placed on the body and the breasts. A normalised vertical displacement curve shows that the breast points followed the body movement. The effect was stronger with a bra than without a bra. It is unclear whether the displayed curves correspond to numerical simulations or experimental data. Liu et al. (2008) modelled a female human while jumping. The jumping motion was synchronised using a metronome with 110 bpm. The human body is vertically displayed based on the motion data. Liang et al. (2019) applied a two-degree-of-freedom rigid body movement to apply running movement on a rigid body.

Harrison et al. (2018) can animate its avatar to running movements by using VitalBody (VitalMechanics 2024). The dynamic breast-bra interaction is analysed through pressure distribution. Information on validation methods or how the avatars are animated is not described.

Zhang et al. (2020) modelled the movement of bra straps. A rigid human body model is used. The rigid human body is moved to model shoulder movements. The movement is captured by the motion capture system Eagle Motion Analysis Corporation, USA by retro-reflective markers. Li et al. (2003) apply a velocity in various translational directions to imitate walking and the breast bouncing. The movements are later validated based on data from the literature.

2.4.4.3 Validation Approaches

Validation methods for estimating the divergence of experimentally measured and numerically determined data of breast-bra interaction models are commonly pressure-based. The pressure distribution between the bra and the skin is considered (Zhang et al. 2020; Hong et al. 2011; Ruixin et al. 2020; Guo et al. 2021; Perez Del Palomar et al. 2008). Liang et al. (2019) discovered breathing to be a noticeable confounding factor for measurements of static poses due to small movements. It is the only source discussing breathing as confounding to the author's knowledge. Zhang et al. (2020) measured the pressure between shoulder straps and shoulder. Guo et al. (2021) compared the pressure between bra cups and the body. Hong et al. (2011) compared a numerically determined pressure with experimentally measured pressures.

Comparisons of movement for the dynamic breast-bra models are investigated in the following studies. Bel-Brunon et al. (2014) captured a running movement with sensors and compared breast movements with and without a bra. A model validation is not discussed. From a single figure, the assumption is made that a rigid body and two bust points with and without bra movements are analysed. The graphs shown do not seem to reflect experimental data. Liang et al. (2019) investigated breast movements while the test subjects are running on a treadmill. Experimental measurements and numerical determined data are compared. The study does not include graphs comparing input (body) versus output (breast) movement profiles of experimental and numerical data.

Zhang et al. (2020) did a displacement comparison. The research aims to analyse the movement behaviour of straps. While a bending movement is modelled, the experiment and numeric determination displacement measurements are compared. Sun et al. (2021) measure the deformation behaviour of breast by measurements of calibrated lines drawn on test subject's chest area.

In summary, the main validation method is pressure comparison. Dynamic breast-bra models perform a movement comparison. However, these movement comparisons are not shown in a degree of detail to be able to evaluate them.

2.4.5 Conclusion

All major literature sources in numerical breast-bra modelling as of December 2023 have been analysed. For each step of the modelling process, the findings are summarised in Table 2.9. Generally, a variety of breast-bra models can be seen. The large variety shows the absence of a standardised modelling technique.

Table 2.9 Key findings of the literature review of breast-bra finite element models

Modelling steps		Key findings
Deformable human body model	Geometry development	*All breast-bra models used a 3D scanner to set up the body geometry* Models, just modelling the breast, also used data from MRI/CT and are used in the medical field
	Meshing	Most breast-bra models modelled the body and breast with volume elements and homogeneous. If the skin is modelled, varies
		The tetrahedral is the most common element type to model the breast and body
		The element sizes vary between 5 and 10 mm
	Material model	*A generalised experimental material determination method does not exists*
		Breast models tend to use ex vivo data
		Breast-bra models tend to use literature values
		A generalised material model calibration method does not exist
		Most models are using hyper-elastic material models
		Generalised, commonly used material parameters cannot be seen
		The breast is modelled incompressible with a Poisson ratio between 0.45 and 0.5
Bra model	Geometry development	*Most bra geometries were developed based on the body's surface*
	Meshing	All bra models were meshed with shell elements
		The element sizes vary between 4 and 6 mm
	Material model	Not many information was given. When described, tensile tests have been performed
		The Mooney-Rivlin, or isotropic, linear-elastic material model, was used when the material models were named
Combined breast-bra model	Dressing method	*Mostly, no model combination method was needed as the bra geometry was the same as the breast*
		The bra is put according to a real dressing process by using the bra closure
		Two researchers extended the human body model to fit into the bra
	Boundary conditions	Analysing the breast-bra interaction with just gravity and without movements
		Modelling jumping or running movements by moving the rigid body core
	Validation	Pressure analysis is performed by most breast-bra models
		Dynamic models, added a movement comparison of the breast point

2.5 Conclusion and Discussion

This literature review presents basic knowledge about scan technologies, especially 3D and 4D scanning, and the development of finite element models (FE models). The scanning technology Sect. 2.1 describes functionality, processing and export formats. Within the FEM section, general modelling steps of FEM are displayed.

The literature is reviewed and focused on breast-bra interaction applications. Thereby, literature published till December 2023 is included.

Only a single research group works publicly on using 4D scan technology for breast-bra interaction analysis. Pei et al. (2021b, c, 2019a) investigate breast movement while wearing a bra without breast support properties. A method for manually selecting and aligning frames within a running movement on the spot has been presented. A method of processing the scanned point cloud data into 100 vertical and horizontal slices with an equal number of points remaining in each slice is used. Pei et al. describe limitations in the automation and repeatability of movements. In conclusion, the following research gaps are discovered:

- **An automated 4D scan data analysis framework is missing**: 4D scan data can be processed and analysed only manually or with the licensed 4D scanner software. The 4D scanner software cannot set individualised analysis areas. The software offers no APIs for programming languages like Python or MATLAB to extend functionality or integrate it in more flexible or complex automation scenarios of scan data analyses.
- **The influence of various bras on the breast deformation while moving has not been analysed**: Analysing the influence of breast movement by acceleration sensors is state of the art. This sensor analysis does not include the breast deformation behaviour. Pei et al. did not compare multiple bra models.
- **A method to retrieve comparable frames of breast deformation in motion for analysis is missing**: The movement profiles of different running cycles recorded in the study of Pei et al. are not compared. As two running recordings of the same person will never be the same in frequency and amplitude, selecting frames manually is insufficient for comparison.

After reviewing the literature regarding finite element breast-bra models, no general modelling setup could be determined. All literature sources present different meshing methods and material calibration methods. The similarity between all breast-bra models is the geometry capturing of the body by 3D scanners and choosing a nearly incompressible Poisson ratio for the breast of 0.4–0.5. The bra models were all modelled with shell elements. However, no bra model is based on a real cut pattern. Most bras are based on the body's surface scan in body-paint style. Having accurate cut pattern representations is necessary for the bra development process. A different method to create a 3D shape of a bra is to use 3D fitting software like Lectra, Optitex, Tukatech, Clo3D, or VStitcher. However, the listed software is not applied in the literature. A detailed description of material testing is often missing. A dressing process is required as normally, a tightly fitted bra is smaller than the body. Only two dressing processes could be determined in the literature. No comparison of the effectiveness between different dressing processes is documented. The extension dressing method shows promising results as the body can fit in a smaller clothing item. However, this process does not accurately reflect real-world conditions, and the dressing process influences the bra fit. Other dressings should be further investigated, and bra models should be based on actual bra cut patterns rather than on the body's surface geometry.

In summary, the following research gaps could be determined:

- **A generalised method to generate FE models for analysing the breast-bra interaction or general tightly fitted-clothing-body interaction is missing**: Most breast-bra interaction models are generated for a specific experiment with minor information on the model generation, material calibration or validation. Different methods are used in every literature source, and a comparison between the methods is missing. Knowledge about model generation and methods from a textile technology point of view for clothing development is missing.
- **Method to calibrate the breast material is missing**: Till now, most sources use values from literature to calibrate the breast models. This is inaccurate as the values are taken in a medical setup by in vivo methods where test subjects are laying on their backs or ex vivo methods where the breast tissue is removed from dead bodies. Methods for tissue measurements of living test subjects who are moving upright must be implemented. Pai et al. (2018) presented a promising device. However, this device is still not available.
- **A realistic, application-near dressing method is required** : Only two methods for dressing are reported. Knowledge about additional dressing methods and their usability in the clothing industry are missing.
- **Cut-pattern-based clothing FE models**: Most FE models in the literature focused on the deformable human body model. The representation of the clothing item, especially bras, is too rudimentary to be a valid model for deriving estimations about the behaviour of the physical counterpart. Original cut patterns or seams are not modelled. For a beneficial digital clothing development process, it is important to model accurate deformable human body models **and** the clothing items. Deformable-Human Body-Clothing FE models focusing on clothing development are missing.
- **Accurate validation methods of comparable body poses, or movements are required**: Most validation methods are based on deformation, pressure or movement comparison. However, only minor attention has been paid to the **comparability** of the measurements and their value for validation.
- **Integration of 4D scan data into a FE human body-clothing model to improve the model development and validation process is missing**: 4D scan data has high potential to analyse body deformation and clothing-body interactions accurately and in a timely manner. They can be used to generate an FE model and select comparable frames in the validation process. Additionally, the deformation behaviour

of a dynamic FE model can be analysed. Combining 4D scan technology with FE modelling promises a disruptive potential for improving clothing development.

References

Artec 3D: Artec Leo. https://www.artec3d.com/de/portable-3d-scanners/artec-leo. (visited on 02/26/2023)

Autodesk Inventor. https://www.autodesk.com/de/products/inventor/overview?term=1-YEAR&tab=subscription. (visited on 07/15/2024)

Azar FS, Metaxas DN, Schnall MD (2000) A finite element model of the breast for predicting mechanical deformations during biopsy procedures. In: Proceedings IEEE workshop on mathematical methods in biomedical image analysis. MMBIA-2000 (Cat. No.PR00737). IEEE Computer Society, 2000, pp 38–45. ISBN: 0-7695-0737-9. https://doi.org/10.1109/MMBIA.2000.852358

Azar FS, Metaxas DN, Schnall MD (2002) Methods for modelling and predicting mechanical deformations of the breast under external perturbations. Med Image Anal 6(1):1–27. ISSN: 1361-8415.. https://doi.org/10.1016/s1361-8415(01)00053-6

Bel-Brunon A et al (2014) Numerical modeling of bra wear during running. In: 11th World congress on computational mechanics. Barcelone, Spain, July 2014. https://hal.science/hal-01864280

Cai Y et al (2018) A piecewise mass-spring-damper model of the human breast. J Biomech 67:137–143. https://doi.org/10.1016/j.jbiomech.2017.11.027

Cai Y, Yu W, Chen L (2016) Finite element modeling of bra fitting. In: Advances in women's intimate apparel technology. Elsevier, pp 147–168. ISBN: 9781782423690. https://doi.org/10.1016/B978-1-78242-369-0.00009-8

Chen L-H et al (2013) A study of breast motion using non-linear dynamic FE analysis. Ergonomics 56(5):868–878. ISSN: 0014-0139. https://doi.org/10.1080/00140139.2013.777798

Cignoni P et al (2008) MeshLab: an open-source mesh processing tool. In: Scarano V, De Chiara R, Erra U (eds) Eurographics Italian chapter conference. The Eurographics Association, 2008, pp 129–136. ISBN:978-3-905673-68-5. 10.2312/LocalChapterEvents/ItalChap/ItalianChapConf2008/

Compression sock adverticement. https://www.artec3d.com/de/learning-center/laser-3d-scanning. (visited on 09/10/2023)

DIANA User Manual (ed) (2015) DIANA-10.0 User's Manual-Theory First ed. Netherlands

Dubuis L et al (2012) Identification of the material parameters of soft tissues in the compressed leg. Comput Methods Biomech Biomed Eng 15(1):3–11. https://doi.org/10.1080/10255842.2011.560666

Eder M et al (2014) Comparison of different material models to simulate 3-d breast deformations using finite element analysis. Ann Biomed Eng 42(4):843–857. https://doi.org/10.1007/s10439-013-0962-8

Guo Z et al (2021) Study on realizing the personalized customization of bra cup by solving the inverse problem of bra cup design. Text Res J 91(23–24):2995–3011. ISSN: 0040-5175. https://doi.org/10.1177/00405175211016560

Haleem A et al (2022) Exploring the potential of 3D scanning in Industry 4.0: an overview. Int J Cogn Comput Eng 3:161–171. ISSN: 2666-3074. https://doi.org/10.1016/j.ijcce.2022.08.003. https://www.sciencedirect.com/science/article/pii/S2666307422000171

Hammad Mian S, Al-Ahmari A (2014) New developments in coordinate measuring machines for manufacturing industries. Int J Metrol Quality Eng 5(1):101. ISSN: 2107-6839. https://doi.org/10.1051/ijmqe/2014001

Han L et al (2012) Development of patient-specific biomechanical models for predicting large breast deformation. Phys Med Biol 57(2):455–472. https://doi.org/10.1088/0031-9155/57/2/455

Harrison D et al (2018) Fitting close-to-body garments with 3D soft body avatars. In: D'Apuzzo N (ed) Proceedings of 3DBODY.TECH 2018-9th international conference and exhibition on 3D body scanning and processing technologies, Lugano, Switzerland, 16–17 Oct 2018. Hometrica Consulting-Dr. Nicola D'Apuzzo, Ascona, Switzerland, pp 184–189. ISBN: 9783033069701. https://doi.org/10.15221/18.184

HerRoom. https://www.herroom.com/breast-separation. (visited on 06/09/2024)

Hong L et al (2011) A study of the relationship between clothing pressure and garment bust strain, and young's modulus of fabric, based on a finite element model. Text Res J 81(13):1307–1319. https://doi.org/10.1177/0040517510399961

Javaid M et al (2021) Industrial perspectives of 3D scanning: features, roles and it's analytical applications. Sens Int 2:100114. ISSN: 2666-3511. https://doi.org/10.1016/j.sintl.2021.100114. https://www.sciencedirect.com/science/article/pii/S2666351121000358

Klein B (2015) FEM. Springer Fachmedien Wiesbaden, Wiesbaden. 978-3-658-06053-4. https://doi.org/10.1007/978-3-658-06054-1

Krouskop TA et al (1998) Elastic moduli of breast and prostate tissues under compression. Ultrason Imaging 20(4):260–274. PMID: 10197347. https://doi.org/10.1177/016173469802000403

Kyosev Y, Tomanova V, Schmidt A-M (2022) Method for automatic analysis of the clothing related body dimension changes during motion using high-speed (4D) body scanning. In: D'Apuzzo N (ed) Proceedings of 3DBODY.TECH 2022-13th international conference and exhibition on 3D body scanning and processing technologies, Lugano, Switzerland, 25–26 Oct 2022. Hometrica Consulting-Dr. Nicola D'Apuzzo, Ascona, Switzerland. ISBN: 9783033095205. https://doi.org/10.15221/22.24

Lanman D, Taubin G (2009) Build your own 3D scanner: 3D photography for beginners. In: ACM SIGGRAPH 2009 courses. SIGGRAPH'09. Association for Computing Machinery, New Orleans, Louisiana. ISBN: 9781450379380. https://doi.org/10.1145/1667239.1667247. https://doi.org/10.1145/1667239.1667247

Li Y, Zhang X, Yeung KW (2003) A 3D biomechanical model for numerical simulation of dynamic mechanical interactions of bra and breast during wear. Fiber 59(1):12–21. https://doi.org/10.2115/fiber.59.12

Liang R et al (2019) Numerical simulation of nonlinear material behaviour: application to sports bra design. Mater Des 183:108177. ISSN: 02641275. https://doi.org/10.1016/j.matdes.2019.108177

Link M (2014) Finite Elemente in der Statik und Dynamik. 4., korrigierte Aufl. Springer Vieweg, Wiesbaden. ISBN: 3658035560

LS-DYNA Keyword User's Manual (2020) Material models, vol 2

Lukas M (2020) Das 3D-Scanner-Praxisbuch: Grundlagen, Nachbau, Nachbearbeitung. 1. Auflage. Heidelberg: dpunkt.verlag. ISBN: 978-3-86490-640-4

Manual of Artec Leo. http://docs.artec-group.com/leo/index.html. (visited on 09/10/2023)

McGhee DE, Steele JR (2011) Breast volume and bra size. Int J Cloth Sci Technol 23(5):351–360. ISSN: 0955-6222. https://doi.org/10.1108/09556221111166284

Move 4D User guide. In: 1.5 (2023)

Myung Y et al (2019) Finite element analysis of long-term changes of the breast after augmentation mammoplasty: implications for implant design. Arch Plast Surg 46(4):386–389. ISSN: 2234-6163. https://doi.org/10.5999/aps.2019.00346

Nguyen MHT, Tran MH (2021) Automatic estimation of body measurements to extract the size and body shape. Int J Eng Adv Technol 11(2):24–28. https://doi.org/10.35940/ijeat.B3262.1211221

Okereke M, Keates S (2018) Finite element applications: a practical guide to the FEM process, 1st ed. Springer tracts in mechanical engineering. Springer International Publishing and Imprint, Springer. ISBN: 9783319671253

Pai DK et al (2018) The human touch. ACM Trans Graph 37(4):1–12. ISSN: 0730-0301. https://doi.org/10.1145/3197517.3201296

Parrilla E et al (2019) MOVE 4D: accurate high-speed 3D body models in motion. In: D'Apuzzo N (ed) Proceedings of 3DBODY.TECH 2019-10th international conference and exhibition on

3D body scanning and processing technologies, Lugano, Switzerland, 22–23 Oct 2019. Hometrica Consulting-Dr. Nicola D'Apuzzo, Ascona, Switzerland, pp 30–32. ISBN: 9783033075283. https://doi.org/10.15221/19.030

Patrick CW (2004) Breast tissue engineering. Ann Rev Biomed Eng 6:109–130. ISSN: 1523-9829. https://doi.org/10.1146/annurev.bioeng.6.040803.140032

Pei J et al (2021a) An exploratory study of bust measurements during running using 4D scanning technology. Int J Fashion Des Technol Educ 14(3):302–313. ISSN: 1754-3266. https://doi.org/10.1080/17543266.2021.1938699

Pei J et al (2021b) Monitoring dynamic breast measurements obtained from 4D body scanning. Int J Cloth Sci Technol 33(5):740–759. ISSN: 0955-6222. https://doi.org/10.1108/IJCST-10-2020-0157

Pei J et al (2021c) The detection of the upper boundary of breasts using 4D scanning technology. Int J Fash Des Technol Educ 14(1):1–11. ISSN: 1754-3266. https://doi.org/10.1080/17543266.2020.1829097

Pei J, Fan J, Ashdown SP (2019a) A novel method to assess breast shape and breast asymmetry. J Text Inst 110(8):1229–1240. ISSN: 0040-5000. https://doi.org/10.1080/00405000.2018.1555876

Pei J, Fan J, Ashdown SP (2019b) Detection and comparison of breast shape variation among different three-dimensional body scan conditions: nude, with a structured bra, and with a soft bra. Text Res J 89(21–22):4595–4606. ISSN: 0040-5175. https://doi.org/10.1177/0040517519839398

Pei J, Fan J, Ashdown SP (2020) A novel optimization approach to minimize aggregate-fit-loss for improved breast sizing. Text Res J 90(15–16):1823–1836. ISSN: 0040-5175. https://doi.org/10.1177/0040517519901318

Perez Del Palomar A et al (2008) A finite element model to accurately predict real deformations of the breast. Med Eng Phys 30(9):1089–1097. ISSN: 1350-4533. https://doi.org/10.1016/j.medengphy.2008.01.005

Rajagopal V (2007) Modelling breast tissue mechanics under gravity loading. PhD thesis. The University of Auckland, Auckland

Rajagopal V et al (2008) Creating individual-specific biomechanical models of the breast for medical image analysis. Acad Radiol 15(11):1425–1436. https://doi.org/10.1016/j.acra.2008.07.017

Rankin SC (2014) MRI of the breast. Br J Radiol 73(872):806–818. ISSN: 0007-1285. https://doi.org/10.1259/bjr.73.872.11026854. eprint: https://academic.oup.com/bjr/article-pdf/73/872/806/54976729/bjr.73.872.11026854.pdf

Roose L et al (2005) Validation of different soft tissue simulation methods for breast augmentation. In: International congress series, vol 1281, pp 485–490. ISSN: 05315131. https://doi.org/10.1016/j.ics.2005.03.126

Ruiter NV, Muller TO, Stotzka R, Gemmeke H, Reichenbach JR, Kaiser WA (2002) Automatic image matching for breast cancer diagnostics by a 3d deformation model of the mamma, vol 47, pp 644–647. https://doi.org/10.1515/bmte.2002.47.s1b.644

Ruiter NV et al (2004) Model-based registration of X-ray mammograms and mr images of the female breast. In: IEEE symposium conference record nuclear science 2004. IEEE, pp 3290–3294. ISBN: 0-7803-8700-7. https://doi.org/10.1109/NSSMIC.2004.1466392

Ruiter NV et al (2006) Model-based registration of X-ray mammograms and MR images of the female breast. IEEE Trans Nucl Sci 53(1):204–211. ISSN: 0018-9499. https://doi.org/10.1109/TNS.2005.862983

Ruixin L et al (2020) Computational modelling methods for sports bra-body interactions. Int J Cloth Sci Technol 32(6):921–934. ISSN: 0955-6222. https://doi.org/10.1108/IJCST-09-2019-0143

Samani A et al (2001) Biomechanical 3-D finite element modeling of the human breast using MRI data. IEEE Trans Med Imag 20(4):271–279. ISSN: 0278-0062. https://doi.org/10.1109/42.921476

Schmidt A-M et al (2023) Automated investigation of the breast-bra interaction using 4D scan data and oscillation analysis. In: Scataglini S et al (eds) Advances in digital human modeling. Lecture notes in networks and systems, vol 744. Springer Nature Switzerland, Cham, pp 52–61. ISBN: 978-3-031-37847-8. https://doi.org/10.1007/978-3-031-37848-5_6

SolidWorks. https://www.solidworks.com/. (visited on 11/13/2023)

Špelic I (2020) The current status on 3D scanning and CAD/CAM applications in textile research. Int J Cloth Sci Technol 32(6):891–907. ISSN:0955-6222. https://doi.org/10.1108/IJCST-07-2018-0094

Stewart ML, Smith LM, Hall N (2011) A numerical investigation of breast compression: a computer-aided design approach for prescribing boundary conditions. IEEE Trans Bio-medical Eng 58(10):2876–2884. ISSN: 0018-9294. https://doi.org/10.1109/TBME.2011.2162063

Sun Y et al (2019b) Optimization method for the determination of Mooney-Rivlin material coefficients of the human breasts in-vivo using static and dynamic finite element models. J Mech Behav Biomed Mater 90:615–625. https://doi.org/10.1016/j.jmbbm.2018.11.016

Sun Y et al (2019a) 3D bra and human interactive modeling using finite element method for bra design. Comput Aided Des 114:13–27. ISSN: 00104485. https://doi.org/10.1016/j.cad.2019.04.006

Sun Y et al (2021) Finite element analysis on contact pressure and 3D breast deformation for application in women's bras. Fibers Polym 22(10):2910–2921. ISSN: 1229-9197. https://doi.org/10.1007/s12221-021-0878-0

Tanner C et al (2006) Factors influencing the accuracy of biomechanical breast models. Med Phys 33(6):1758–1769. ISSN: 0094-2405. https://doi.org/10.1118/1.2198315

Tanner C et al (2003) 3D rezoning for finite element modelling of large breast deformations. University College London, Technical report

Uriel J et al (2022) A methodology to obtain anthropometric measurements from 4D scans. In: Proceedings of the 7th international digital human modeling symposium (DHM 2022) and Iowa Virtual Human Summit 2022-University of Iowa Libraries Publishing. https://doi.org/10.17077/dhm.31758

Vendl Craveiro M, Bittencourt TN, Della Bella JC (2021) Design and verification of reinforced concrete shell elements. Revista IBRACON de Estruturas e Materiais 14(3). https://doi.org/10.1590/s1983-41952021000300005

VitalMechanics. https://www.vitalmechanics.com/ (visited on 04/15/2024)

Wai CM, Rivai A, Bapokutty O (2013) Modelling optimization involving different types of elements in finite element analysis. In: IOP conference series: materials science and engineering, vol 50, p 012036. ISSN: 1757-8981. https://doi.org/10.1088/1757-899X/50/1/012036

Yiqing CAI (2016) Nonlinear dynamic analysis of bra fitting using finite element models. PhD Thesis. The Hong Kong Polytechnic University, 2016

Zain-Ul-Abdein M et al (2013) Numerical simulation of breast deformation under static conditions. Comput Methods Biomech Biomed Eng 16(Suppl 1):50–51. https://doi.org/10.1080/10255842.2013.815857

Zhang J et al (2022) Non-linear finite element model established on pectoralis major muscle to investigate large breast motions of senior women for bra design. Text Res J 92(19–20):3511–3521. ISSN: 0040-5175. https://doi.org/10.1177/00405175221075049

Zhang S et al (2020) Finite-element modelling of elastic woven tapes for bra design applications. J Text Inst 111(10):1470–1480. ISSN: 0040-5000. https://doi.org/10.1080/00405000.2019.1704962

Zhang Y et al (2007) 3D finite element modeling of nonrigid breast deformation for feature registration in -ray and MR images. In: 2007 IEEE workshop on applications of computer vision (WACV'07), pp 38–38. https://doi.org/10.1109/WACV.2007.2

Zienkiewicz OC, Taylor Robert L, Jianzhong Z (2005) The finite element method, vol 1, 6th edn. Elsevier Butterworth-Heinemann, Amsterdam. 0 7506 6320 0

Part II
Experimental Methodologies to Analyse Clothing-Body Interactions Using 4D Scan Data

Chapter 3
Introduction and Objectives

Abstract As an introduction to Part 2, this chapter focuses on methods for the experimental analysis of clothing-body interactions using 4D scan data. These methods are developed and applied to the example of breast-bra interaction.

Keywords 4D scanning · Dynamic movement analysis · Automated scan data analysis · Body deformations

The following part comprises the experimental analysis of body deformations and clothing deformable-body interactions. An examination of how the development of tightly fitted clothing can be improved experimentally using 4D scan data is presented. This part structure can be seen in Fig. 3.1.

Tightly fitted clothing interacts with the body to fulfil a specific function, such as distributing pressure, changing body shapes or restricting movements. Bodies with the same body shape but different soft tissue and muscle composition will deform differently in motion and interact differently with the clothing. Subsequently, fundamental knowledge about bodies in motion and deformations is essential in clothing development.

Additionally, it is important to test and evaluate clothing's functionality. When testing clothing on real humans, it is impossible to reproduce exactly the same poses and motions consistently. Methods are required to help select comparable poses and movements from varying datasets.

4D scanners capture body surfaces in motion. Information about body deformations or to track specific body parts' motion can be extracted from the recorded 4D scan data. The interaction between the body and clothing items can be captured, even when wearing tightly fitted clothing. However, scanning objects in 4D generates huge amounts of data. For example, 6 sec at a low frequency of 30 fps results in more than 3 GB of raw data. Higher frequencies (up to 178 fps) produce proportionally more data as the point density per frame stays constant. Methods for an automated analysis of such large datasets are necessary to save time during the evaluation process. Additionally, the application fields in clothing development are vast. A modularised, specialised analysis toolbox is required to adapt to the ever-changing purposes of the analysis. When analysing the influence of tightly fitted clothing on a human body, the challenge lies in the confounding factor of humans not being able to perform

A.-M. Schmidt, *Novel Methodologies for Digital Investigations of Clothing-Body-Interactions*, Lecture Notes in Bioengineering,
https://doi.org/10.1007/978-3-031-91794-3_3

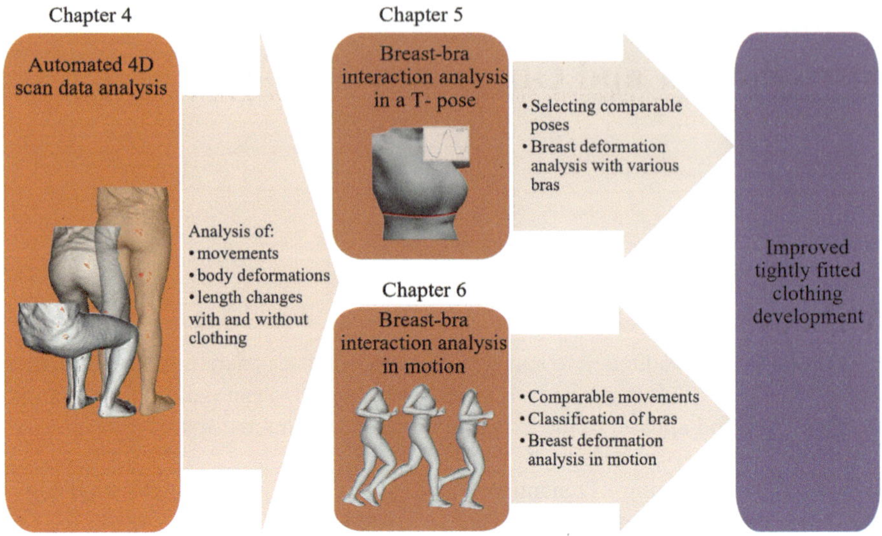

Fig. 3.1 Chapter overview of Part II

the same pose or movement repeatably. Therefore, methods are required to extract comparable poses and movements to increase the accuracy and validity of clothing evaluations. To fill the described research gaps, the following topics are addressed in this work:

- **A 4D scan data analysis framework is developed with with**
  - **an automated analysis process**: 4D scan data results in millions of points representing 3D surfaces of objects. A manual analysis process is too time consuming to be applicable in the clothing development process.
  - **selectable regions of interest (ROI) for analyses**: Different body regions require to be investigated for the clothing development of tightly fitted pants, socks, bras or T-shirts. Pre-set measurement areas are not flexible enough to cover all required ROI in clothing development.
  - **simple extension of analysis tools**: To increase the usability and automation of the analysis process of 4D scan data, it is beneficial when simple analysis methods, such as deformations, shape or movement analyses, can be extended and interchangeably used for application-specific analyses. For example, to select comparable body poses. An advanced programming interface (API) is required.

- **Methods to select comparable poses, movements and frames to evaluate developed clothing**. When comparing the functionality of a clothing item experimentally the challenge presenting itself is to get comparable body poses and movements. Methods to improve the selection of data points from 4D scan data for increased experimental comparability are developed. Proving comparability in the static and dynamic breast-bra interaction serves as an example.

- **The application of the developed 4D scan data analyses framework to improve the analyses of the static and dynamic breast-bra interaction**: The usability of the developed 4D scan data analyses framework is presented in two chapters for a static and dynamic breast-bra interaction analysis. The transferability is shown by analysing two test subjects, and the extension possibilities are presented by developing the breast-bra interaction-specific analysis methods.

Chapter 4
Developing BodyTex4D: A Framework for Automated 4D Data Analysis

Abstract This chapter presents the development of an automated Python framework called BodyTex4D, which enables the (semi-)automatic analysis of 4D scan data. Various analysis methods are developed and implemented, with a particular focus on their usability for analysing body deformations and clothing-body interactions. The fundamental analysis methods include motion analysis, length analysis and shape analysis. Additionally, selected application examples demonstrate the extendability of BodyTex4D for specific use cases. Among other features, a method for classifying 4D scan data is developed and implemented.

Keywords 4D scanning · Dynamic movement analysis · Automated scan data analysis · Body deformations

4.1 Introduction

This chapter investigates how 4D scan data can be analysed automatically to improve the development of tightly fitted clothing. A 4D scanner captures the body's surface in motion by generating a sequence of 3D models at equidistant time intervals. This enables the scanning of the body both with and without tightly fitted clothing. However, analyzing, evaluating, and extracting meaningful data from these scans for clothing development remains a significant challenge. A huge amount of data is produced within one dynamic movement scan of a single movement. A 6 s scan with 30 fps with 50.000 data points per frame results in 3 GB data. Analysing vast amounts of data manually is time-consuming and, therefore, not practical for clothing development.

For different clothing applications, specific body regions must be analyzed individually. The scan data should therefore be evaluated with a focus on application-specific body areas, which require separate measurement and assessment. As the fields of application vary widely, simple extensions of the analysis methods must be provided to adapt to the desired application individually. The only semi-automated analysis option is to use the Move4D internal software. This software, however, is limited in the analyses of circumference and length. Individualisation of the analysis

A.-M. Schmidt, *Novel Methodologies for Digital Investigations of Clothing-Body-Interactions*, Lecture Notes in Bioengineering,
https://doi.org/10.1007/978-3-031-91794-3_4

areas is not possible, and an interface for allowing extensions by the users to add additional analysis methods is not intended till now, which restricts the automation to a small set of intended applications.

In the following section, BodyTex4D is developed in Python to improve clothing development by giving methods to analyse body deformations and clothing-body interactions in motion. The 4D scan data framework can be applied to analyse body lengths, movements and shape changes in motion. The influence of tightly fitted clothing on the body can be investigated. Additionally, application-specific methods are developed to improve the comparability of poses and motions.

For this chapter, knowledge about homologous meshes, scan data export formats (*.obj), scan series, dynamic movement scans and basic programming knowledge is presumed. Further explanations about homologous meshes can be seen in Sect. 2.1.3.2 and about scan data export formats in Sect. 2.1.4.

4.2 Workflow of BodyTex4D

The workflow of BodyTex4D is divided into four main steps, resulting in 4D scan data analysis methods. As the framework is specially developed for application in clothing development, the developed analysis methods focus on investigating body deformations and the deformable body-clothing interaction.

Figure 4.1 shows the basic structure of BodyTex4D. Section 4.3 provides a more detailed description of the programming environment. After capturing, processing and exporting 4D scan data, they are consumed by BodyTex4D and saved in tabular form (Sect. 4.4). Next, filter conditions are defined. Several filter methods categorised by filter conditions are implemented and can be selected (Sect. 4.5). The scan data saved in an array are filtered (Sect. 4.6). After choosing an analysis parameter from which the filter condition is derived, the filtered vertices-IDs are processed to measure the properties of interest (Sect. 4.7). Finally, 4D scan data analysis methods are available to evaluate and interpret the measured properties dataset (Sect. 4.7).

4.3 Programme Environment

BodyTex4D is implemented in Python 3.9. Six open-source libraries are used. The open-source python libraries are displayed in Listing 4.1.

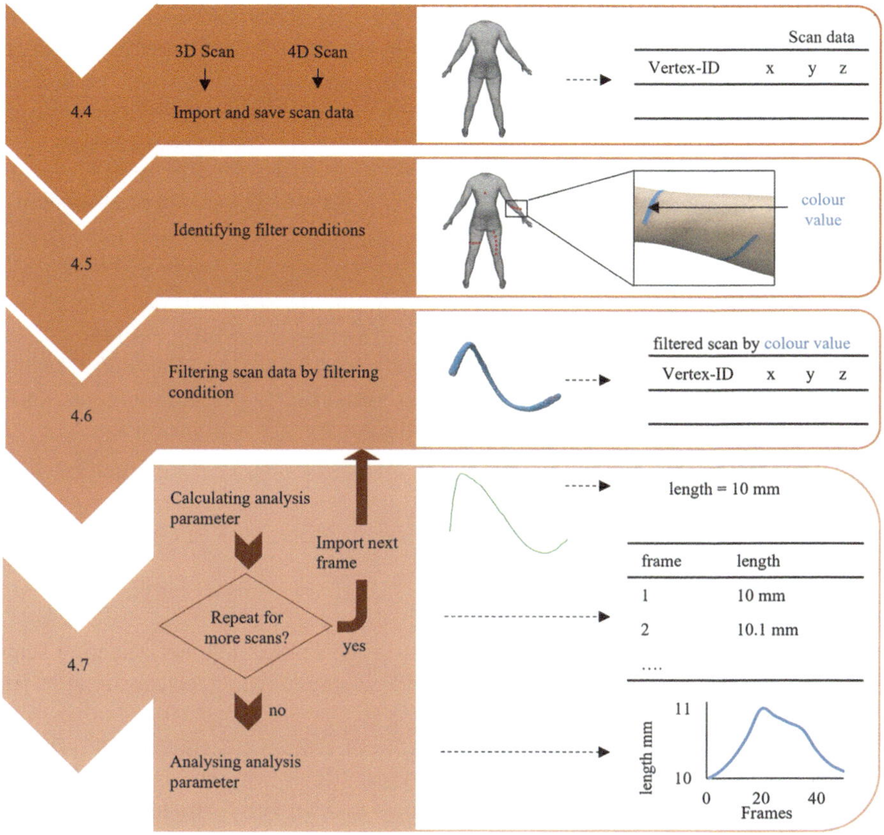

Fig. 4.1 Structure of the 4D scan data analysis framework—BodyTex4D

Listing 4.1 Open-source python libaries used in BodyTex4D

```
python == 3.11.4
numpy == 1.26.3
open3d == 0.18.0
pandas == 2.1.4
pyvista == 0.43.1
matplotlib == 3.8.2
```

4.4 Importing and Saving Scan Data

4D scan data can be imported into BodyTex4D with the file formats *.obj, *.mtl and *.png (see Sect. 2.1.4.2), *.ply or *.xyzrgb. Python's Standard Library, pyVista, Open3D, and NumPy are used to import the scanned data. As most 3D scan data can be exported to the listed output files types, 3D scans can also be imported. The implementation of importing scan data as an *.obj-file and saving it with its coordinate values without using a specific 3D point cloud library can be seen in Listing 4.2. Implementing the obj-reader by hand gives BodyTex4D control over the vertex-IDs, which is beneficial for the homologous meshes. Other libraries read *.obj files with inconsistent vertex-IDs from frame to frame of a 4D dataset, making it hard to track vertices over time. In the developed function to read on *.obj-file, the *.obj-file is opened as a text file using its file path. Then, the first line starting with "v" (Listing 4.2 line 7) is looked up. If the line is found, the three coordinates of that line are saved in a vertex array, further called `vertex_coordinates`. `get_homologous_vertices_from_obj` iterates over all lines saving each `new_vertex` in the correct order to `vertex_coordinates`. The used 4D scanner keeps the order of the vertex the same when exporting a homologous mesh. Subsequently, vertex on line X represents always the same point on the scanned object.

Listing 4.2 Reading and saving scan data (reader.py)

```
1  import numpy as np
2
3  def get_homologous_vertices_from_obj(obj_file_path: str) -> np.ndarray:
4      vertex_coordinates = []
5          with open(obj_file_path, 'r') as f:
6              for line in f:
7                  if line.startswith('v '):
8                      new_vertex = np.array([float(x) for x in line.split()[1:]])
9                      vertex_coordinates.append(new_vertex)
10     return np.asarray(vertex_coordinates)
```

If colour values are required in the analysis, an additional array is generated with the colour values.

Alternatively, the Open3D or pyVista library can read the meshes with the function `open3D.io.read_triangle_mesh(file_path, True)` for Open3D or `pyvista.read(file_path)` for pyVista. However, the vertex arrays created by these libraries do not match the order of the vertex-IDs, which is important for homologous mesh-based analysis. For colour-/texture-based analysis, these libraries are suitable as the texture array extraction is simple. All import methods save the data as NumPy datatypes, which results in compatibility between all libraries used.

In summary, two arrays are created as NumPy datatype (Fig. 4.2). One array contains the vertex-IDs of the 3D meshes or point clouds and the corresponding spatial coordinates. The other array has the corresponding colour values for the same vertex-IDs. The spatial coordinates and colour values are stored independently, as the colour values are not always required for the following methods.

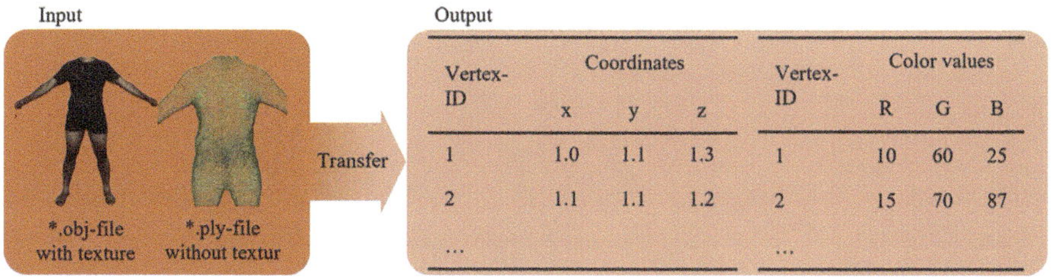

Fig. 4.2 BodyTex4D: importing and saving scan data

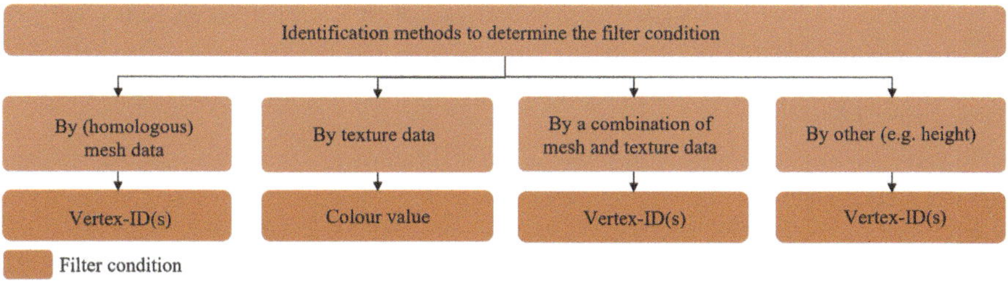

Fig. 4.3 BodyTex4D: identification methods to identify filter conditions

4.5 Identification of Filter Conditions

The next step is identifying and defining the filter conditions for which the vertex and texture arrays should be filtered. This identification step can be performed using mesh data of a homologous mesh, texture data or a combination of both (Fig. 4.3). The output of this identification step is to identify the filter conditions, which can be one or several colour value(s) or vertex-ID(s).

4.5.1 Identification Method by Homologous Mesh Data

The identification method by mesh data is usable for scanning data with a homologous mesh over all frames. Vertex-IDs are selected and used in the filtering process described in the following. If the mesh is not homologous, the vertex-IDs differ for each frame and the filter process by vertex-ID(s) is not possible. A single vertex-ID or a list of vertex-IDs can be selected for filtering. For tracking movement, a single vertex-ID is picked. A list of vertex-IDs can be picked by performing analyses based on lines (Fig. 4.4).

The vertex-ID(s) can be selected by different methods. One method is to use key-vertex-ID(s). An example of key-vertex-IDs of the homologous mesh of the Move4D scanner by IBV can be seen in Fig. 2.7. These key-vertex-ID(s) are generally known on which body area they are. Using this method, the manual work is neglectable.

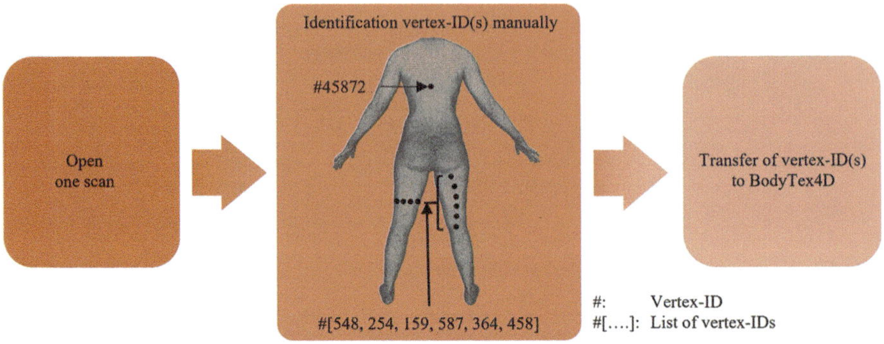

Fig. 4.4 BodyTex4D: choosing filtering conditions by using the identification method by homologous mesh data

If a different vertex should be filtered than the key-vertex-ID, a vertex-ID can be selected manual per scan series or per dynamic movement scan. The first frame of a scan serie or a dynamic movement is opened for example, by Meshlab, pyMeshlab, or the plotter of PyVista and an individual vertex-ID (list) can be picked. Selecting the vertex-ID(s) per dynamic movement scan is more accurate as the manual control is higher. Whether the selection process is performed by key-vertex-IDs manual selection depends on the use case and body area. The single vertex-ID or a vertex-ID list is transferred to the next step, the filtering process.

4.5.2 *Identification Method by Texture*

Using the identification method by texture, the captured scans are filtered by colour values. This method is only applicable if texture data is available. This method is particularly suitable for scan data without a homogeneous mesh.

The colour for filtering the scan data can be pre-set or selected manually. In Fig. 4.5, a manual selection process by colour is shown. Using the open-source library PolyScope (Sharp et al. 2019), a scanned object is visualised, and a filter colour can be selected manually. The colour values are displayed in red (R)—green (G)—blue (B) RGB values and are transferred to the scan data analysis script for the next step. This RGB value is the determined filter condition.

4.5.3 *Identification Method by Mesh and Texture*

Using the mesh and texture identification method, a vertex-ID or list of vertex-IDs is picked based on a texture marker (Fig. 4.6). These vertex-ID(s) perform the

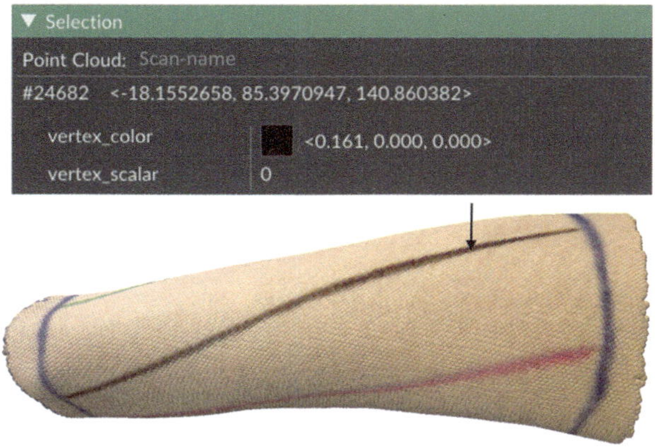

Fig. 4.5 BodyTex4D: choosing filtering conditions by using the identification method by texture data

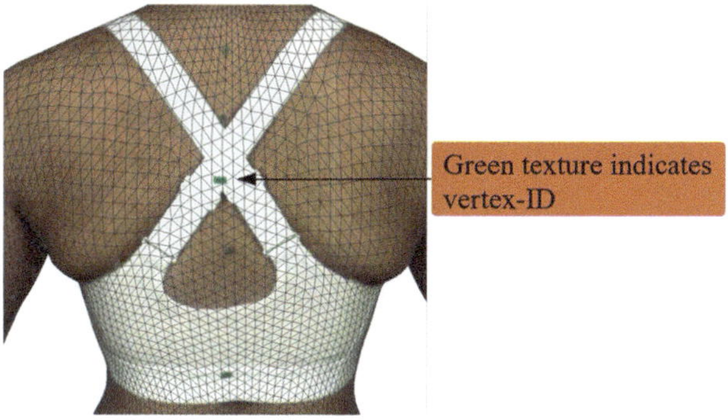

Fig. 4.6 BodyTex4D: choosing filtering conditions by using the identification method by mesh and texture data

filtering process using the method of filtering scan data by vertex-ID(s) (Sect. 4.6.1). A homologous mesh and texture data are necessary. This identification method is mainly used when markers are required to identify the area to investigate. Markers for tracking certain body parts are applied in cases where a textile layer covers reference body markers like the belly button or bust point. The advantage is the high accuracy in the selection process.

4.5.4 Identification Method by Other

There are many more methods to identify filter conditions. One method is by height and cross-sections. A certain height accompanied by a certain margin can be selected for one scan, and all vertex-IDs within this cross-section are selected for the filtering process. Figure 4.7 shows an example of this method. This method by cross-section has also been implemented in the BodyTex4D framework. Other methods could also be single-shape detection or looking for a colour in a specific region. Further methods

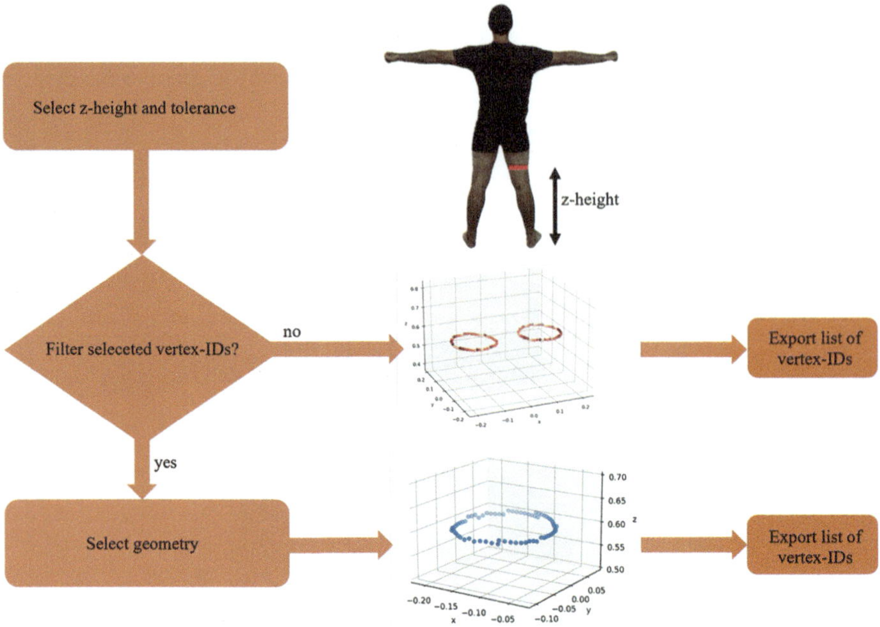

Fig. 4.7 BodyTex4D: choosing filtering conditions by using the identification method by height

can be implemented to extend BodyTex4D. Compatibility is preserved by ensuring the output data format to be the same colour values or vertex-ID(s) layout.

4.6 Filtering Scan Data by the Filter Conditions

In the following, the filter by vertex-ID(s) (Sect. 4.6.1) and by colour values (Sect. 4.6.2) is presented. The application fields of the filtering methods and their advantages and disadvantages are discussed and compared. A combination of both filters is possible.

4.6.1 Filtering Scan Data by Vertex-ID(s)

The filtering process by vertex-ID(s) require as input the `vertex_coordinates` of Sect. 4.4 and the filter conditions of Sect. 4.5. Figure 4.8 shows an overview of the filtering process.

The structure of the `vertex_coordinates` is such that the row number corresponds to the vertex-ID. To extract the coordinate of a single vertex-ID, defined in

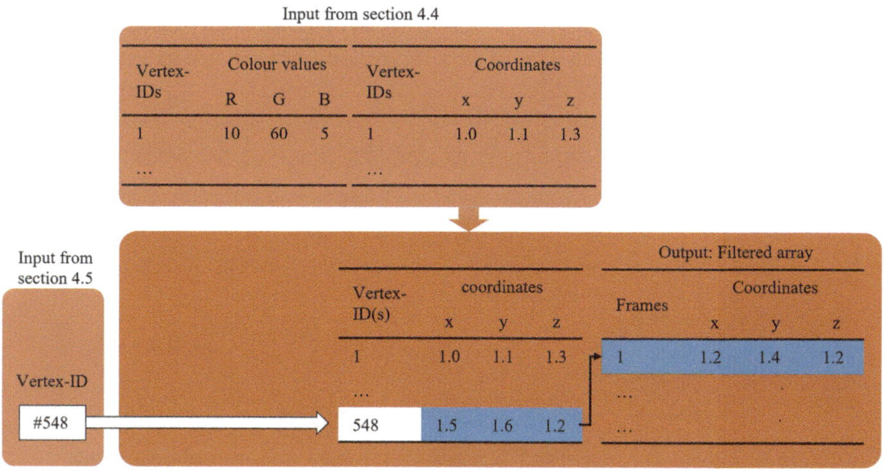

Fig. 4.8 Filtering scan data by selected vertex-ID(s)

Sect. 4.5, the coordinates in the line equal the vertex-ID are filtered and saved. This results in the code displayed in Listing 4.3.

Listing 4.3 Getting the coordinates of the filter-vertex-ID of one scan (homologous-mesh-utils.py)

```
import numpy as np

def get_vertex_coordinates_by_VertexID_from_frame(
    vertex_id: int, frame: np.ndarray) -> np.ndarray:
     return frame[vertex_id]
```

This filtering process is repeated for all desired frames of a movement. If a whole scan series with several dynamic movements of a 4D scan should be filtered, a for-loop is looping through each frame (Listing 4.4). The output of this process is an array of coordinates of the filtered vertex-IDs per dynamic movement.

Listing 4.4 Filtering several frames by one vertex-ID of a scan series (homologous-mesh-utils.py)

```
import numpy as np

def get_vertex_coordinates_by_VertexID_for_several_frames(
    frame_paths: list[str], vertex_id: int] -> np.ndarray:

    frames = [get_homologous_vertices_from_obj(frame_path) for
    frame_path in frame_paths]

    filtered_vertex_coordinates = [frame[vertex_id] for frame
    in frames]

    return np.asarray(filtered_vertex_coordinates)
```

If a vertex-ID list is the filter condition, the same filtering process is started. Only the output array differs. The coordinates of several vertex-IDs are filtered, and these coordinates are saved in one array. The result is one coordinates array per frame (Listing 4.5).

Listing 4.5 Filtering several frames by a list of vertex-IDs of a scan series (homologous-mesh-utils.py)

```
import numpy as np

def get_vertex_coordinates_by_VertexIDs_for_several_frames(dir_path: str, frames_of_interest: list[str], vertex_ids: list[int]) -> np.ndarray:
    frame_paths = [dir_path + '/' + name + ".obj" for name in frames_of_interest]

    frames = [get_homologous_vertices_from_obj(frame_path) for frame_path in frame_paths]

    vertices_coordinates = [frame[vertex_ids] for frame in frames]

    return np.asarray(vertices_coordinates)
```

The requirement to use this filtering method is to have a homologous mesh over several frames. The advantage of this filtering method is that the process can track movements and deformation easily. Since the filtering is based on the homologous mesh data, any vertex-ID can also be chosen and analysed after scanning. This method has a high level of flexibility. Additionally, it is not very error-prone during the filtering process as each vertex-ID has defined spatial coordinates. A disadvantage of this method is that a homologous mesh is required. Due to the refitting process of the homologous mesh in each frame, the vertex-ID to filter may be shifted a bit to its original place. An inaccuracy of the position can occur.

4.6.2 Filtering Scan Data by Colour Values

The filtering process by colour values is similar to the filtering process by vertex-ID(s). The `texture_array` is required in addition to the vertex coordinates imported by Sect. 4.4. Figure 4.9 shows an overview of the filtering process.

Additionally to the selected colour value, a margin is required. The margin is needed as the colour of a fixed area on the scanned object can differ over many frames due to lighting and other distortions during the scan process. The `texture_array` is filtered by a simple Euclidean distance approach, which proofed to be sufficient (Eq. 4.1). The `texture_array` is looped over each colour value and is evaluated for being within the set margin. A more accurate approach for calculating the colour difference is by calculating the colour distance ΔE after converting the RGB-values into CIELab values. Information about the CIELab colour room and the colour distance ΔE can be found in Gomez-Polo et al. (2016). A Python implementation can be found in Reddy (2024).

$$\Delta colour_{Euclidean} = \sqrt{(R_{target} - R_{actual})^2 + (G_{target} - G_{actual})^2 + (B_{target} - B_{actual})^2} \tag{4.1}$$

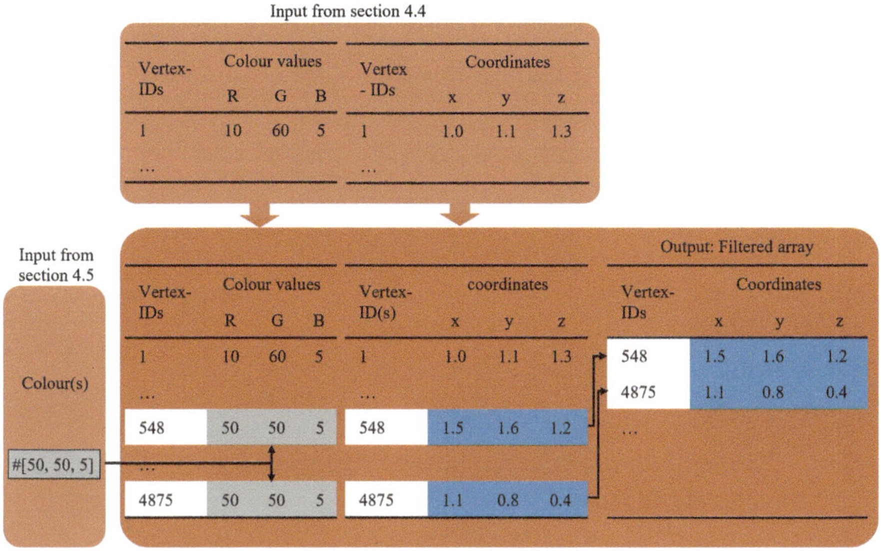

Fig. 4.9 Filtering scan data by selected colour values

Preliminary tests show sufficient accurate results for a colour margin of 0.15. If the colour margin seems to be too small in the filtering process, it can be increased. A higher margin is required when the colour distribution of the area to filter is less homogeneous.

After selecting a suitable margin, all vertex-IDs within this margin and corresponding coordinates are saved in a new array (Fig. 4.9). A new array is generated for each frame if several frames are analysed. The output is the same as, when filtering by a vertex-ID list.

To apply this identification method, scan data with texture is required. During the scan process, the texture of the filtering area must be as homogeneous as possible. Light distortions or a scanned object not adequately illuminated can lead to strongly diverging colour values of the same area over time. To get precise results, colour markers on the region of interest (ROI) should be done using striking colours. The colour of the ROI should be chosen in the analysis area to distinguish it from the rest of the object. Selecting a different analysis area based on texture data after scanning is impossible by colour, if it has not been marked before. The element size of the mesh has to be smaller than the coloured object to be filtered. In Fig. 4.10, a comparison between two mesh element sizes can be seen. The red line in the figure is the object to be filtered. In Fig. 4.10b, at least three red coloured elements are within this red line, which is an appropriate number for the filtering process of the red colour. In Fig. 4.10c, the red line is placed on a third of a mesh element. This mesh in Fig. 4.10c is too coarse to filter the red line during filtering. If a bigger coloured area, such as a t-shirt, would be chosen to be filtered, the mesh size from Fig. 4.10c is dense enough.

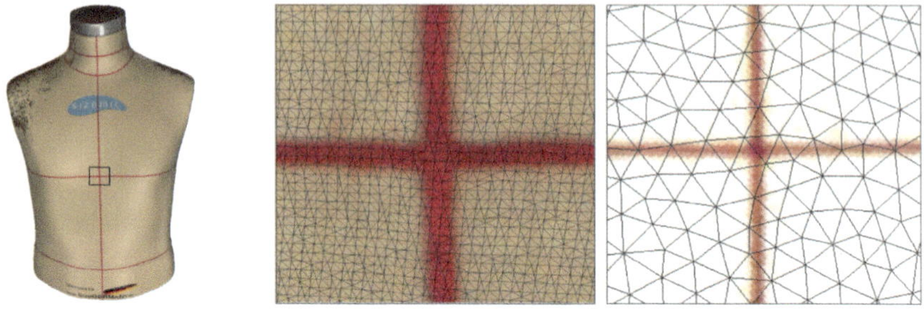

(a) Red lines on a Manikin, which should be filtered by colour **(b)** Mesh element small enough to filter the red line **(c)** Mesh element too coarse to filter the red line

Fig. 4.10 Mesh density of two scanner with the purpose to filter the red line

4.6.3 Comparison and Discussion of Filtering by Vertex-ID(s) and Colour Value

The methods to filter by vertex-IDs and/or colour value is compared in the following section. The comparison focuses on usability in clothing development. The advantages and disadvantages of filtering by colour value and filtering by vertex-ID(s) are discussed. The requirements for filtering by colour or vertex-ID(s) differ. Texture data is required to filter by colour. Homologous meshes are required to filter by mesh data.

The following evaluation criteria are defined and analysed to provide guidance on selecting appropriate filtering methods for various use cases:

- **Flexibility to change analysis area after the scan process**: Filtering by vertex-ID(s) enables selection of the region of interest after the scan process due to the homologous mesh property of the 4D scan data. Therefore, flexibility of analysis area selection is high. Filtering by colour requires the areas to be defined before the scan process. Therefore, flexibility of analysis area selection is low.
- **Dependency on illumination and colouring of clothing**: Illumination is tightly linked to the quality of texture data. If high-quality texture data is required, high levels of illumination is required. For example, shadows cause colour gradients artificially. Reflective material properties are unsuitable for filtering by texture. An example of the challenge of filtering texture of a reflective blue band is shown in Fig. 4.11. Regions turned white due to the reflecting properties of the band (Fig. 4.11a), subsequently, white regions are omitted by the filter process (Fig. 4.11b). The colour value-based filtering method requires high contrast between marker and the rest scan object. Similar effects can be seen in Matthys et al. (2023) in the author's published paper. If the texture of interest is the colour of the clothing, more planning of the scan process is required. The process is prone to error. It is advised against introducing dependency on colours for clothing development as the colour of the clothing to be tested may not be changeable.

(a) 3D scanned blue strip before the filtering process **(b)** Blue strip filtered with missing areas

Fig. 4.11 Colour filtering of a bad illuminated, reflective blue strip

Table 4.1 Evaluation of filtering methods (Orange colour indicates high usability in clothing development)

Evaluation criteria	Filtering by colour(s)	Filtering by vertex-ID(s)
Flexibility to change analysis area after scan process	low	high
Dependency on illumination	high	low
Dependency on wearers clothing colour	high	low
Accuracy in filtering process	medium	high
Accuracy in filtering results	high	medium

- **Accuracy of filtering process**: Filtering by vertex-IDs is highly accurate as every vertex-ID has a defined position on the scanned body due to the homologous mesh property. Filtering by colour is less accurate as the filtering process is highly dependent on the quality of texture data.
- **Accuracy of filtering results**: Filtering by colour results in a higher accuracy compared to filtering by vertex-IDs. Filtering by colour is based on markers. The marked areas filtered represents the same area on the body in each frame by design. Filtering by vertex-ID(s) requires homologous meshes. Homologous mesh is refitted for each frame based on key-vertex-IDs causing non-key-vertex-IDs to vary relative to the body. The deviations are depended on the fitting algorithm, are minor, but should be kept in mind.

In conclusion, filtering by vertex-ID(s) is less prone to errors by test setup. Filtering by vertex-ID(s) is highly usable for clothing development due to texture independence and flexibility in the analysis area after the scan. Colour filtering is more accurate, if high quality texture data is obtainable. When human body scan data without homologous mesh should be analysed by methods based on vertex-IDs, they can be fitted with an open-source homologous mesh, for example, the open-source tool Skinned Multi-Person Linear model (SMPL) (Loper et al. 2015). The SMPL generates homologous meshes in a post-processing step. It is specialised to identify human body poses in scan data. A summary of the comparison of the filtering methods by vertex ID(s) and colour values can be seen in Table 4.1. The evaluation criteria that are more beneficial for clothing development are marked in orange.

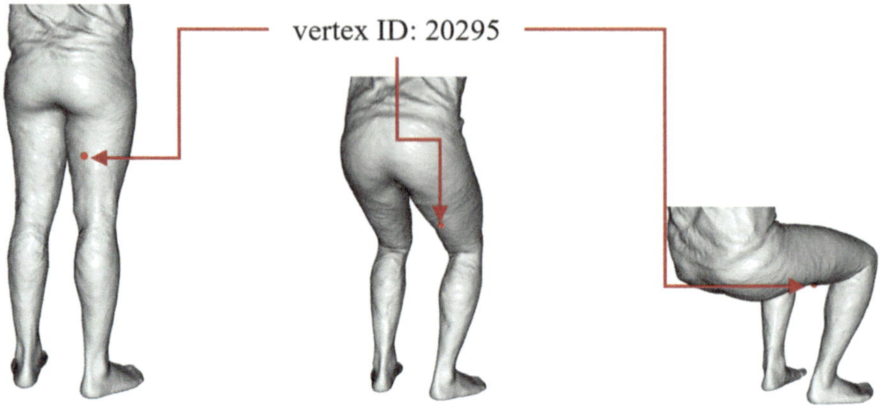

Fig. 4.12 Scan data analysis method: movement analysis

4.7 Scan Data Analysis Methods

Different scan data analysis methods can be applied after the filtering step. Generalised analysis methods are developed and implemented into BodyTex4D: "Movement analysis by single point tracking", "Length Analysis" and "Shape Analysis". Selected analysis methods focus on body movement, body deformation and clothing-body interactions.

4.7.1 Movement Analysis by Single Point Tracking

Tracking the movement of a single point captures the x-, y- and z-coordinates of a single vertex over the span of a dynamic movement. This method is based on filtering by single vertex-ID (Sect. 4.6.1). A vertex ID to track is identified, and its coordinates are filtered and saved in a new array. A tracked vertex while a test subject is sitting down is visualised in Fig. 4.12. All positional components (x, y, z) or a custom set of components can be analysed and visualised depending on the use case. Tracking single points on the body is useful to understand body movements, compare body movements while wearing a clothing item or see which influences clothing has on the movement of a body part of interest. Inaccuracies must be kept in mind during the analysis, as discussed in Sect. 4.6.3. Tracking single points by homologous mesh is less complex compared to tracking by acceleration sensors. Acceleration sensors must be placed at the point of interest before the scan. The number of measurement points is limited to the amount of available sensors, which is generally less than vertices of a mesh. Movement analysis based on 4D scan data with homologous mesh property can be performed on any vertex.

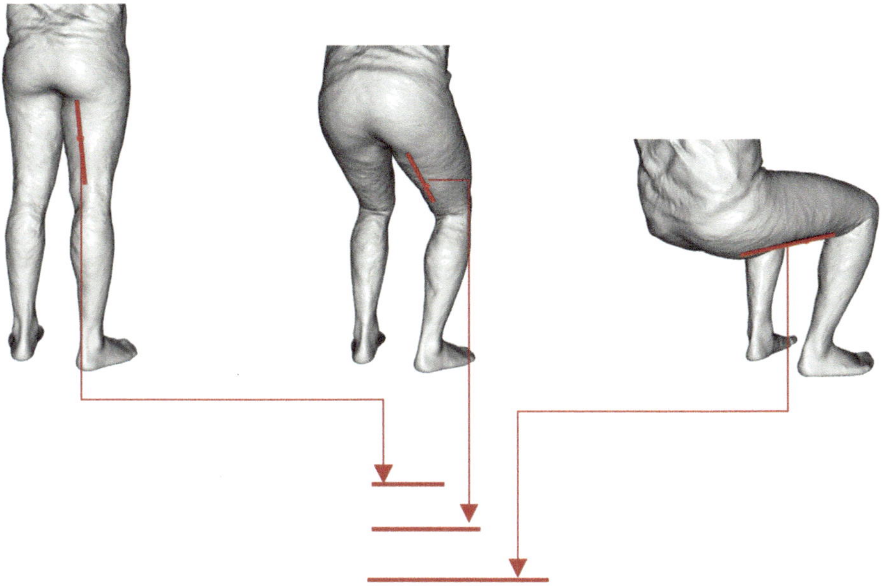

Fig. 4.13 Scan data analysis method: length analysis

4.7.2 Length Analysis

The length analysis aims to determine length changes over the course of a movement (Fig. 4.13). The skin stretches due to deforming body parts and due to muscle activation over the course of a movement. Knowledge of the magnitude of change in length is beneficial for clothing development to design cut patterns based on movements and the dynamic changes of the body or to choose adequate materials.

The length analysis requires adaption on a case-by-case basis. Several methods to analyse length are explored and implemented in BodyTex4D. Only the methods used within this work are described in more detail. In general, two categories distinct length measuring cases are determined: open line length measurement and closed (looped) line length measurement. The open line is defined by unequal coordinates of start and end points. The closed or looped line is defined by equal coordinates of start and end points. Applications of analysis by length measurement of an open line include, measuring the length of the upper leg or the elongation of the skin on parts of the finger when grabbing an object. Applications for measurements of closed lines are, a leg or underbust circumference. Possible methods to measure the line length depending on the application are listed below. The methods are described in detail after the lists. The method description includes fields of application as well. Implementations follow in Sect. 4.8 of selected methods

Fig. 4.14 Polyline (orange line) and optimised fitting with a smoothed spline/curve (blue line)

The main length analysis methods determined and explored are:

- calculating the length of an open line
  - by two vertex-IDs and a straight line
  - by two vertex-IDs and a surface line
  - by a sorted point cloud and a polyline
  - by a point cloud and a polynomial algorithm
  - by a point cloud and a spline algorithm
- calculating the length of a closed line (circumference)
  - by a single vertex-ID and a normal
  - by a point cloud and a polyline
  - by a point cloud and a projection and a polyline
  - by a point cloud and a shape approximation

The high number of methods shows that it depends on the use case, which length analysis fits best.

Calculation the length of a straight line between two vertex-IDs is straight forward by Euclidean distance. Use cases are calculating, for example, shoulder width or height from head to toe. Two vertex-IDs are chosen as input parameters. All frames of a dynamic movement are filtered by chosen vertex-IDs. The distance is calculated frame-wise by the Euclidean equation Eq. 4.2.

$$distance = \sqrt{(x_1 - x_2)^2 + (y_1 - y_2)^2 + (z_1 - z_2)^2} \tag{4.2}$$

Calculating the length of a surface line by two vertex-IDs is based on the shortest distance between these two vertex-IDs. For example, the approach can be used to calculate the change in length of the back during movements. Two vertex-IDs are the input parameter, and the shortest path connecting the vertices by intermediate vertices is calculated based on the Dykstra algorithm and Euclidean distance of each path segment (Helmberg 2020). The resulting line is a so-called polyline. Generally, polylines have the disadvantage of adding extra length due to the deviations from the shortest distance. An example of a polyline is presented in Fig. 4.14 in orange. An optimisation is to smooth the polyline or use other fitting methods such as curve or spline fittings (Fig. 4.14 blue line).

Calculating line length by combining sorted point cloud data with polyline. The method can be used when filtering by a vertex-ID list or by colour. The polyline connects the vertex-IDs frame-wise.

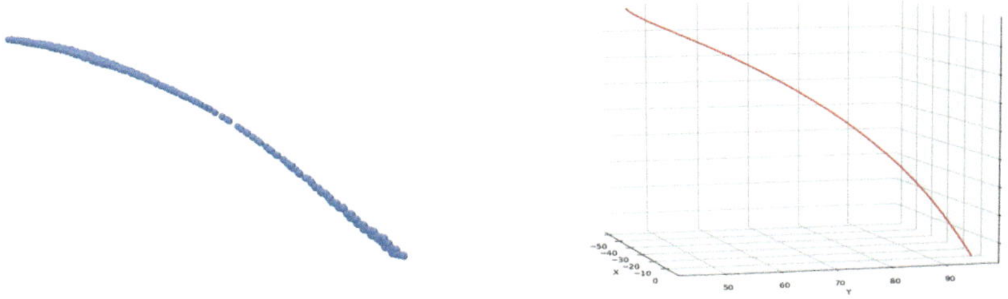

(a) Filtered pointcloud **(b)** Fitted curve of the point cloud in (a) with a polynomial function

Fig. 4.15 Implemented curve measuring algorithms: polynomial fitting

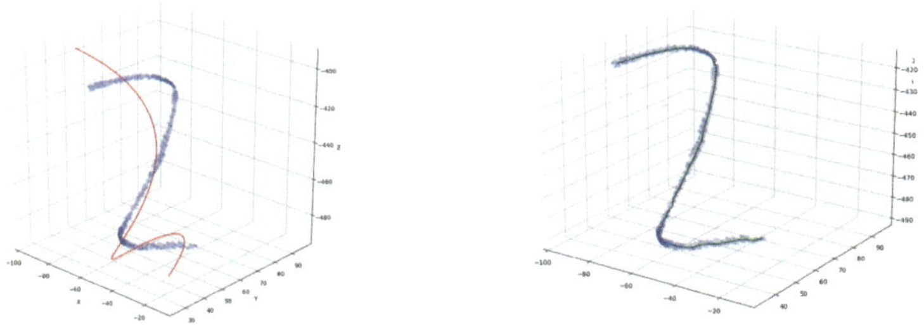

(a) Difficulties fitting a complex point cloud with the polynomial fitting approach **(b)** Fitting a complex point cloud with the L1-medial approach

Fig. 4.16 Fitting a complex-shaped point cloud with the **a** polynomial fitting approach and the **b** spline fitting approach

Another approach is to use polynomial or spline algorithms. Polynomial or spline algorithms increase the automation of line length measurement, especially, if the scan data does not have homologous mesh properties.

Polynomial-based length calculation is defined by the order of the polynomial equation. The constants of the polynomial are commonly fit until a deviation threshold or the max number of iterations is reached. This approach is fast and accurate for simple line shapes (Fig. 4.15).

Inaccuracies at the line edges when using polynomials are observed. Furthermore, complexity increases and stability of the fitting process decreases for 3D lines. Figure 4.16 displays a point cloud with a complex shape. Fitting a polynomial showed insufficient results (Fig. 4.16a).

The **L1-medial-skeleton based polyline approach** calculates central points geometrically of an object and connects the central points by a polyline (Huang et al. 2013). The method is implemented into the scan data analysis framework. Figure 4.17 shows an example of the L1-medial-skeleton-spline algorithm. Figure 4.17b shows the centre point line and Fig. 4.17c measures the curve along the centre points. Further algorithms to find central points or the skeleton of an object are found in Tagliasacchi et al. (2025) and Mei et al. (2017).

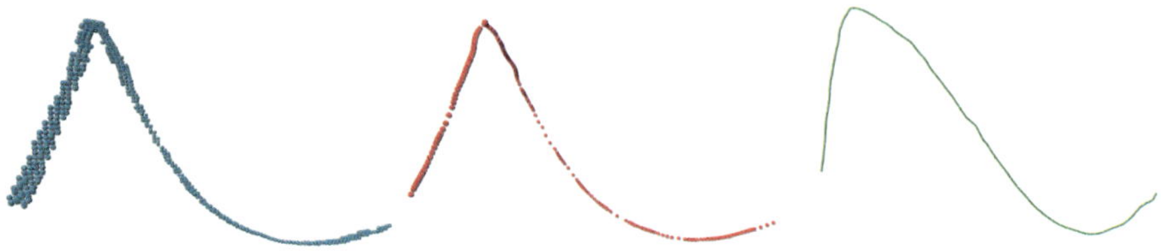

(a) Filtered point cloud **(b)** Centred point cloud **(c)** Curve (L1 medial skeleton algorithm)

Fig. 4.17 Implemented spline fitting using the L1 medial skeleton algorithm **a** filtered point cloud, **b** centred points placed by the algorithm, **c** resulting spline

Calculating a circumference using a closed polyline. In the following, the approach of using a polyline is displayed. Depending on the input data, various methods can be applied.

If the input data is a single vertex-ID, a corresponding normal vector is required. The vertex and the normal vector define the slicing plane. Slicing the mesh by the plane creates a cross-section represented in points. The points are connected by a polyline. The length of the polyline is calculated segment-wise by Euclidean distance. The calculated lengths are accumulated resulting the circumference of the cross-section. The method is fully automated. A more detailed description of the implementation of the is presented in Sect. 4.8.2.

If the input data is a slice of a point cloud, the points are sorted by nearest neighbour approach and connected to a closed polyline (Fig. 4.18a). The sorting algorithm is explained in detail in Sect. 4.8.2. The method is more inaccurate because the points are dispersed around a plane, which adds an extra length (Fig. 4.18a). To reduce the inaccuracy, the points can be projected onto a plane.

Another approach is to fit a geometric shape into the point cloud. For example, elliptical geometries (Fig. 4.18b). The shapes have to be predefined, making it difficult to describe complex closed curves.

Another approach to measure the length is to split the closed curve into open curves and define boundary conditions for the end points. Subsequently, open curve algorithms are applicable. However, this method has not been investigated within this work.

Overall, the framework implements many different length-measuring methods interchangeably. The method to choose depends on the application and the present input data. For circumference measurements, the method by a single vertex and normal vector is determined to be fast, accurate, consequently, highly useful in analysing body deformation.

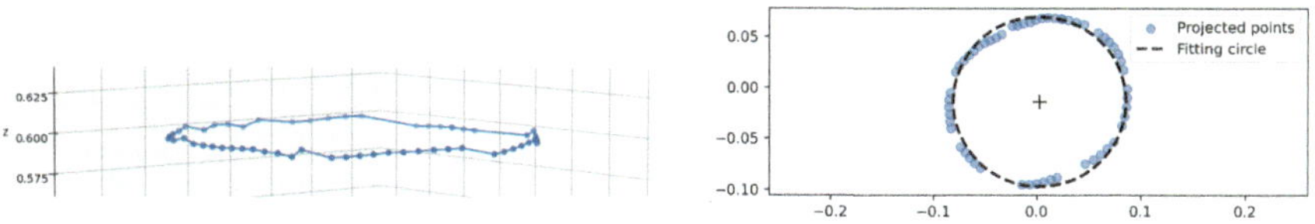

(a) Polyline **(b)** Geometric shape approximation

Fig. 4.18 Methods to measure the circumference of a closed point cloud

4.7.3 Shape Analysis

Shape analysis is used to investigate body deformations caused by movement, muscle activation or the deformable body-clothing interaction. Body measurements like distance or circumferences with the same values can differ by shape. Analysing and understanding of the body dynamics is required for clothing development.

Input parameters of the shape analysis are an array of filtered vertex-IDs, single vertex-ID per scan or an array produced by filtering point clouds by colour.

Point clouds, vertex list or arrays in general can be easily visualised by using the Python libraries pyVista or Matplotlib (Fig. 4.19). Using a single vertex-ID, a normal vector has to be defined to create a slice. Multiple approaches to define a normal vector for slicing by single vertex-ID exist. The vertex defines the support vector of the slicing plane. The normal can be a global axis, constant for all analysed scans. Another approach is based on two vertex-IDs of a homologous mesh. For example, the cross-section of a leg is investigated for deformation during sitting down (Fig. 4.19). Two vertices on the leg's surface approximate the longitudinal direction of the femur are chosen to calculate the normal vector. One of the two vertices are chosen as support vector of the slicing plane. Subsequently, the normal vector and the support vector of the slicing plane are bound to the pose of the upper leg. Consequently, the same scanned leg in every frame is sliced in the same manner when a homologous mesh is used. Consequently, every scanned leg in every frame is sliced similarly when a homologous mesh is used. Figure 4.19 shows the leg deformation due to muscle activation while sitting down. The method is implemented in Python (Listing 4.6)

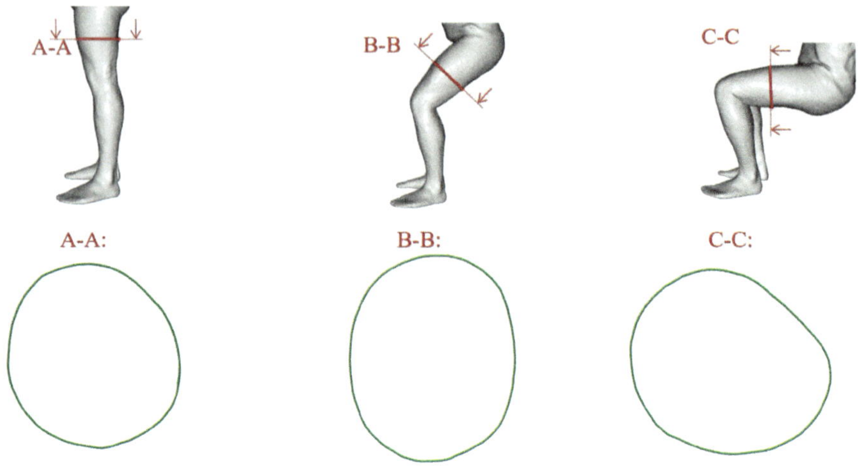

Fig. 4.19 Scan data analysis method: shape analysis

Listing 4.6 Shape analysis - track cross-section during motion via two vertex-IDs (cross-section.py)

```python
import numpy as np
import pyvista as pv

def calculate_normal_vector(
    homologous_mesh: np.ndarray,
    normal_origin_vertex_id: int,
    normal_target_vertex_id: int) -> tuple[float, float, float]:

    origin = homologous_mesh[normal_origin_vertex_id]
    target = homologous_mesh[normal_target_vertex_id]

    normal_vector = target - origin

    return normal_vector

def get_cross_section(
    mesh: np.ndarray,
    slice_plane_normal_origin: tuple[float, float, float],
    slice_plane_normal_target: tuple[float, float, float],
) -> np.ndarray:
    # calculate the cross section of the mesh is using pyvista's slice method
    pv_mesh = pv.PolyData()
    pv_mesh.point_data["homologous mesh"] = mesh
```

```python

    cross_section = pv_mesh.slice(
        normal=slice_plane_normal_target, origin=slice_plane_normal_origin)
    return cross_section.points

def get_cross_section_by_vertex_ids(
    homologous_mesh: np.ndarray,
    origin_vertex_id: int,
    target_vertex_id: int,
) -> np.ndarray:

    origin = homologous_mesh[origin_vertex_id]
    normal = calculate_normal_vector(
        homologous_mesh, origin_vertex_id, target_vertex_id)

    cross_section = get_cross_section(homologous_mesh, origin, normal)
    return cross_section

def track_cross_section_during_motion(
    homologous_meshes: list[np.ndarray],
    origin_vertex_id: int,
    target_vertex_id: int,
) -> list[np.ndarray]:
    cross_sections: list[np.ndarray] = []

    for mesh in homologous_meshes:
        cross_section = get_cross_section_by_vertex_ids(
            mesh, origin_vertex_id, target_vertex_id
        )

        cross_sections.append(cross_section)

    return cross_sections
```

4.8 Advanced Analysis Methods of BodyTex4D for Clothing-Body Interactions

Clothing-body interaction analyses methods are developed based on the methods described in Sect. 4.7. A method to classify bras by support through movement analysis by enabling comparability of motions is developed in Sect. 4.8.1. Underbust circumference calculation-based closed curve length calculation is derived, enabling subtraction of textile thickness influencing the chest circumference measurements in Sect. 4.8.2.

4.8.1 *Advanced Movement Analysis: Amplification Factor Analysis for Breast-Bra Interactions in Motion*

The amplification factor analysis provides comparability of motions. The amplification factor analysis is based on Fast Fourier Transformation (FFT). Target of the amplification factor analysis is to enable classification of the support property of bras, restricting breast movement during intense exercises. The main functionality of bras is to restrict breast movement. Lorentzen and Lawson (1987) showed that vertical, relative movement between breast and body causes pain and discomfort. Restricting breast movement is especially useful and necessary for exercising women with larger breasts. Most sources in the literature use treadmills or metronomes to achieve comparable movements. Preliminary investigations of this work show inconsistent amplitudes and frequencies in movements recorded using a treadmill or a metronome. Comparability is not given as long as the body movement is not considered during the analysis.

In the amplification analysis, breast and body movement are measured to generate data for the bra support classification by amplification factor analysis. The first step of the amplification factor analysis is to extract the body and breast motion curves from experimental 4D scan data. Therefore, a vertex-ID representing the body movement and one vertex-ID representing the breast movement are selected (Fig 4.20a).

Movement analysis by single point tracking described in Sect. 4.7.1 is performed. The outputs are two arrays with the coordinates of the vertex-IDs over all captured 4D scans, tacking the motion. The vertical motion of breasts and body are analysed. The FFT is a fast form of the discrete FT with relatively low complexity of $O(N \cdot logN)$, which can be derived from the Fourier Series. The FT decomposes time-based signals into a representation, where signal intensity is a function of frequency. The x-axis represents frequency values and the y-axis represents intensity of given frequency

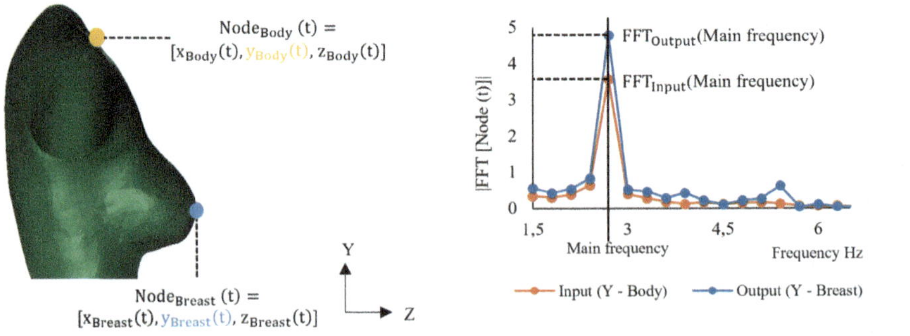

(a) Vertices / nodes to perform the movement analyses **(b)** Overview Fast Fourier Transformation and main frequency

Fig. 4.20 Overview of relevant method parameter for using automated amplification factor analysis (figures made by author, published in Schmidt et al. 2023)

(amplitude level) (Fig. 4.20b) (Follinger 2021; Waller and Schmidt 1989). The FFT is performed using the open-source Python package NumPy. Used functions are displayed in Listing 4.7. The frequency is calculated based on the number of frames captured per second (fps) during the 4D scan.

Listing 4.7 Python script for an advanced movement analysis: FFT analysis of motion curves

```
import numpy as np

fft_result = np.fft.fft(
    displacement_array_of_vertexID)
fft_result_abs = np.abs(fft_result)
fft_frequency = np.fft.fftfreq(number_frames, d=1.0
    / fps)
```

An example result of FFT applied to motion curves is presented in Fig. 4.20b. One frequency produces a significantly higher amplitude compared to all other measured frequencies. The frequency has higher amplitudes, subsequently, is more intense in the movement. The frequency with the highest amplitude is called main frequency (Fig. 4.20b). The amplification factor is calculated based on the main frequencies of breast and body motion. The amplification factor is defined as the quotient of FFT values of the main frequencies of breast (output) and body (input) (Eq. 4.3).

$$amplification\ factor = \frac{FFT_{Output\_main\_frequency}}{FFT_{Input\_main\_frequency}} \tag{4.3}$$

The definition assumes a linear amplification. Motions of test subjects are made comparable by the amplification factor, which is unprecedented in the literature. The amplification factor describes the breast movement in relation to the body movement. Low amplification factors indicate small relative breast motions while high amplification factors indicates higher relative breast motions. Bra functionalities for test subjects running on treadmills at different speeds with various bras are made comparable. How the body movement changes and whether the same main frequencies dominate at the same treadmill speed can be analysed. Bras can be classified based on their support by using the amplification factor as metric. The method is applied in Sect. 6.5.3. Additionally, this method is published by the author in the paper (Schmidt et al. 2023).

4.8.2 Advanced Length Analysis: Circumference Calculation and Textile Thickness Subtraction

The following algorithms extend the circumference calculation of body parts by subtracting the textile thickness on the basis of methods introduces in Sect. 4.7.2. The target is to retrieve accurate circumference measurements of the body to compare different body states under the influence of different clothing items. Influences of clothing items on body measurements must be factored out. Specifically, it is important to subtract the textile thickness to decouple body circumference values from the clothing geometries. Decoupling achieves comparability between different clothing items interacting with the body, for example, to investigate body deformations.

This method is based on a circumference calculation by a single vertex-ID and a normal vector aligned with a global axis. The cross-section is calculated based on PyVista's method `mesh.slice()` by the function `get_cross_section` in Listing 4.6. In Listing 4.8, the normal vector is parallel to the global y-axis is chosen as the test subject is standing upright, cutting through the chest horizontally. The coordinates of the single vertex-ID define the origin, height relative to body. The coordinates of the vertex-ID are extracted by the method described in Sect. 4.6.1. The output is an unsorted NumPy array of Cartesian coordinates (Listing 4.8).

Listing 4.8 Advanced length analysis: Cross-section points relative to global axis

```
from cross-section import get_cross_section
from reader import get_homologous_vertices_from_obj

if __name__ == '__main__':
    # Import mesh from obj
    obj_file_path = "/path/to/file"
    homologous_mesh =
   get_homologous_vertices_from_obj(obj_file_path)

    # Create a cross-section of the mesh along
   slicing plane
    global_y_axis = (0, 1, 0)
    vertex_ID = <see text>
    height = homologous_mesh[vertex_ID]

    cross_section = get_cross_section(
   homologous_mesh, height, global_y_axis)
```

Next, the circumference of the cross-section is calculated and the textile thickness is subtracted. The function `subtract_textile_thickness_from_cross_section` is used. The implementation is presented in Listing 4.9. The cross-section points array from Listing 4.8 and the textile thickness are the input parameters. The cross-section is approximated as ellipses. Preliminary tests indicate that the textile thickness can be approximated by reducing the ellipse's axis by the textile's thickness. Subsequently, width and height of the ellipses is required. Width and height of the ellipses is calculated by fitting a plane to the cross-section and rotating the cross-section to align its normal with the z-axis. Minimal and maximal x- and y-values are calculated, width and height is derived. `PolyData.scale()` from PyVista is used to move the cross-sections points evenly. The detailed algorithm is presented in Listing 4.9 by the function `subtract_textile_thickness_from_cross_section()`. In Fig. 4.21, a cross-section reduced by a thickness of 2 mm can be seen.

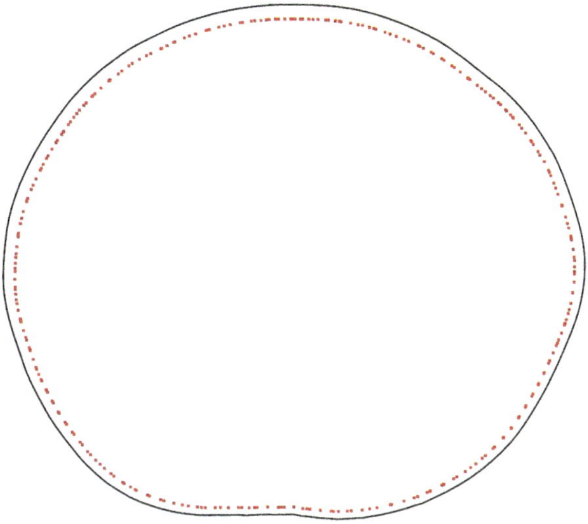

Fig. 4.21 Textile thickness subtraction using the script in Listing 4.9. The black line represents the original cross-section, while the red line depicts the modified cross-section, shifted 2 mm inward

Listing 4.9 Preparing cross-section point cloud for circumference calculation (cross-section.py)

```
import pyvista as pv
import numpy as np
import scipy.spatial.transform as transform

def subtract_textile_thickness_from_cross_section(cross_section: np.ndarray, textile_thickness_m: float) -> np.ndarray:

    # Find best fitting xy-plane of cross section to calculate the normal vector
    centroid = np.mean(cross_section, axis=0)
    _, _, Vt = np.linalg.svd(cross_section - centroid)
    normal = Vt[2, :]

    # Rotate the points so that the normal vector is aligned with the z-axis
    rotation_axis = np.cross(normal, [0, 0, 1])
    rotation_angle = np.arccos(
        np.dot(normal, [0, 0, 1]) / np.linalg.norm(normal))
    rotation = transform.Rotation.from_rotvec(rotation_axis * rotation_angle)
    rotated_points = rotation.apply(cross_section - centroid) + centroid

    # Now the width and height can be calculated as the range of x and y coordinates
    min_point, max_point = np.min(
```

```
            rotated_points, axis=0), np.max(rotated_points, axis=0)
    width = max_point[0] - min_point[0]
    height = max_point[1] - min_point[1]

    # Cross section circumference is approximated as ellipses
    scale_width = (width - textile_thickness_m * 2) / width
    scale_height = (height - textile_thickness_m * 2) / height

    # Scale the original points
    pv_cross_section = pv.PolyData(cross_section)
    pv_scaled_cross_section = pv_cross_section.scale(
        [scale_width, 1, scale_height])

    return pv_scaled_cross_section.points
```

The cross-section represents a looped curve. The cross-section points are sorted to introduce a looped polyline for the circumference measurement. Therefore, an algorithm to sort the cross-section points by nearest neighbour is implemented as Python function `sort_points_by_nearest_neighbor` (Listing A.1).

Finally, the cross-section circumference is calculated using the algorithm shown in Listing A.2.

The presented functions are combined into a high-level module for BodyTex4D to calculate the circumference of the underbust while factoring out textile thickness. The input arguments are the path to the *.obj file, the vertex-ID cross-section support vector, the normal of the cross-section and the textile thickness. Listing 4.10 presents the module applied for a random example.

Listing 4.10 Python script of an example main script for the circumference calculation subtracting a textile thickness

```
from homologous-mesh-utils import get_homologous_vertices_from_obj
from circumference import calculate_circumference_of_points
from cross-section import (
    calculate_normal_vector,
    get_cross_section,
    subtract_textile_thickness_from_cross_section,
    )


def calculate_cross_section_circumference_of_obj(
        obj_path: str,
        cross_section_origin_vertex_id: int,
        cross_section_normal: int | tuple[float, float, float],
        textile_thickness_m: float,
) -> float:
```

```python
    homologous_mesh = get_homologous_vertices_from_obj(obj_path)

    slice_plane_origin = tuple(homologous_mesh[cross_section_origin_vertex_id])

    slice_plane_normal: tuple[float, float, float]
    if isinstance(cross_section_normal, tuple):
        slice_plane_normal = cross_section_normal
    else:
        slice_plane_normal = calculate_normal_vector(
            homologous_mesh, cross_section_origin_vertex_id, cross_section_normal)

    cross_section_points = get_cross_section(
        homologous_mesh, slice_plane_origin, slice_plane_normal)

    underbust_cross_section_points = subtract_textile_thickness_from_cross_section(
        cross_section_points, textile_thickness_m)

    return calculate_circumference_of_points(underbust_cross_section_points)

if __name__ == '__main__':
    # random example data
    obj = 'data/3D_scan.obj'
    cross_section_origin_vertex_id = 31586
    cross_section_normal = (0, 1, 0)
    textile_thickness_m = 0.002
    circumference = calculate_cross_section_circumference_of_obj(
        obj, cross_section_origin_vertex_id, cross_section_normal, textile_thickness_m)

    print(f'The circumference of {obj} is {circumference:.2f} m')
```

4.9 Conclusion and Discussion

This chapter introduces the BodyTex4D framework, designed for the automated processing and analysis of 4D scan data. It presents a selection of Python module implementations within BodyTex4D, with an emphasis on methods that are applied in the subsequent chapters. These implementations enhance the automation of analyzing body movements and clothing-body interactions. In addition, the methods contribute to improved comparability of body postures and motions captured using 3D and 4D scanning technologies.

Currently, 4D scan data must be analysed manually. Software of 4D scanner manufactures exist, but is too limited to be applied in research and development. The software is limited by measurements options being pre-set. For example, pre-set locations of length measurements on the human body. Adaptation within the software or extension by APIs (Advanced Programming Interfaces) is not possible. The limitation inhibit data processing automation for specific use cases such as breast-bra interaction analysis. Through the developed 4D scan data analysis framework, the following research gaps addressed in Part I are closed successfully:

1. 4D scan data can be analysed semi-automatically
2. 4D scan data can be analysed independent of scanner software
3. 4D scan data analysis methods are explored, described and in Python
4. Individualised body regions can be analysed, concerning movement, length, and shape changes
5. Extension of the analysis framework by custom methods is easy due to the simplicity, accessibility and level of distribution of Python

The main findings of BodyTex4D are summarised in the following:

1. **The framework allows for use of 3D and 4D scan data**: Presented methods differ depending on the availability of homologous meshes. Currently, 4D scan data is mostly analysed manual or by the scanner software of Move4D with limited feature sets. The novel developed framework improves 4D scan data handling by automation.
2. **BodyTex4D is based on vertex coordinate and vertex colour arrays**: A simple *.obj-file parser for keeping the homologous property is presented.
3. **Two filter conditions to extract data from scan data are presented—filtering by colour values and filtering by vertex-IDs**: Data properties and task requirements govern the identification of the appropriate filter condition. The filter condition identification process considers, for example, homologous mesh property and texture data quality, among others.
4. **Filtering by vertex-IDs is more suitable for the clothing development process**: The process is highly automated. The filtering process is accurate. Regions of interest can be changed after the scanning process. No coloured markers or similar preparation steps have to be taken. The dependency on illumination and the wearer's clothing colour is negligible. The only disadvantage are slight inaccuracies of non-key-vertex positions occur because of the refitting of the homologous mesh for each frame. The vertex may not be on the exact same body surface position. The inaccuracies depend on the fitting algorithm and must be evaluated on a case-by-case basis.
5. **1D, 2D and 3D general analysis methods are determined—single point tracking, change of length and change of shape**: Single point tracking is realised by extracting vertex coordinates of vertex-ID of a dynamic movement. The homologous mesh property ensures that the extracted coordinates describe the point of

interest on the body. The change of length analysis tracks changes in the length due to movements by two or more vertex-IDs analogous to single point tracking. The shape analysis compares body deformations due to muscle activation, soft-tissue distribution or the influence of tightly fitted clothing. All methods are highly usable in clothing development. The methods are all generalised and can easily be modified for specific use cases. Different implementations and application examples are presented for all analysis methods to help future users decide on the methods which fits their use case best.

6. **An advanced movement comparison metric to classify bras by support is developed**: The dependency of body and breast movement is characterised by an amplification factor calculated based on Fast Fourier Transformation (FFT). Main frequencies of the movement curves are determined which represent the intensity of the motion. Higher amplification factors indicate lower bra support. Lower amplification factors indicate higher bra support. Classifications of the bras functionality can be made.

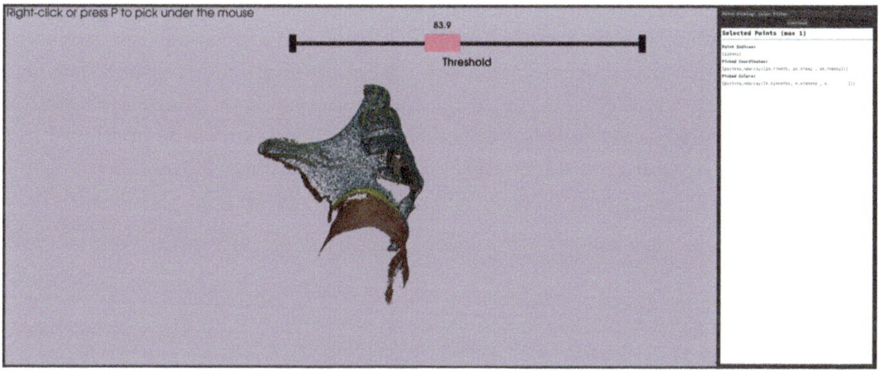

(a) Colour filter with a higher colour tolerance

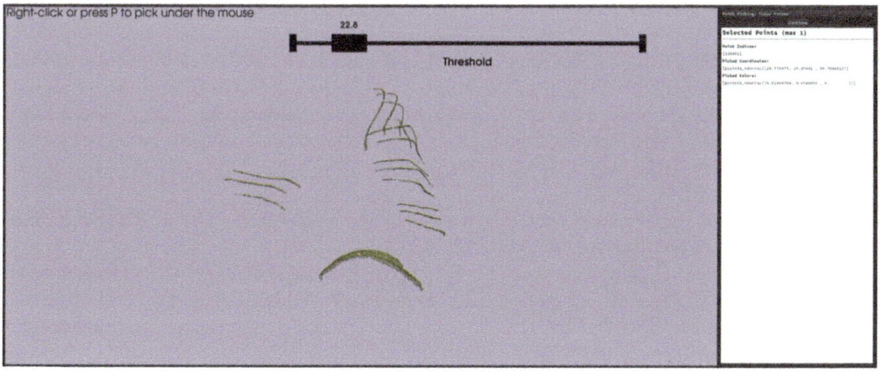

(b) Colour filter with a lower colour tolerance

Fig. 4.22 Graphical user interface featuring a slider for setting the filtering tolerance of the filter colour value

7. **An advanced circumference calculation is presented**: The presented method improves the comparability of circumferences of test subjects wearing clothing with different thicknesses by factoring out the influence of the textile's thickness. Factoring out the influence of the textile's thickness enhances the comparability of different clothing items that perform the same functionality but are made of different materials.

Following research steps should improve BodyTex4D usability. People with little programming knowledge could profit from a graphical user interface (GUI). Figure 4.22 shows the first concepts of implementing a simple GUI. The feature set should be extended in regards to applications that would profit from 4D scan data. Applications that would profit from 4D scan data are all applications analysing motion of objects. The main advantage of using 4D scan data is the homologous mesh property. Homologous meshes enable tracking of points, lines or shapes on an object over the course of a movement by design. Using sensors limits the amount of analysis points and repeatability, because points of measurement must be selected before the experiment. However, acceleration sensors are specially designed to provide accurate movement data. Acceleration sensors stay at the point which should be measured. Sensors' disadvantage is that body deformations cannot be captured. In follow-up research work, parallel measurements by acceleration sensors and homologous meshes should be performed to compare the accuracy of the movement analysis by the 4D scan data analysis framework. Nevertheless, depending on the application, the operator has to decide if higher flexibility in the analysis or a higher accuracy is required.

References

Beyer G, Schmidt A-M, Kyosev Y (2022) Entwicklung von Auswertungsalgorithmen fur die Linienmessung von sich farblich abgrenzenden Bereichen von 3D gescannten Objekten. Dresden

Follinger O (2021) Laplace-, Fourier- und z-Transformation. 11., uberarbeitete Auflage. Berlin: VDE VERLAG GMBH, 2021. ISBN: 9783800753710

Gomez-Polo C et al (2016) Comparison of the CIELab and CIEDE2000 color difference formulas. J Prosthet Dent 115(1):65–70. https://doi.org/10.1016/j.prosdent.2015.07.001

Helmberg G (2020) Optimale Wege-der Dijkstra-Algorithmus. In: Glaeser G (ed) 77-mal Mathematik fur zwischendurch: Unterhaltsame Kuriositaten und unorthodoxe Anwendungen. Springer, Berlin, Heidelberg, 2020, pp 265–267. ISBN: 978-3-662-61766-3. https://doi.org/10.1007/978-3-662-61766-3_70

Huang H et al (2013) L1-medial skeleton of point cloud. ACM Trans Graph 32(4):1–8. ISSN: 0730-0301. https://doi.org/10.1145/2461912.2461913

Kyosev Y, Tomanova V, Schmidt A-M (2022) Method for automatic analysis of the clothing related body dimension changes during motion using high-speed (4D) body scanning. In: D'Apuzzo N (ed) Proceedings of 3DBODY.TECH 2022-13th international conference and exhibition on 3D body scanning and processing technologies, Lugano, Switzerland, 25–26 Oct 2022. Hometrica Consulting-Dr. Nicola D'Apuzzo, Ascona, Switzerland . ISBN: 9783033095205. https://doi.org/10.15221/22.24

Loper M et al (2015) SMPL: a skinned multi-person linear model. ACM Trans Graph 34(6):1–16. ISSN: 0730-0301. https://doi.org/10.1145/2816795.2818013

Lorentzen D, Lawson L (1987) Selected sports bras: a biomechanical analysis of breast motion while jogging. Phys Sports Med 15(5):128–139. ISSN: 0091-3847. https://doi.org/10.1080/00913847.1987.11709355

Matthys D et al (2023) Dynamic scanning of cyclists: techniques and applications. In: D'Apuzzo N (ed) Proceedings of 3DBODY.TECH 2023-14th international conference and exhibition on 3D body scanning and processing technologies, Lugano, Switzerland, 17–18 Oct 2023. Hometrica Consulting-Dr. Nicola D'Apuzzo, Ascona, Switzerland, 2023. ISBN: 9783033101371. https://doi.org/10.15221/23.52

Mei J et al (2017) 3D tree modeling from incomplete point clouds via optimization and L1-MST. Int J Geograph Inf Sci 31(5):999–1021. ISSN: 1365-8816. https://doi.org/10.1080/13658816.2016.1264075

Reddy T (2024) Difference between 2 colours using python. https://dev.to/tejeshreddy/color-difference-between-2-colours-using-python-182b. (visited on 04/20/2024)

Schmidt A-M et al (2023) Automated investigation of the breast-bra interaction using 4D scan data and oscillation analysis. In: Scataglini S et al (eds) Advances in digital human modeling, vol 744. Lecture notes in networks and systems. Springer Nature Switzerland, Cham 2023, pp 52–61. ISBN: 978-3-031-37847-8. https://doi.org/10.1007/978-3-031-37848-5_6

Sharp N et al (2019) Polyscope. www.polyscope.run. (visited on 02/27/2023)

Tagliasacchi A, Zhang H, Cohen-Or D (2025) Curve skeleton extraction from incomplete point cloud. In: Funkhouser T, Hoppe H (eds) ACM SIGGRAPH 2009 papers. ACM, New York, NY, USA 7272009, pp 1–9. ISBN: 9781605587264. https://doi.org/10.1145/1576246.1531377

Waller H, Schmidt R (1989) Schwingungslehre fur Ingenieure: Theorie, Simulation, Anwendungen. Mannheim,Wien, and Zurich: BI-Wiss.-Verl. ISBN: 3184014843

Chapter 5
Analysing the Breast-Bra Interaction in a T-Pose Using 4D Scan Data and BodyTex4D

Abstract The 4D scan data evaluation methods developed in Chap. 4 are applied to the static analysis of breast-bra interaction. Multiple bras differing in design, size and material are tested on two subjects and compared with each other. During 4D scans, the subjects are recorded while breathing, and comparable body positions are filtered and analysed using the automated evaluation methods of BodyTex4D. This chapter demonstrates how 4D scan data can enhance the analysis of static clothing-body interactions

Keywords 4D Scanning · Clothing development · Automated scan data analysis · Body deformations · Breast-bra

5.1 Introduction and Objectives

This chapter analyses if the experimental comparison of various bras on test subjects in static poses can be improved by 4D scan data and BodyTex4D (Chap. 4).

Bras support and shape women's breasts daily. Bras are worn during sporting activities, pregnancy, healing after an illness and many more. The bra interacts with the breasts and body. The many forms of breasts and bras result in an incalculable number of individual fittings. When new bras are developed, it is important to analyse the influences on the human body of parameters such as seam placements, materials or cut patterns. One method to analyse these influences of various bras is comparing the breast deformation. For an accurate breast deformation analyse, the test subjects have to be in the same body state repeatedly, which is the main challenge encountered during evaluation.

State-of-the-art to achieve comparable body poses is to get test subjects into similar leg and arm positions. Pre-marked foot positions or constructions to grab onto are used as helpers (Simmons 2001). In EN ISO 20685-1, it is recommended to breathe normally and calmly (DIN EN ISO 2018). However, scans done while a test subject is breathing in and while breathing out cause breast deformations to differ greatly, especially when comparing the fit of bras (Fig. 5.1 and Sect. 5.2). This chapter used automated 4D scan data analysis to improve the selection of comparable body states,

A.-M. Schmidt, *Novel Methodologies for Digital Investigations of Clothing-Body-Interactions*, Lecture Notes in Bioengineering,
https://doi.org/10.1007/978-3-031-91794-3_5

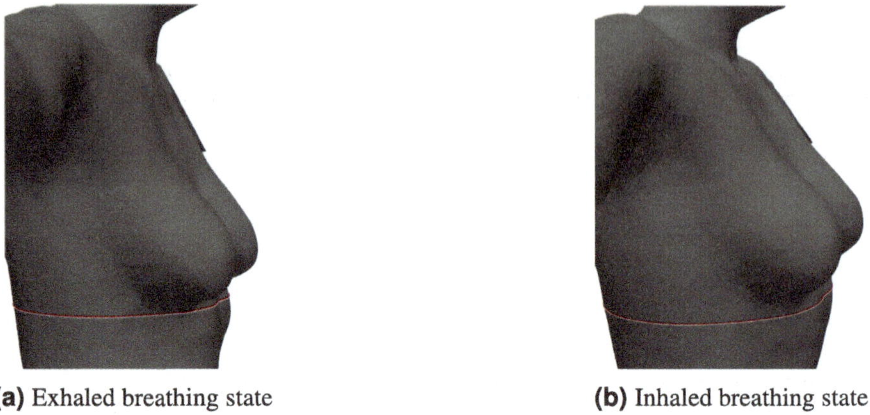

Fig. 5.1 Chest difference between **a** an exhaled and **b** an inhaled breathing state

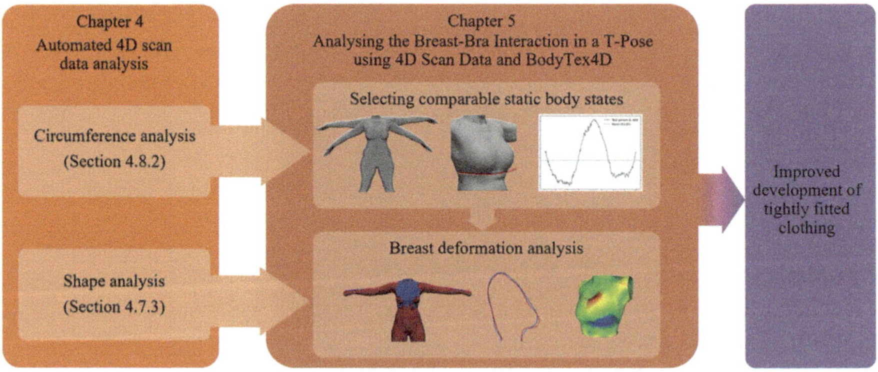

Fig. 5.2 Overview of the experimental investigation of the breast-bra interaction in a T-pose

focusing on similar breathing states. Comparable body states are analysed by shape analysis to detect the influence of various bras on breast deformation.

Figure 5.2 displays the chapter structure and the dependency to Chap. 4. In Sect. 5.2 the preliminary examination is discussed. The most significant criteria for the comparability of static poses are determined. Firstly, methods for selecting comparable body states are described. Secondly, methods for analysing breast deformations are described. Results are shown and discussed in Sect. 5.5. Lastly, in Sect. 5.6, the methods used are discussed within the context of usability for tightly fitted clothing development.

The primary focus of the following chapter is to develop a method for individualised clothing-body analysis based on 4D scan data in an automated way to improve its accessibility for a larger group of researchers and the textile industry. The presented methods can further be used to perform studies with more test subjects or at other body deformation applications. Statements about the wearing behaviour, such as the comfort of bras on a broad mass of women, are out of the scope of this research work. The research work presents the groundwork for individualising tightly fitted clothing for many people.

5.2 Preliminary Examinations

The preliminary examinations investigate parameters for body deformation analysis in bra development and fitting applications. The influence of the body pose (Sect. 5.2.1) and breathing state (Sect. 5.2.2) are investigated. Section 5.2.3 deals with the influence of the bra-band-thickness on the underbust circumference.

5.2.1 Analysing the Influence of Body Poses

Different body poses result in different body deformations. Recognisable pose differences are change in position of extremities or rotated body parts (e.g. leaning forward). Activation of muscle and gravity acting on soft tissue changes the shape of particular body parts. Various poses have been investigated to analyse the breast-bra interaction. Example poses are shown in Fig. 5.3.

Figure 5.4 shows the influence of the arm position on the breast's shape. The arm position is important for the pose comparability in the breast-bra interaction. Comparing the A- (Fig. 5.3a) and T-pose (Fig. 5.3c) indicates that minor arm movements effects the breast deformation (Fig. 5.4a). In Fig. 5.4, a breast lift occurs due to lifting of the arms from the A- to T-pose. The influence on the breast's deformation is higher without a bra (Fig. 5.4a). Wearing the bra lifts the breast, reducing the influence of lifting the arms (Fig. 5.4b).

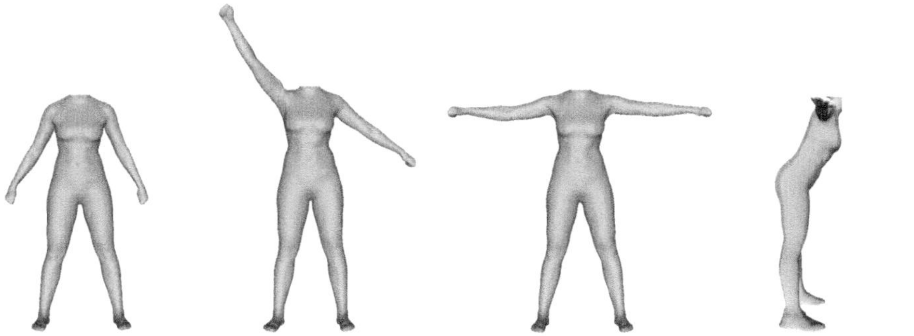

(a) A-pose **(b)** Diagonal pose **(c)** T-pose **(d)** Leaning forward pose

Fig. 5.3 Static poses scanned as a pre-investigation

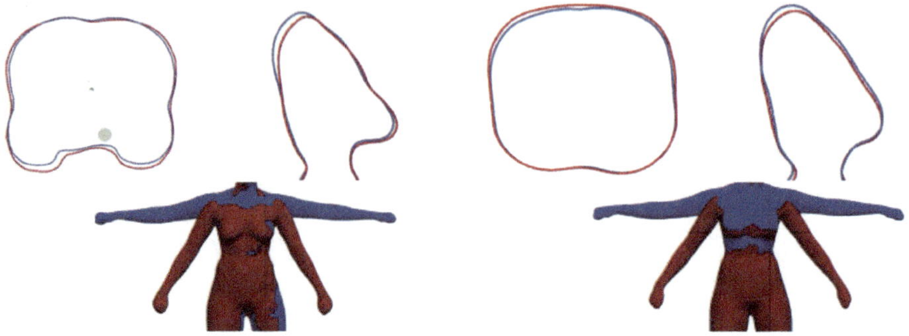

(a) Test subject without a bra **(b)** Test subject with a bra

Fig. 5.4 Cross-section comparison of the influence of the body pose on breast deformation

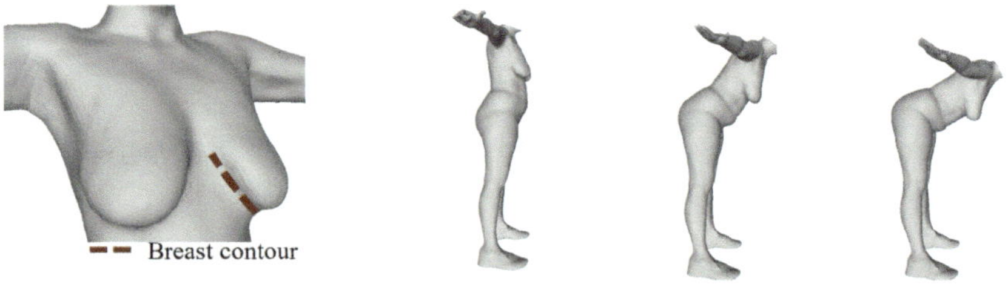

(a) Breast capturing challenges for test subjects with larger breasts (red dashed line: inframammary fold)

(b) Leaning forward 4D scan data

Fig. 5.5 T-pose variation to capture the breast surface of test subjects with larger breasts and inframammary fold (dashed red line)

The T-pose has proven to deliver the best results during experimental investigations and finite element analyses (Chap. 9). The pose is symmetrical, and all body areas are visible for 3D and 4D scanners. For the surface capturing of larger breasts, the standing up right T-pose can lead to inaccuracies. Larger breasts tend to form an inframammary fold (Fig. 5.5a). 3D and 4D surface scans are fusing the inframammary fold to the thorax, resulting in a wrong model of the breast. To circumvent the fusing of the inframammary fold, the test subject can lean forward (Fig. 5.5b). Capturing surfaces of naked breasts is not focused on further by this chapter and the inframammary fold does not occur when wearing a bra. Subsequently, this research work has not chosen the leaning forward technique for further experimental breast deformation analysis. The leaning forward technique is documented to be used in future studies.

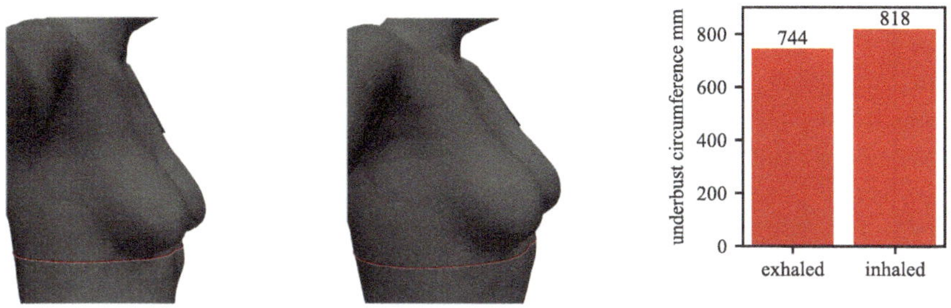

(a) Breathing state: exhaled **(b)** Breathing state: inhaled **(c)** Underbust circumference of two breathing states

Fig. 5.6 Comparison of the circumference of an exhaled and inhaled breathing state of test subject-A without a bra in a T-Pose

5.2.2 Analysing the Influence of Breathing States

Breathing causes the thorax to expand and shrink. The chest lifts up and lowers down the breast. The breast shape and thorax changes during the breathing process. In Fig. 5.6, a test subject has been scanned in the exhalation breathing state (Fig. 5.6a) and the inhalation breathing state (Fig. 5.6b). During inhalation, the breast is rounder and lifted up (Fig. 5.6b). If bras and their interaction with the breast are compared, the breathing state significantly influences it.

The underbust circumference has been identified to be linked to the breathing state. The underbust circumference presents itself as parameter to improve the selection of comparable breathing states. For test subjects of this study a maximum of 70 mm difference in underbust circumference was measured (Fig. 5.6c). Section 5.4.1 develops a method to identify comparable breathing states to improve the comparability of T-pose scans while using the 4D scan technology and BodyTex4D by identifying similar breathing states.

5.2.3 Analysing the Influence of Underbust Circumference Measurements Without Bra-Band Thickness

As described in Sect. 5.2.2, the breathing state is a crucial parameter for selecting comparable T-poses. The bra-band-thickness influences the measurement of the underbust circumference. The bra-band-thickness adds to the underbust circumference an extra length, leading to wrong comparison results. In the following work, when calculating the underbust circumference of a test subject wearing a bra, the bra-band-thickness is always subtracted from the underbust circumference by the method presented in Sect. 4.8.2. An example of underbust measurements with and

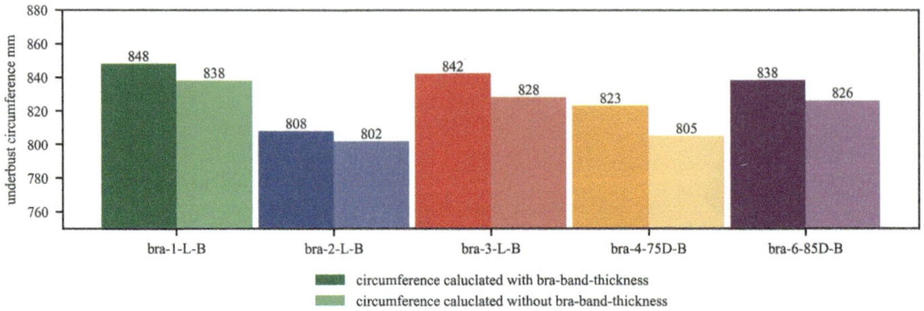

Fig. 5.7 Underbust circumference with and without subtracting the bra-band thickness

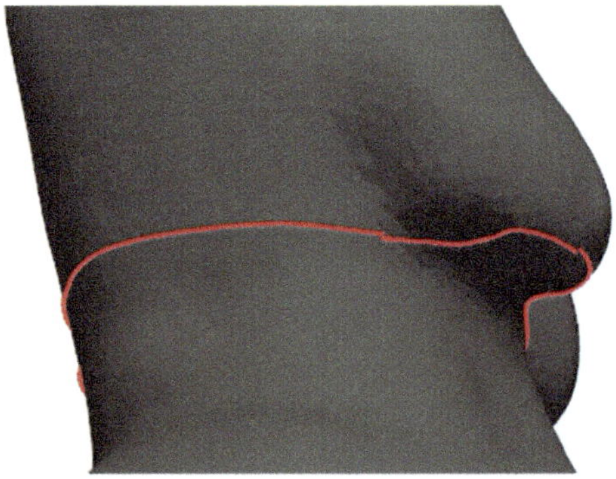

Fig. 5.8 Challenge in the underbust measurement of larger breasts: The underbust circumference is measured on top of the breast

without the bra-band subtraction can be seen in Fig. 5.7. Bra-2 and bra-4 had a difference of 15 mm before the bra-band thickness reduction. After the bra-band thickness reduction, the circumference difference decreases to 3 mm.

For people with larger breasts, the underbust measurement must be taken at a lower measurement point than for people with smaller breasts because the breast hangs relatively lower (Fig. 5.8). An approximation is to choose the area where the inframammary fold ends. The approximation did not affect the accuracy of determining the breathing state.

5.3 Materials

The following section describes the materials used for further analyses, including the test equipment (Sect. 5.3.1), test subjects (Sect. 5.3.2) and bras (Sect. 5.3.3).

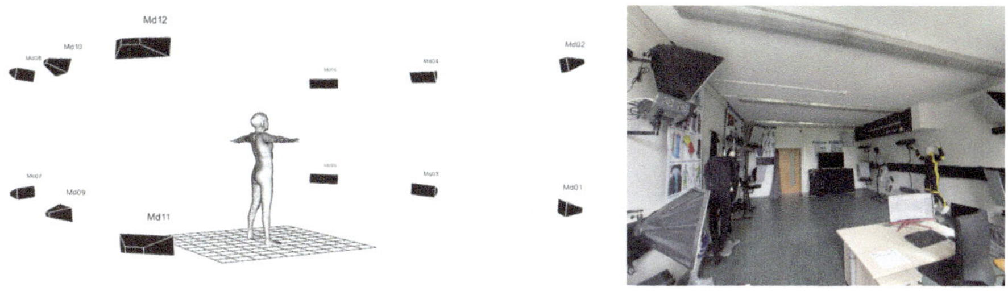

(a) Conceptual sketch of the scanner room

(b) Scanner room

Fig. 5.9 4D scanner used in the following experimental work

5.3.1 Test Equipment

The Move4D Scanner by IBV is used to capture the breast-bra interaction. Section 2.1.3 describes its principle and relevant terminology. Before each scan series, the scanner is calibrated. The scan volume is 2 m x 2 m x 2.5 m. The test setup of the Move4D Scanner consists of 12 cameras. Six cameras are in the front of the scan area, and six cameras are in the back of the scan area. On each front/ back side, three cameras are placed on a lower and three on a higher level (Fig. 5.9). The test subject must be placed in the middle of the scanner area (Fig. 5.9a).

5.3.2 Test Subjects

Two test subjects have been selected. Test subject-A is a 30-year-old woman with a 75C bra size and is 168 cm in height (Fig. 5.10a). Test subject-A's upper body clothing size is 36 (EU). She has an athletic body type. Test subject B is a 60-year-old woman with an 85D bra size (Fig. 5.10b). She is 172 cm in height. Test subject-B's upper body clothing size is 42 (EU). Test subject-B has a higher body fat percentage than test subject-A.

5.3.3 Bras

Six different bras are tested experimentally. A description of cut patterns (Sect. 5.3.3.1) and material properties (Sect. 5.3.3.2) are given.

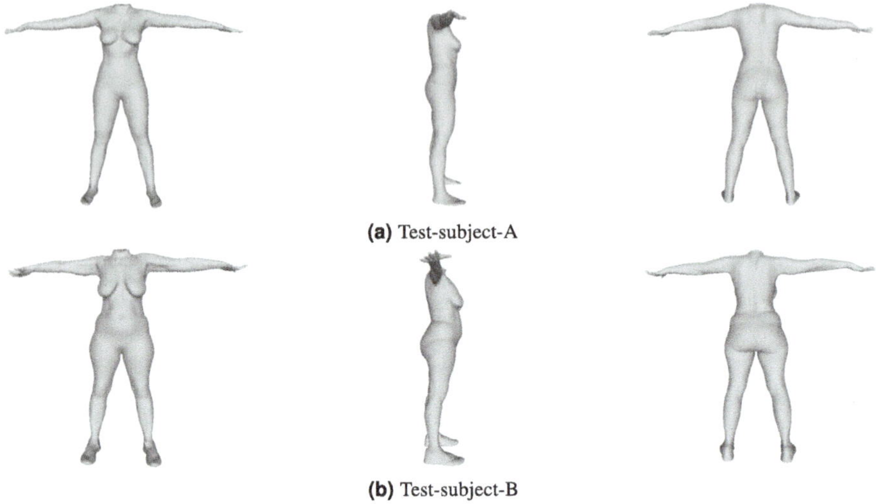

(a) Test-subject-A

(b) Test-subject-B

Fig. 5.10 Test subjects

5.3.3.1 Cut Patterns

Figure 5.11 shows all tested bras front and back views. The bras are further divided and described in the following parts: "front part", "back part" and "bra-band". The front and back part is seamed together at the sides of the bra right under the armpits. Bra-1 to bra-6 are sorted in the order of the expected support. Bra-5 and bra-6 are the test subjects' own bras, which they use for high-impact activities like jogging. The test subjects characterise the bras as well fitted and providing a high level of support. With the exception of bra-2, all bras were purchased for the purpose of testing and evaluating commercially available products. sstab Bra-2 has been recreated from

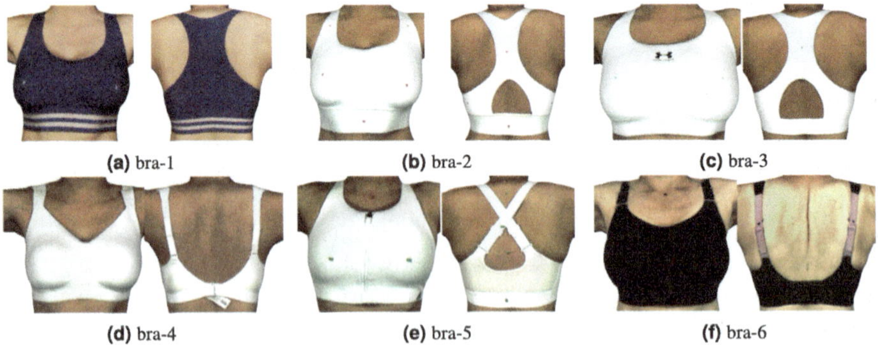

(a) bra-1 (b) bra-2 (c) bra-3

(d) bra-4 (e) bra-5 (f) bra-6

Fig. 5.11 Tested bras and defined names

Table 5.1 Cut pattern details of the tested bras

Bra parts		Layers	Cut pattern details	Support (purposed)
Bra-1	Front	2	Moulded cups, elastic edge seams	Light
	Back	1	No closure, elastic edge seams	
	Band	1		
Bra-2	Front/back	1	Same cut pattern as bra-3, no seams	Light
	Band	1	longer, wider bra-band as bra-3	
Bra-3	Front	2	Elastic edge seams	Medium
	Back	2	Elastic edge seams	
	Band	3	Elastic band encased by front material	
Bra-4	Front	2	Moulded cups, wider bra straps	Medium
	Back	1	with closture	
	Band	2		
Bra-5	Front	2	Moulded cups with zipper (stiff)	High
	Back	2	Crossed, adjustable straps	
	Band	2		
Bra-6	Front	2	Moulded cups (stiff)	High
	Back	2	With closture	
	Band	2		

bra-3 to have the original cut pattern and material properties. The original cut pattern from all other bras is not known and can just be approximated or tested on the bra.

Table 5.1 summarises details about the bras cut patter including number of layers, seam information and by the manufacturer purposed support. Bra-1 has a rounded front part and a crossed back part. The front part consists of two layers of the same fabric and has moulded cups (Fig. 5.12a). The back part has the same fabric as the front and is single-layered. The bra has seams on all edges and no closure. The brand declares the bra to be low-supportive.

(a) Bra-1 with a moulded front part

(b) Bra-3 with a not-moulded front part

Fig. 5.12 Difference between a moulded and not moulded bra front part

Bra-2 is a self-made bra. The cut pattern from the front and back are the as bra-3. The bra-band is 20 mm shorter and 15 mm wider than bra-3 for bra size M and S to provide a better underbust-fit. The front and back parts are single-layered and are made out of the same fabric. The edges are seamless. The bra has no closure.

Bra-3 has a rounded front part and a crossed back part. The front and back parts are double-layered, and the bra-band is triple-layered. The outer front fabric encloses an elastic band. The front and back parts outer fabrics are the same. The inner fabrics differ. The bra has no closure. The brand declares the bra to be a medium-supportive sports bra.

Bra-4 differs from bra-1 to bra-3 as follows. Bra-4 has straight straps at the back and an adjustable closure. The straps are adjustable. The straps are wider at the shoulders. The cups are moulded and are partially covered by a stiff lace. The brand declares the bra to be an everyday bra for breasts that require high supportive properties for daily activities.

Bra-5 is from test subject-A. Bra-5 has high supportive properties. The front part is double-layered, has stiffer cups and has a zipper. The back-straps are crossed and adjustable.

Bra-6 has a similar cut to bra-4. Bra-6 has a closure and adjustable straps. The cups are moulded and stiffer, similar to bra-5. Bra-6 has been declared by test subject-B to be a high-supportive sports bra.

In summary, the differences of the bras are:

- Bra-1 to bra-3 have no adjustable elements.
- Bra-5 and bra-6 have moulded, stiff cups.
- Bra-2 has the same front and back cut pattern as bra-3
- Bra-2 is one-layered and has no seams at the edges.
- Bra-4 and bra-6 have a closure at the back and no cross-back.

A nomenclature to distinguish the bras by following pattern is introduced: bra-*number-size-testsubject*. The size given in EU size. For example, if test subject-B has worn bra-2 of size L, its ID is derived as bra-2-L-B.

5.3.3.2 Material Characterisation

The Table B.1 describes each bra's material composition. As all bras have been purchased except bra-2, the manufacturers provide not all material parameters. However,

(a) Clamp positions for bra-1 to bra-3 on the front part

(b) Clamp positions for bra-1 to bra-3 on the back part

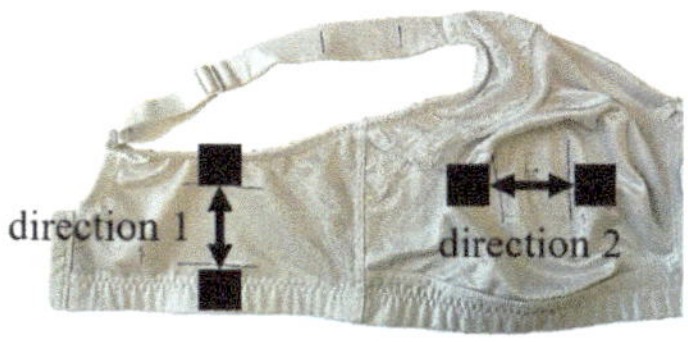

(c) Clamp positions for bra-4 on the left part

(d) Clamp positions for bra-4 on the right part

Fig. 5.13 Clamp positions and testing areas of the cyclic tensile tests for the material characterisation from bra-1-bra-4

the bras are evaluated against each other, cancelling out missing information by cross-referencing. Cyclic tensile tests, thickness measurements and density measurements were performed to characterise the bras. The tests are performed on bra-1 to bra-4 as those could be destroyed. Bra-5 and bra-6 are the test subjects' bras and a characterisation was not possible. As this chapter focuses on the scan data analysis a characterisation is not mandatory. However, characterisations are useful to get an impression and intuition of the material. Additional, the results of this material characterisation are used the finite element analysis of the breast-bra interaction in the following part III.

The methods to perform a cyclic tensile test, the thickness and density measurements are in the following described.

A cyclic tensile test is performed in reference to DIN 13934-2 at the tensile testing machine Z2.5 "Zwicki Junior" with a load cell of 2.5 kN (Fig. 5.14a). The clamps have a clamping surface of 30 mm x 30 mm to ensure the clamps fit all bra parts. The clamping length is set to 60 mm for the front and back parts. The front and back parts are tested in vertical and horizontal directions. Vertical and horizontal directions are called direction 1 (dir1) and direction 2 (dir2). Figure 5.13 shows the clamp positions and testing directions. On bra-1 to bra-3 the clamps are positioned as displayed in Fig. 5.13a and b. The clamp positions of bra-4 can be seen in Fig. 5.13c and d. The bra-band has been tested in the direction of length.

The thickness of the bra parts is measured accordingly to ISO 5084 with 1 kPa and a pressure area of 20 mm^2 (Fig. 5.14b). Five different regions of each bra part are measured. The five thickness measurements are averaged.

For the density measurements (Fig. 5.14c) five samples of the size of 100 mm^2 have been cut out of each bra part. The area of the samples is measured after the cut-out again to determine the exact area. The weight of the samples is determined.

(a) Tensile testing

(b) Thickness testing

(c) Density measurement

Fig. 5.14 Material characterisation tests of the bra parts

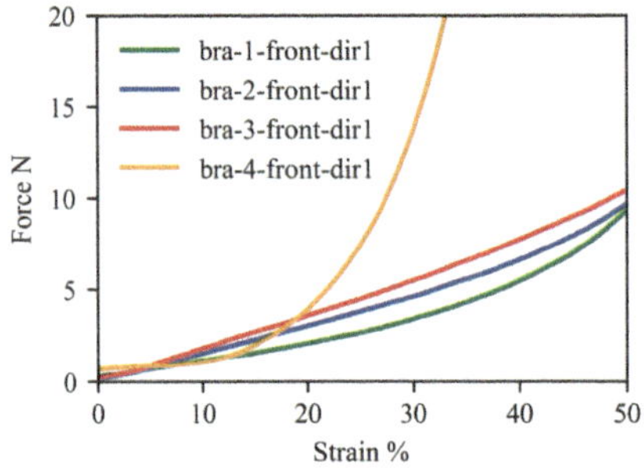

(a) Front side - direction 1

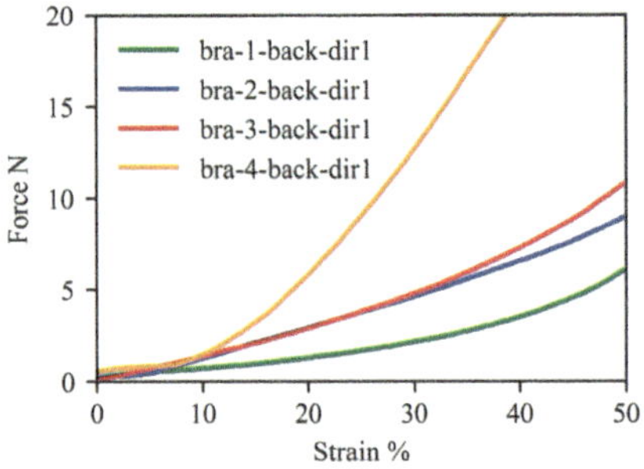

(b) Back side - direction 1

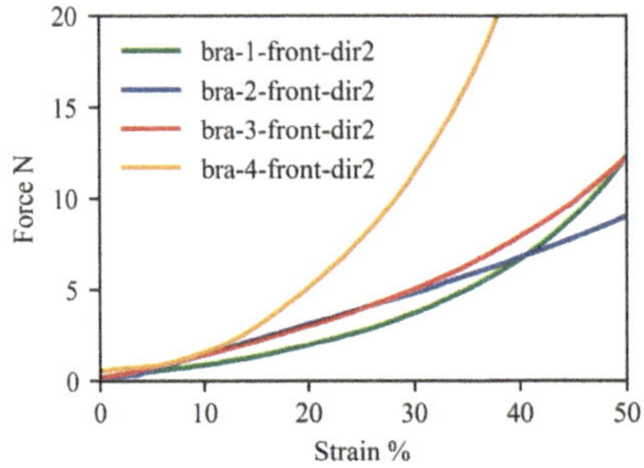

(c) Front side - direction 2

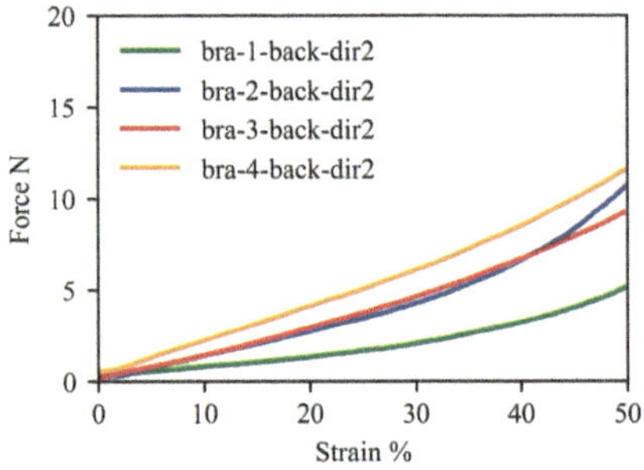

(d) Back side - direction 2

Fig. 5.15 Tensile test results of the loading phase

The density ρ is calculated based on the area A, weight m and thickness t with $\rho = m/(A * t)$.

The results of the cyclic tensile tests for the 5th cycle are displayed in Figs. 5.15 and B.1. The cyclic tensile tests are sectioned into the loading and unloading. For all loading sections, bra-4 showed the stiffest material behaviour. For bra-2's and bra-3's back part similar tensile properties in both directions can be determined (blue and red curves in Figs. 5.15b, d and B.1b, d). The material behaviour of bra-1 shows a

Table 5.2 Thickness and density of bra-1 to bra-4

Bra part	Thickness mm						Density E-10 t^3 mm^{-1}			
	Bra-1	Bra-2	Bra-3	Bra-4	Bra-5	Bra-6	Bra-1	Bra-2	Bra-3	Bra-4
Front	0.9	0.482	1.31	1.01	1.72	1.65	4.137	5.162	3.35	4.21
Back	0.48	0.482	1.13	0.59	1.4	1.02	4.03	5.162	3.178	5.03
Band	1.78	0.79	2.0	2.73	1.24	1.26	2.97	6.288	3.758	5.94

lower tensile strength in the loading and unloading section in both testing directions. The loading (Fig. 5.15a) and unloading section (Fig. B.1a) of the front part of bra-4 in direction 1 shows the highest tensile strength, followed by bra-3, bra-2 and bra-1.

In summary, the key-findings from the cyclic tensile test are:

- Bra-4 has the highest tensile strength.
- Bra-1 has the lowest tensile strength.
- Bra-2 and Bra-3 have similar tensile strengths except the front part in direction 1, where bra-3 shows slightly higher tensile strength.

Results of the thickness and density measurements are shown in Table 5.2. The results are used in part III.

5.4 Methods

The breast deformation due to various bras in a T-pose are compared. Firstly, methods to select comparable body states by analysing the breathing states (Sect. 5.4.1) are presented. Secondly, methods for a breast deformation analyse are shown.

5.4.1 Selection of Comparable Poses of the Body

The scan procedure to capture 4D scan data is described in Sect. 5.4.1.1. Comparability criteria are defined to select comparable states of the body using the 4D scan data (Sect. 5.4.1.2). Methods to determine the fulfilment for each comparability criteria are described based on the scan procedure (Sects. 5.4.1.3–5.4.1.5).

5.4.1.1 Scan Procedure

4D scans of the test subjects wearing various bra variations are captured. For each bra variation one T-pose scan (static scan) and one scan while breathing (dynamic

Table 5.3 Scan plan of test subject-A and test subject-B for the deformation analysis in a T-pose

Bras	Static scan		Dynamic movement scan		colours
	T-Pose		breathing in T-Pose		
	test subject A	test subject B	test subject A	test subject B	
without-bra	x	x	x	x	
bra-1	M	L	M	L	
	S		S		
bra-2	M	L	M	L	
	S		S		
bra-3	M	L	M	L	
	S		S		
bra-4	75C	75D	75C	75D	
bra-5	M		M		
bra-6		85D		85D	

movement scan) is performed. The Move4D scanner by IBV (Sect. 5.3.1) is utilised. The test subjects described in Sect. 5.3.2 volunteered for the data collection. Table 5.3 shows the scan procedures for test subject-A and test subject-B. All scans were captured in one day to keep them comparable. The breast volume can change during the female cycle, by weight gain/loss and more confounding factors yet to be determined.

Test subject-A has been captured without a bra, wearing bra-1 to bra-3 in two different sizes each and wearing bra-4 and bra-5 in a single size. Test subject-B has been captured without a bra, bra-1 to bra-4 and bra-6 in a single size. The focus of the analysis of test subject-A is to compare the influence of various bras and the influence of variation in the sizes of the same bra. Test subject-B focuses on analysing the influence of various bras. As test subject-B has larger breasts, only a single size of the bra fits her.

Both test subjects are scanned in the T-pose (Fig. 5.3c). To identify the same breathing state, a dynamic movement scan has been captured of the test subjects while deeply breathing in and out while holding the T-pose. All dynamic scans are captured in 30 fps. Test subject-A's dynamic movement consists of 400 frames. Test subject-B dynamic movement consists of 200 frames. The scan data is processed as a homologous mesh (Sect. 2.1.3.2) and exported as *.obj-, *.mtl- and *.png- files (Sect. 2.1.4).

5.4.1.2 Comparability Criteria

Three comparability criteria have been indicated in preliminary examinations to improve the comparability of body states. The three defined criteria are:

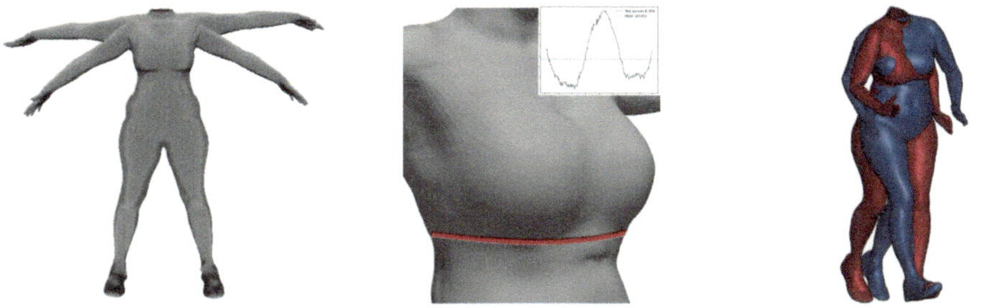

(a) Body pose (b) Breathing state (c) Force similarity

Fig. 5.16 Comparability parameters to select comparable body states

- body pose (Fig. 5.16a)
- breathing state (Fig. 5.16b)
- force similarity (Fig. 5.16c)

The first criteria is the body pose. The body pose is influenced by rotation and translation of the body parts. In the application of breast-bra interaction, the upper body position influences breast deformation. Poses should be similar, especially the arm position and upper body rotation.

The second criterion is the breathing state. Due to breathing, the breast is moving up and downwards causing the breast to change shape, furthermore, the thorax volume increases and decreases. Frames are comparable if a similar state of breathing is selected.

The third criterion is the force similarity. Force similarity describes the forces acting on the body. Especially during movements, the forces change dramatically due to accelerations over time. The force similarity criteria are important for dynamic scans. If a test subject runs with a higher amplitude and frequency, the breast experiences higher accelerations and, therefore, stronger deformations. The breasts deform less if the test subject moves with a lower frequency and a smaller amplitude.

To evaluate the comparability criteria, BodyTex4D of Chap. 4 is utilised.

5.4.1.3 Selection of Comparable Body Poses

The first comparability criterion of having comparable body poses is fulfilled for all scans, as for all scans, the test subjects are in a T-pose. Little pose changes due to the arm height in the T-pose are neglected in this analysis.

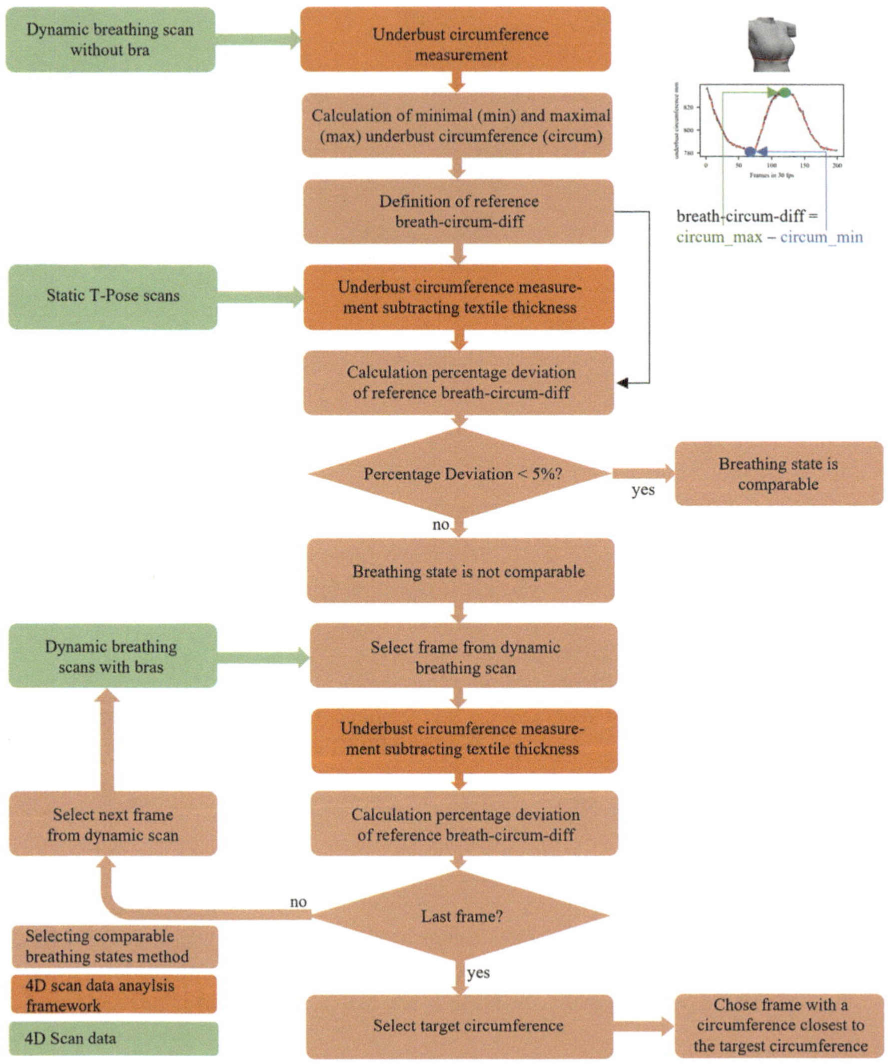

Fig. 5.17 Method to select a comparable breathing state

5.4.1.4 Selection of Comparable Breathing States

The second comparability criteria is based on 4D scans of the test subjects breathing while being the T-pose. A method to select comparable breathing states has been developed and is displayed in Fig. 5.17 through a flowchart.

The breathing state is identified by tracking the test subject's underbust circumference (circum) over time. If the test subject wears a bra, the bra-band-thickness

(a) Key-vertex-ID 9593 **(b)** Vertex-ID 9590

Fig. 5.18 Vertex-IDs for the underbust circumference measurement

is subtracted by the method presented in Sect. 4.8.2 to keep the measurements comparable.

The first step is to calculate the maximum circumference difference during a breathing cycle ($\Delta breath\_circum$) to get a reference. Step one is performed without bra. The dynamic breathing scans (Table 5.3) are used as input data. The circumference measurement method (Sect. 4.8.2) is based on homologous mesh data and uses the identification method by a key-vertex-ID for test subject-A and by manual selection of vertex-IDs for test subject-B (Sect. 4.5.1). As test subject-B has larger breasts, the underbust key-vertex-ID of the Move4D scanners happens to be part of the breast geometry (Fig. 5.8). The underbust key-vertex-ID is 9593 (Move 4D User guide 2023) and is visualised in Fig. 5.18a. For test subject-B the vertex-ID 9590 is chosen (Fig. 5.18b). An additional input parameter for the circumference measurement (Sect. 4.8.2) is a normal vector for the cross-section plane. This normal vector is set to the vertical direction to [0, 1, 0] to achieve a horizontal slice.

The result of the circumference measurements for each frame of the dynamic breathing scan is a curve showing the circumference changes throughout a breathing cycle. The minimal and maximal circumference is determined and the difference of the circumference during breathing is calculated. This breathing-circumference-difference is defined as the reference parameter $\Delta breath\_circum$ (Eq. 5.1).

$$\Delta breath\_circum = |circum_{without\_max} - circum_{without\_min}| \tag{5.1}$$

The next step is to measure the underbust circumferences of the T-pose scan without breathing instruction. This step aims to analyse the effect of the breathing state regarding comparability if no breathing instructions have been given. The 4D scan data of the static T-pose (Table 5.3) is used for the analysis. The method to calculate the circumference without bra-band thickness Sect. 4.8.2 is applied again. The bra-band-thickness is subtracted in this method to improve the comparability. The bra-band-thicknesses can be seen in Table 5.2. When calculating the circumference without bra-band-thickness, a thickness change of the underbust material due to the elongation is neglected.

The calculated underbust circumference of the scans with the bra variations $circum_{bra}$ is compared to the underbust circumference measurement without a bra $circum_{without}$. The $percentage\_deviation$ is a metric for the influence of breathing of the state of the body. The $percentage\_deviation$ metric is defined as

$$percentage\_deviation = \frac{|circum_{without} - circum_{bra}|}{\Delta breath\_circum} < 5\,\% \quad (5.2)$$

Preliminary tests show that a $percentage\_deviation$ from the reference circumference of more than 5 % renders two body states non-comparable to each other due to the geometric deviation. A range of 1 % to 2 % $percentage\_deviation$ is due to measurement inaccuracies of 4D scan data and the scan data evaluation process. The underbust circumference is measured for all breathing scans. For each bra variations the key-vertex-ID 9593 (Fig. 5.18a) is used. The next step is to find a reference breathing state for all scans that should be compared. Subsequently, all scans must include the reference breathing state. The underbust circumference reflects the reference breathing state and must be measured during the same breathing phase (either inhalation or exhalation) for all scans being compared. For example, if a specific underbust circumference value appears in all scans during inhalation but occurs during exhalation in only one scan, it is discarded as the reference underbust circumference. The use of two simple boundary conditions to identify a comparable breathing state across multiple scan series enables the development of an automated breathing state detection system.

In the last step, the frames closest to the reference circumference for all bra variations, including the results of the first circumference breathing analysis without a bra, are selected. Frames from the inhalation section are selected only. Listing B.1 shows the Python script for this automated selection process.

5.4.1.5 Selection of Similar Force States

All scans are made in an upright position, and differences in breathing result in trivial adhering to the force similarity criteria.

5.4.2 *Analysis of Breast Deformations*

Using the selected, comparable T-pose scans, two methods are used to perform the analysis of breast deformations.

1. 3D comparison of two scan meshes (Sect. 5.4.2.1).
2. Analysis of breast deformations based on 2D cross-sections (Sect. 5.4.2.2).

The selected comparable frames are aligned for both analyses of the breast deformations. The open-source software CloudCompare (CloudCompare 2023) is used. Meshes to be compared are imported as *.obj - files. They are cropped to the relevant section, the thorax. The **I**terative **C**losest **P**oint registration procedure (**ICP**) aligns the meshes. Similar cropping is a prerequisite for accurate alignment results.

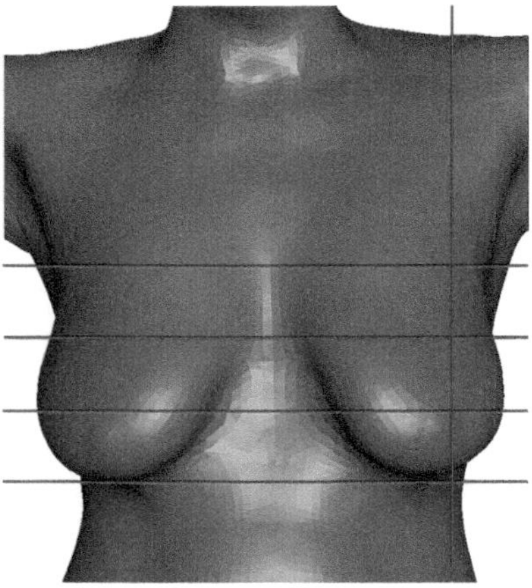

Fig. 5.19 Positions (red lines) of the horizontal and vertical cross-sections for the cross-section analysis

5.4.2.1 3D Mesh Comparison

The 3D mesh comparison is used to evaluate different bra variations by 3D meshes. The open-source software CloudCompare (CloudCompare 2023) is used. The mesh distances are computed by calculating the shortest mesh distances with the **C**loud-**to**-**M**esh distance (**C2M distance**) function. Meshes of scans without a bra are set as a reference frame. Meshes of test subjects wearing bras are set compared to the reference mesh. Colours are visualised on the compared meshes. Areas coloured blue indicate a negative distance; the compared mesh is smaller than the reference mesh. Areas coloured in red indicate a positive value; the compared mesh is larger than the reference mesh. A chart indicates how many vertices are located in which distance area. The value of mesh distance represents the metric, whereby the greater the distance, the greater the difference of the meshes. For test subjects A and B, the 4D scan without a bra serves as the reference mesh. All other scans with different bra variations are compared against this reference mesh.

5.4.2.2 Analysis of Cross-sections

The BodyTex4D framework's cross-section analysis is used. The selected, cropped and registered 3D meshes are imported to BodyTex4D. Four horizontal slices spacing 50 mm apart from each other are cut through the breast area and visualised (Fig. 5.19). The lowest slice is made exactly where the underbust measurements are made using the same vertex-ID. For test subject-B, this is vertex-ID 9590. One vertical slice is made through the left bust point of the test subject while wearing a bra. Shape differences are compared. Conducting the analysis is mainly an example for the use of BodyTex4D, subsequently, the analysis is performed for test subject-B only.

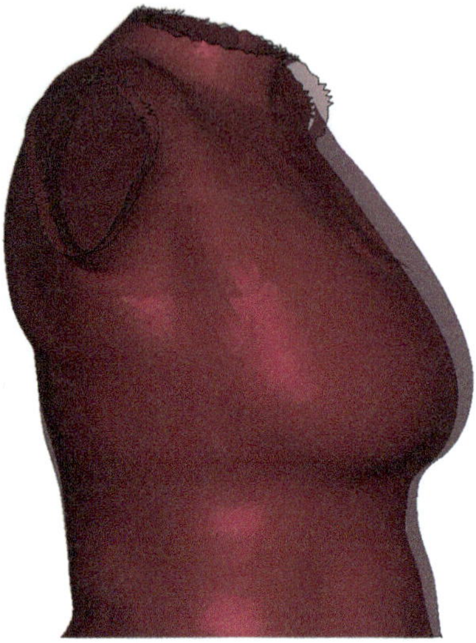

Fig. 5.20 Two breathing states captured in the dynamic breathing scans: The transparent part represents the inhaling state

5.5 Results

This section presents the breast-bra deformation analysis results in a T-pose after selecting comparable breathing states using BodyTex4D. Results of selecting comparable body states (Sect. 5.5.1) and results of comparing breast deformations due to various bras (Sect. 5.5.2) are discussed separately.

5.5.1 Selection of Comparable Breathing States

Figure 5.20 shows two aligned frames of dynamic breathing scans of test subject-B wearing bra-3. Differences in chest volume and height can be seen due to different breathing states. The need to find comparable breathing states is illustrated in this figure.

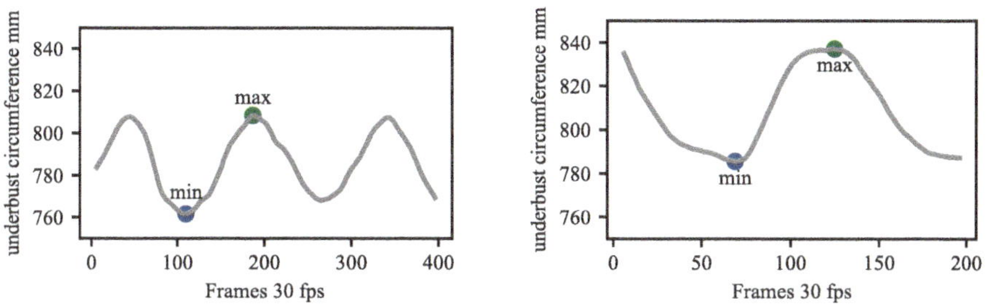

(a) Test-subject-A **(b)** Test-subject-B

Fig. 5.21 Underbust circumference changes during breathing

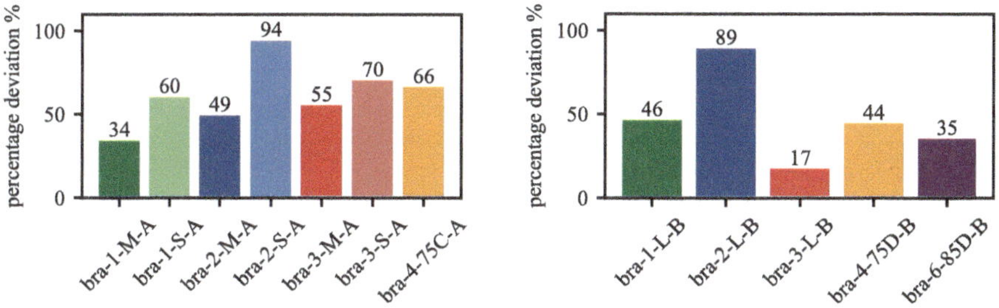

(a) Test-subject-A **(b)** Test-subject-B

Fig. 5.22 Percentage deviation of bra variations compared to the without-bra circumference at a random breathing state in a T-pose

Figure 5.21 presents the circumference changes of test subject-A (Fig. 5.21a) and of test subject-B (Fig. 5.21b) while wearing no bra and breathing.

The breathing-circumference difference for both test subjects is calculated based on the change in underbust circumference throughout the breathing cycle.

- Test subject-A's $\Delta breath\_circum\_A$ is 47 mm.
- Test subject-B's $\Delta breath\_circum\_B$ is 52 mm.

The average underbust circumferences are 787 mm and 814 mm for test subject-A and test subject-B, respectively. Based on these breathing differences $\Delta breath\_circum$ of the underbust circumference, the deviation percentage of the underbust circumferences with bra compared to without a bra are calculated for the static T-pose scans at a random breathing state. Results can be seen in Fig. 5.22. The $percentage\_deviation$ for all bra variations of both test subjects is higher than 5 %. For both test subjects, the breathing-similarity criteria are not fulfilled; thus, the breathing states are not comparable. This analysis shows that performing T-poses without breathing instructions is error-prone due to the significant influence of breathing. To obtain comparable data points the method presented in Sect. 5.4.1 is conducted.

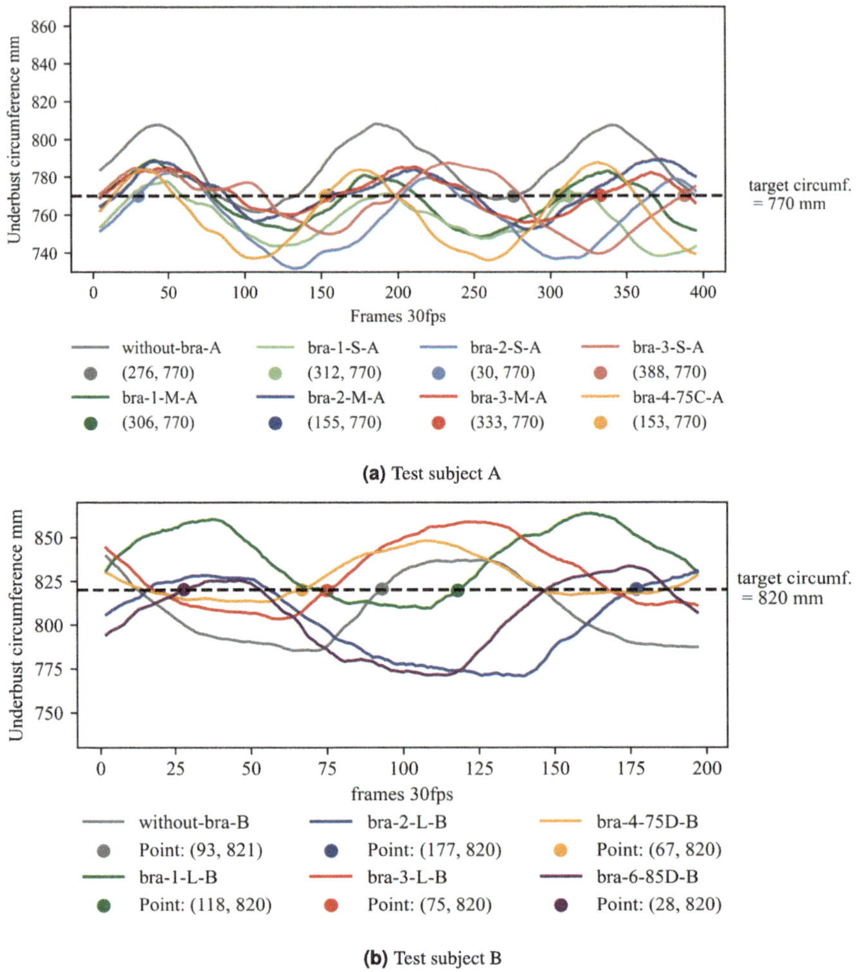

Fig. 5.23 Changes in underbust circumference during breathing and selected comparable frames (coloured dots) with an similar underbust measurement

The results of the underbust circumferences changes while breathing are shown in Fig. 5.23a for test subject-A and in Fig. 5.23b for test subject-B as curves.

For test subject-A, a noticeable trend can be seen: the average underbust circumference without bra is higher than the average underbust circumferences while wearing a bra. The mean breathing circumference of the bra-variation without-bra is 787 mm. The underbust circumference while wearing all other bra variations is smaller than this value. Additionally, the smaller sizes (size S) of bra-1, bra-2 and bra-3 always lead to a lower underbust circumference. The reason could be that the bra compromises the soft tissue around the thorax or the breathing is restricted due

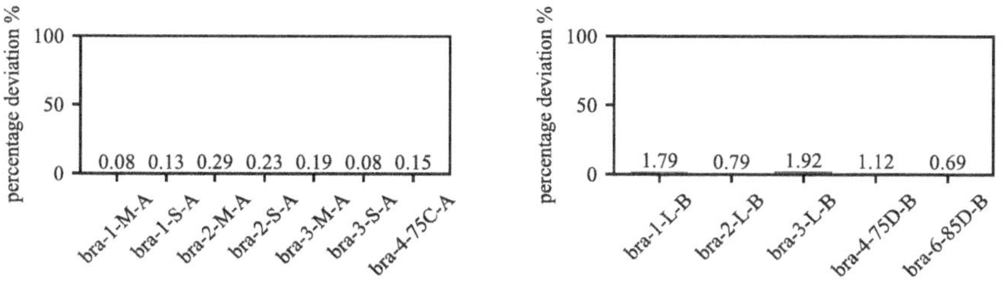

(a) Test-subject-A: Frames selected closest to 770 mm **(b)** Test-subject-B: Framse selected closest to 820 mm

Fig. 5.24 Percentage deviation of bra variations compared to the without-bra circumference after applying methods to identify comparable breathing states

to the bra-band pressure. Restriction breathing leads to discomfort and reduction of performance in sports. The impact of breathing on discomfort by bras needs to be investigated more thoroughly to build up a valid metric to optimise bras for in the future. For test subject-B, this trend could not be identified that clearly. The breathing circumference without a bra is not the highest. The reason is that the measurement of the underbust circumference has been performed below the breast and not under the breast at the original underbust measurement line. Measuring at a lower point leads to a smaller circumference as the thorax slows towards the waist due to test subject-B's body shape. Nevertheless, it can be seen that bra-2 and bra-6 lead to the breathing-circumference curves with smaller amplitude.

To identify comparable breathing frames, the next step is to select the reference circumference at the intersection of all breathing-circumference curves across different bra variations. Determined reference breathing-circumferences are 770 mm and 820 mm for test subject A and test subject b, respectively (Fig. 5.23 dashed line).

The results of the selection of frames closest to the reference circumference while inhaling can be seen in Fig. 5.23 marked by a point.

The circumference differences between the selected frames of the test subjects without a bra and with a bra are calculated again and the deviation percentage is calculated. Results can be seen in Fig. 5.24.

For all bras, the $percentage\_deviation$ is smaller than 5 % after the selection of frames close to reference circumference. The T-Pose scan in a random breathing state without breathing instructions showing deviations in the circumference of up to 89 %. Selecting by comparable breathing state reduces the error in deviation to 0.2 and 1.8 % for test subject-A and test subject-B, respectively. These frames with comparable breathing states are used in the following analysis of breast deformation. It could be shown that using the BodyTex4D framework automates and improves the selection of comparable body states.

5.5.2 Analysis of Breast Deformations

This section presents the results of the breast deformation analysis after selecting comparable frames of a 4D scan. Results of comparing breast deformation by 3D meshes are presented in Sect. 5.5.2.1 and of analysing cross-sections in Sect. 5.5.2.2. Test subject-A's analysis focuses on comparing the influence of various bra sizes. Test subject-B's analysis focuses on the influence of different bra models with different cut patterns.

5.5.2.1 3D Mesh Comparison

Figure 5.25 shows the mesh differences of test subject-A wearing three different bras (bra-1 to bra-3) in two sizes and bra-4-75C-A. The blue indicates that the bra-variation's geometry is smaller than without-bra, and the red indicates an outwards bulking of the bra's geometry. All bra variations show regions coloured in red and in blue simultaneously leading to the conclusion that the breast is compressed in some regions and lifted in other regions.

All figures from Fig. 5.25b–h show a darker red at the upper side of test subject-A's right breast compared to the left breast. The natural asymmetry of the breasts is observed. The right breast is bigger than the left breast. Bra-1-M-A leads to the least breast deformations, and bra-3-S-A to the highest breast deformations. The cut pattern and material properties influence breast deformation. The tensile strength of bra-3 is higher in the front part, which increases the breast deformation. Additional, bra-1 has moulded cups, which results in a smaller breast deformation.

Looking at the bra variations in size M (Fig. 5.25b, d and f), a smaller breast deformation compared to the smaller sizes (Fig. 5.25c, e and g) is noticeable. The size S bras have darker blue and red colouring, indicating a higher compression level and more breast lift. Differences in the influence of the sizes are observed. The size difference between bra-2-M and bra-2-S is smaller than that of bra-1 and bra-3, even though the cut patterns of the front and back parts of the bra from bra-2 and bra-3 are the same. A possible explanation is that bra-2 has a single-layered front part while bra-3 is double-layered. Additionally, bra-3 has seams placed at the part edges, which stiffen up the material. The bra material in the vertical direction of bra-3 also has higher tensile strengths. As the bras mainly experience tensional loading, tensile strength strongly influences the bras' behaviour. It is assumed, that lower tensile strength and seamless edges cause bra-2 to elongate more resulting in overall less breast deformation and smaller size differences.

In conclusion, seam, bra materials and cut patterns influence breast deformation. A more detailed analysis is done in further research work. However, the 3D mesh comparison showed good and plausible results in investigating influencing parameters on the fit of a bra, which helps further bra developments and evaluations. The following findings are achieved by the developed method:

- Influences of the bra material parameter can be visualised.

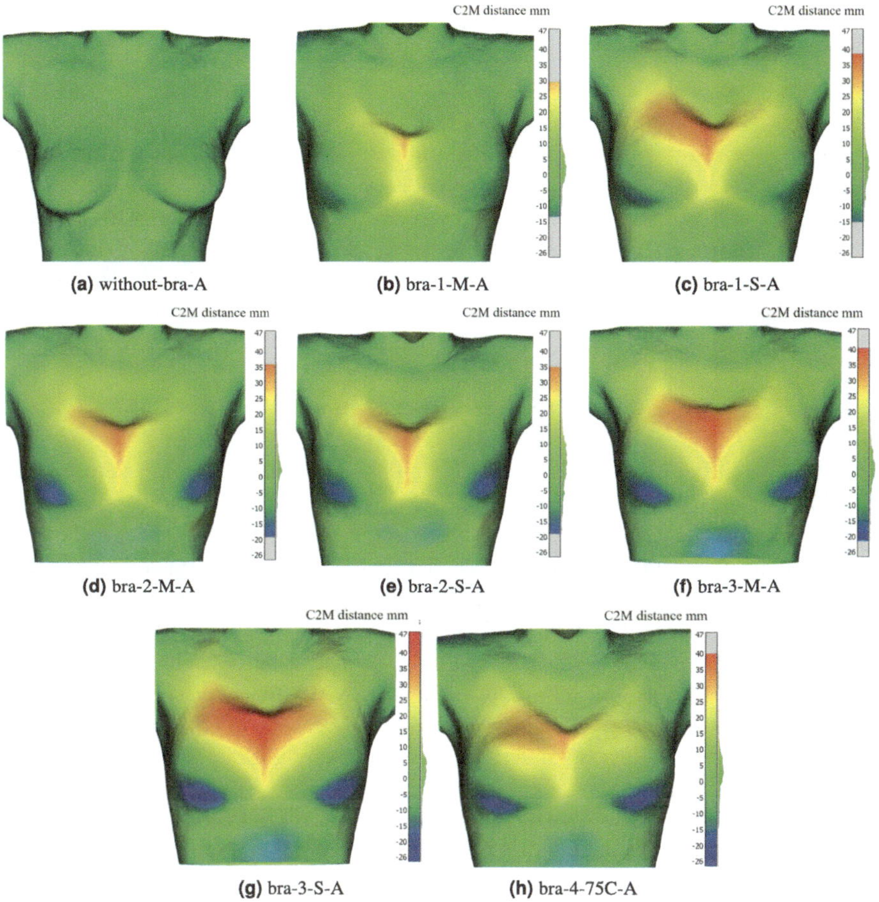

(a) without-bra-A **(b)** bra-1-M-A **(c)** bra-1-S-A
(d) bra-2-M-A **(e)** bra-2-S-A **(f)** bra-3-M-A
(g) bra-3-S-A **(h)** bra-4-75C-A

Fig. 5.25 3D mesh comparison of bra-1 to bra-3 in two sizes (M and S) and bra-4-75-C-A in reference to without-bra-A of test subject-A in a comparable breathing state

- Size difference of bras can be visualised.
- Cut pattern influences can be visualised.

The 3D mesh analysis results of test subject-B are shown in Fig. 5.26. Based on the colour differences of Fig. 5.26b–f, the different breast deformation due to various bras are displayed. Wearing bra-3-L-B, bra-4-75D-B and bra-6-85D-B leads to a higher lift of the breast (blue colour at the lower part of the breasts in Fig. 5.26d–f). Bra-1-L-B has the lowest influence on breast deformation. The breast is lifted slightly. Most areas in the 3D mesh comparison are green (Fig. 5.26b), which indicates only a slight difference from the scan without a bra. Bra-1 is supposed to be a comfortable, low-support bra. The textile is elastic and has low strength, resulting in minor deformation. Bra-2 and bra-3 have the same cut but different materials and

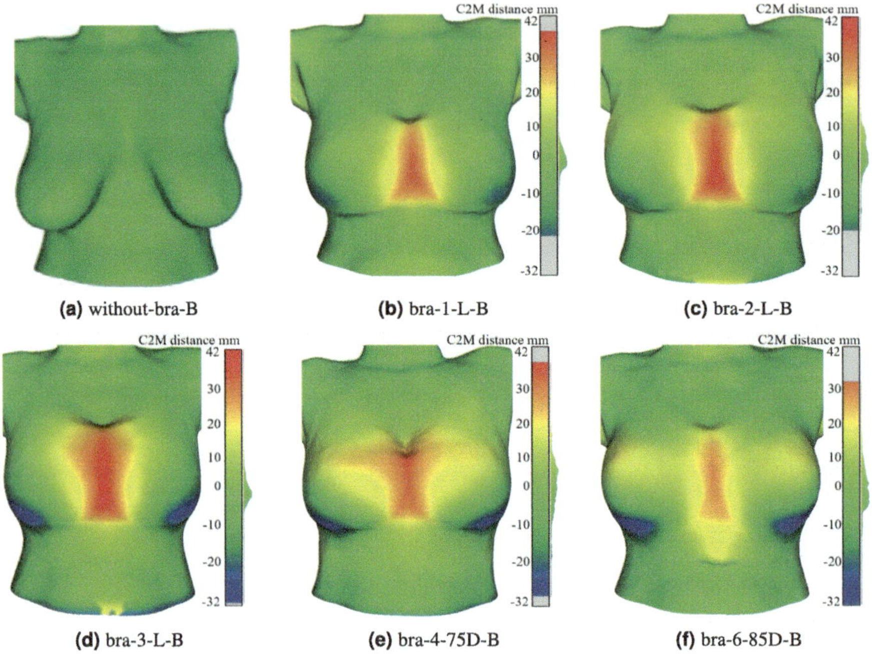

Fig. 5.26 3D mesh comparison of bra-1 to bra-4 and bra-6-85D-B in reference to without-bra-B of test subject-B in a comparable breathing state

seams. The different influences of materials can be seen in Fig. 5.26c and d. As for test subject-A bra-2 causes a smaller breast deformation than bra-3. The assumption, that this difference is caused by the higher material stiffness and strengthened seams of bra-3 is strengthened. Bra-6 is a cupped bra. The breast deformation is similar to the preformed bra.

5.5.2.2 Analysis of Cross-sections

The results of the cross-section analysis can be seen in Fig. 5.27. Slice 4 of Fig. 5.27 shows accurate alignment for the same breathing state of two frames, indicated in Sect. 5.5.1.

The cross-section slices differ slightly because the original scan data with the bra-band-thickness is visualised, which is different for each bra. Slices 1–3 show that the breasts are differently shaped due to the variation in bras. In slice 3 (Fig. 5.27c), the breast shape without a bra (grey line) is smaller and the breasts are more separate than the bra variation with the bras worn. Compared to Fig. 5.27d, the breast shape without a bra is the outer line. This indicates that the breast is lifted up by the bras. Bra-4-75D-B keeps the breasts wider separated than bra-3-L-B (Fig. 5.27d). The back soft

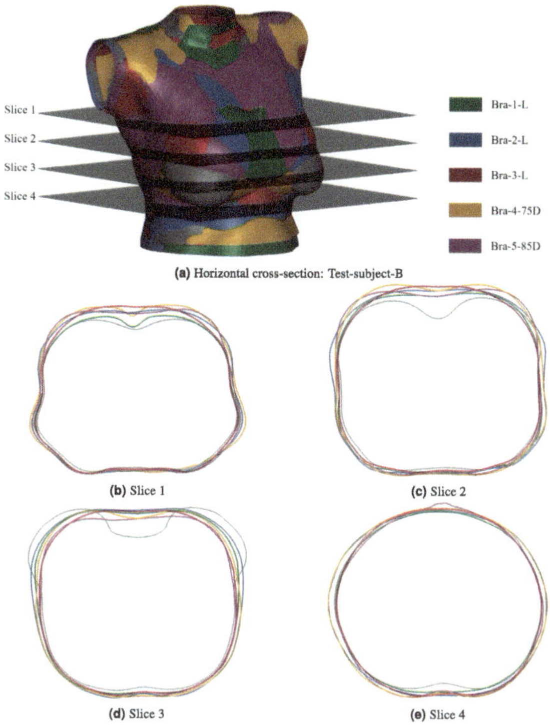

Fig. 5.27 Horizontal cross-section analysis of without-bra-B, bra-1 to bra-4 and bra-6-85D-B of test subject-B.

tissue is visualised in slice 1 (Fig. 5.27b). The most outer line of the cross-sections with bra is of bra-4 (orange), which indicates a tight fit. The soft tissue is pushed to the outside. This bra may be less comfortable.

Further analyses are made by looking at the vertical cross-sections (Fig. 5.28b). The good alignment due to BodyTex4D and the breathing state analysis of the frames can be confirmed. All bras lift the breast, subsequently offering support. Bra-4 did lift the breast the least. Even though bra-6 has a stiff cup, the breast shape is similar to bra-3. For further analysis, more cross-sections have to be made.

5.6 Conclusion and Discussion

Methods and results have been presented to select comparable body states and analyse breast-bra interactions. Bras interact with women's breast. This interaction is influenced by the bra's materials, seams and cut pattern but also by the woman's breast shape and composition. Capturing and understanding the influences of this breast-bra interaction is crucial for clothing development. The main challenging factor when experiments with humans is to create comparable results. Especially when comparing the fit of various bras, the breathing state greatly influences the outcome.

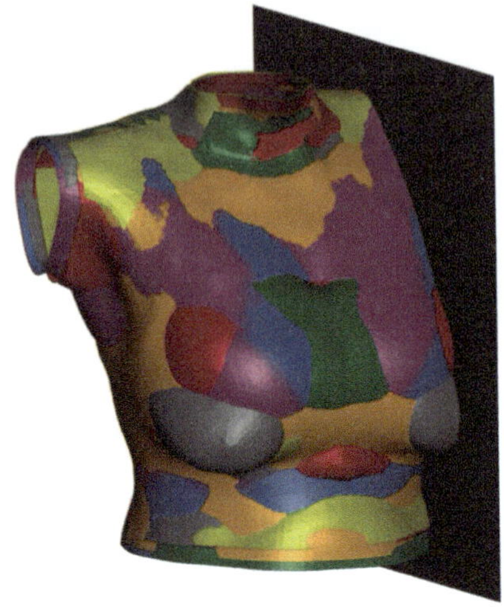

(a) Vertical slice position

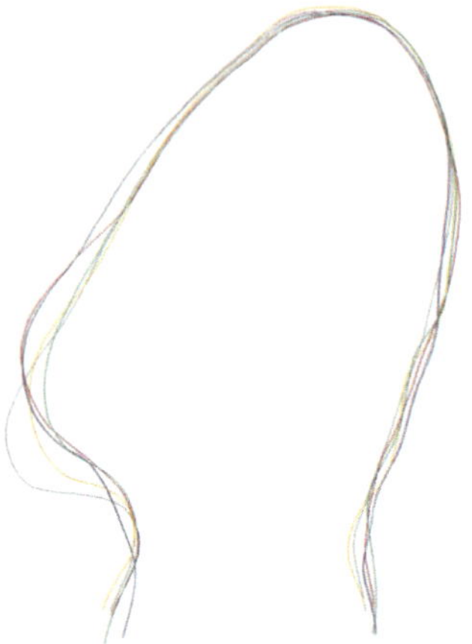

(b) Vertical cross-sections

Fig. 5.28 Vertical cross-section analysis of without-bra-B, bra-1 to bra-4 and bra-6-85D-B of test subject-B

This chapter presented predominantly a method to automatically select comparable breathing states using 4D scan data and the automated 4D scan data analysis framework BodyTex4D (Chap. 4). After performing 4D scans of two test subjects a methodology to select comparable breathing states was developed and performed.

The following key findings have been observed in this chapter:

- **4D scan data utilised by BodyTex4D improves the accuracy of bra comparisons in a T-Pose**: Through analysis of 4D scan data during inhalation and exhalation, comparable frames of similar breathing states are automatically selected. The selection of comparable breathing states increases the accuracy of bra comparisons. A comparison is prone to error when picking a not-defined breathing state (Fig. 5.20).
- **BodyTex4D's usability is confirmed in an automated 4D scan data analysis**: An application-specific method has been implemented to subtract each frame's bra-band-thickness. BodyTex4D can identify the underbust circumference closest to a reference circumference selected before.
- **Influences of various bra cut patterns, materials, seams, sizes on two test subjects with different body shapes is investigated by the improved breast deformation analysis**: Bra-1 showed the smallest breast deformation, followed by bra-2, bra-4, bra-3 and bra-6. The stiffer bra front material of bra-3 and the seams seem to lead to higher breast deformation due to the higher material stiffness.

For completion, many more influences on breathing exist (chest, belly breathing). The presented method focused on the influence on breast movement, but more influences need to be evaluated to increase comparability parameters and automation further in the future. Too tight bra-bands lead to partial suppression of breathing movements. Future research includes how bras of varying structures and materials affect breathing.

Filtering frames for comparable breathing states is automatically done; the 3D mesh comparison is made using CloudCompare. An implementation of a 3D mesh comparison into BodyTex4D is made in further steps.

Analysis of breast-bra interaction in motion is essential to evaluate bras' functionality further. The following Chap. 6 develops methods to enable comparability of movements to evaluate the bra functionality. The breast deformation during movements is analysed using BodyTex4D.

References

CloudCompare. https://www.cloudcompare.org/. (visited on 02/15/2023)

DIN EN ISO 20685-1:2019-03, 3D-Scanverfahren fur international kompatible anthropometrische Datenbanken- Teil 1: Prufprotokoll fur aus 3D-Scans extrahierte Korpermase (ISO 20685-1:2018); Deutsche Fassung EN ISO 20685-1:2018. Berlin. https://doi.org/10.31030/3021568

Move 4D User guide. In: 1.5 (2023)

Simmons K (2001) Body measurement techniques: a comparison of 3-D body scanning and physical anthropometric methods. PhD thesis. North Carolina State University, North Carolina. https://www.dbpia.co.kr/journal/articledetail?nodeid=node09570041

Chapter 6
Analysing the Breast-Bra Interaction in Motion Using 4D Scan Data and BodyTex4D

Abstract The 4D scan data evaluation methods from Chap. 4 are extended and applied to dynamic analyses of breast-bra interaction. Two test subjects run on a treadmill at varying speeds while wearing different bras. BodyTex4D is used to conduct a motion analysis, which provides insight into the comparability of running movements. Additionally, an extended BodyTex4D method selects 3D images of the running movement that meet predefined comparability criteria, allowing for a detailed examination of breast deformations. This comparison makes it possible to assess the influence of bra design, materials and seams on breast deformation. Furthermore, a newly developed method for bra support classification is applied to both test subjects, demonstrating high usability and repeatability. Differences in bra support based on design, size and treadmill speed become clearly visible. Overall, the results confirm that 4D scan data and BodyTex4D are highly effective for analysing and evaluating clothing-body interactions.

Keywords 4D scanning · Dynamic movement analysis · Automated scan data analysis · Body deformations · Breast-bra

6.1 Introduction and Objectives

This chapter discusses how 4D scan data improve the analyses of breast-bra interactions in motion. Many women experience breast pain, back pain and discomfort during activities in general. Lorentzen and Lawson (1987) determined higher vertical breast movements relative to body movements as reason for discomfort. Relative breast movement increases, especially while running because of increased body accelerations. A bra designed for the woman's needs is necessary for her comfort.

To analyse the breast-bra interaction a technology to capture the whole chest area in terms of movements and deformation is beneficial. The 4D scanner is able to capture body surfaces fully. The deformation of every body part during motions can be analysed, and dependencies between body movement, breast deformation and bras are detectable. To detect those dependencies, comparable movements are necessary. As in Chap. 5, non-repeatability of movements by test subjects is a major challenge.

A.-M. Schmidt, *Novel Methodologies for Digital Investigations of Clothing-Body-Interactions*, Lecture Notes in Bioengineering,
https://doi.org/10.1007/978-3-031-91794-3_6

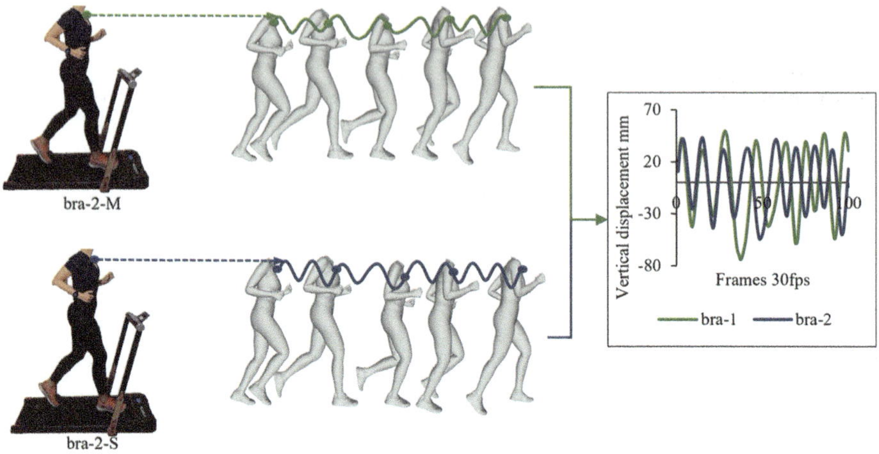

Fig. 6.1 A test subject running at the same treadmill velocities in different bra variations. The body's movement differs and is not comparable

Comparisons over multiple scans is not trivial. Figure 6.1 displays how two times running on the treadmill at the same speed lead to different body movement curves. Methods have to be developed to make those movements comparable.

Analysing tools to evaluate the breast-bra interaction in motion using 4D scan data and BodyTex4D is presented in the following. Running is the investigated movement on which the analysing method are developed upon. One methods investigates the similarity of body movement curves while running and selects comparable body curve segments for a breast deformation comparison. The other developed method is based on an amplification factor analysis. The method classifies bras by support properties. The method to find comparable frames of a running movement is developed as a prerequisite to compare the breast deformation in motion. Finding comparable frames and executing the amplification factor analysis is done by BodyTex4D based on the same 4D scan dataset. These methods improve the comparability of motions and increase the understanding of the breast-bra interactions and further support the improvement of bra development processes. Figure 6.2 gives an overview over dependencies of Chap. 6 to Chap. 4 and results of applying BodyTex4D for preliminary studies are displayed.

6.2 Preliminary Examinations

The 4D scanner is used as a measurement tool to analyse the breast-bra interaction in motion. 4D scanner settings to capture a running movement are determined in Sect. 6.2.1. The preliminary study identifies a test setup that achieves the closest

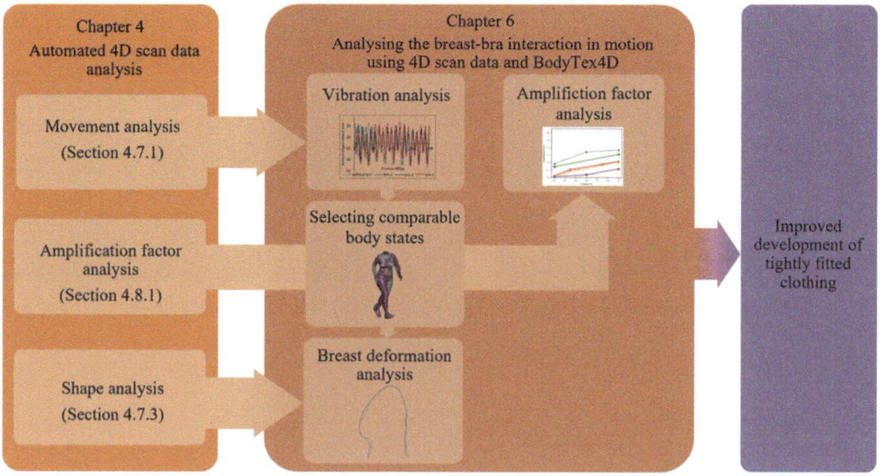

Fig. 6.2 Overview of Chap. 6

possibility of repeatability of movements for comparing breast-bra interactions in motion (Sect. 6.2.2).

6.2.1 4D Scanner Settings

The Move4D scanner captures running motions on the treadmill for following analyses. The treadmill position in the scanner area (Sect. 6.2.1.1) and capturing frame rates (Sect. 6.2.1.2) have been investigated.

6.2.1.1 Treadmill Position

The treadmill position for further experimental investigation has been determined to be the best in a 45° angle (Fig. 6.3) within the scan volume. Rotating the treadmill leads to a better capturing angle for the cameras of the used scanner setup as the used 4D scanner does not have cameras on the left and right sides.

6.2.1.2 Capturing Frame Rate

Setting the sampling rate high enough not to influence and falsify the measurement result to capture oscillating motions is important. Figure 6.4a displays data with an excessively coarse sampling rate, rendering the dataset insufficient. Figure 6.4b

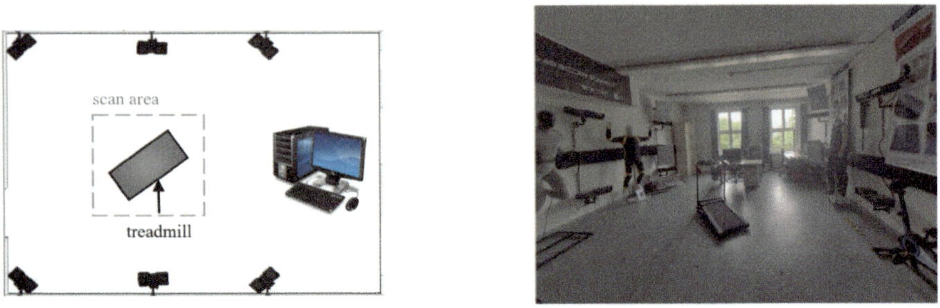

(a) Outlined treadmill position **(b)** Original scanner room with treadmill

Fig. 6.3 Treadmill position and rotation of 45° angle in the scanner area

Fig. 6.4 Two sampling rate examples to capture an oscillating movement (figure adapted from Salvatella and Redo-Sanchez 2010)

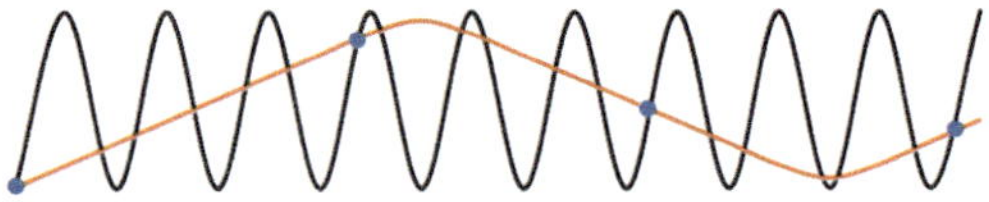

(a) Sampling rate not fulfilling the Nquist-Shannon Theorem

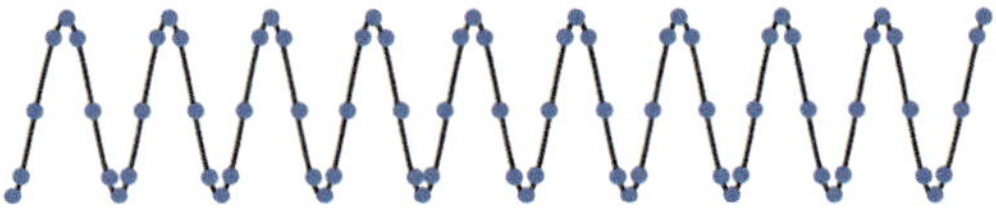

(b) Sampling rate fulfilling the Nquist-Shannon Theorem

displays data sampled at a sufficient rate, capturing the actual oscillation of the motion.

To define a sampling rate high enough to capture the actual oscillation, the Nyquist-Shannon sampling theorem (Salvatella and Redo-Sanchez 2010) is applied. The Nyquist-Shannon sampling theorem defines that the sampling frequency $f_{sampling\_rate}$ should be more than double the maximal frequency f_{max} of the measured signal (Fig. 6.2.1.2).

$$f_{sampling\_rate} \succ 2 \cdot f_{max} \tag{6.1}$$

For the 4D scanner, the sampling rate equals the frame rate. The frequency to be captured is the expected frequency of the running person. A cadence around 3 Hz is expected for a person running at velocity higher than 5 km h^{-1} (Hanley and Mohan 2014). This expected frequency result in a frame rate $f_{sampling\_rate}$ higher than 6 Hz. In the following experimental investigations, the frame rate is set to 30 fps or 60 fps. The sampling frequency fulfils the Nyquist-Shannon sampling theorem and can be used in further analyses. Both frame rates have been tested regarding breast deformation analyses and movement analyses and no difference in the accuracy of the results could be determined.

6.2.2 *Analysis of the Movement Repeatability*

The motion should be similar to measure the influence of various bras in motion. Faster movements lead to higher accelerations of the breast and higher forces impacting the body. The motion directly influences the breast behaviour and changes the boundary conditions of the breast-bra interaction.

Three different movements have been compared in terms of their repeatability:

- Running in place without any instructions
- Running in place with a metronome to set a step frequency. A similar technique was used by Liu et al. (2023)
- Running on a treadmill at a constant velocity of 7.5 km h^{-1}

For motions performed in one spot the test subjects put their hands on the hips to reduce the influence of the upper body and arm movement. All three motion types are performed with four bra variations and without a bra for comparison. The bras are described in Sect. 5.3.3. The motions are captured using the Move4D Scanner by IBV (Sect. 5.3.1) at 60 fps for 200 frames. From each motion and bra variation, a body movement analysis is performed using the movement analysis presented in Sect. 4.7.1 and the key-vertex-ID 13905 at the throat between both collarbones.

Results of the movement analysis are shown in Fig. 6.5. For each motion type, the curves are phase-shifted to overlap to retrieve a matching frame for comparison. Running on the spot resulted in the highest differences in oscillation between the curves. Frequency and amplitudes differ. The differences are shown in Fig. 6.5a at frames 0–40 and 150–200. Using a metronome leads to less fluctuation in frequency (Fig. 6.5b). The metronome has no measurable effect on amplitudes (red, blue curve in Fig. 6.5b). Running on a treadmill (Fig. 6.5c) shows the most consistent results for frequencies and amplitudes. The results using a treadmill are consistent with using a metronome to guide running motions in place, while treadmill running is more closely resembling real sports activity.

In conclusion, performing a running motion in place without any guidelines results in the least consistent data. Utilising a metronome or a treadmill results in more similar body movement curves, but did not result in trivial harmonic motions of the breast-bra interactions. In the following work, treadmill running is chosen for the breast-bra interaction analysis as this motion is used in many daily activities and sports. However, if no treadmill is available, repeatability improves by performing running motions in place conducted with a metronome.

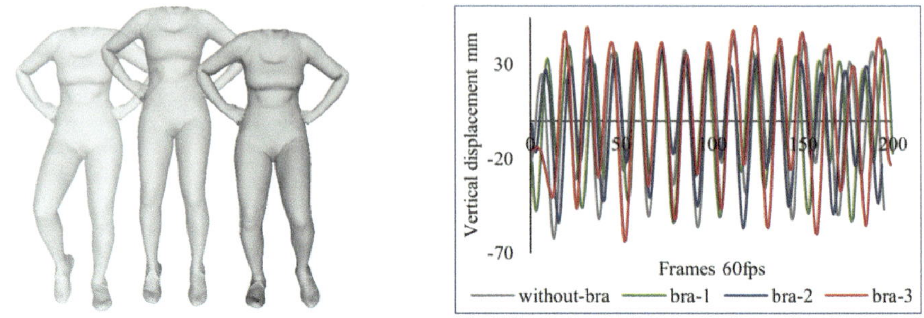

(a) Running in place intuitively, without instructions

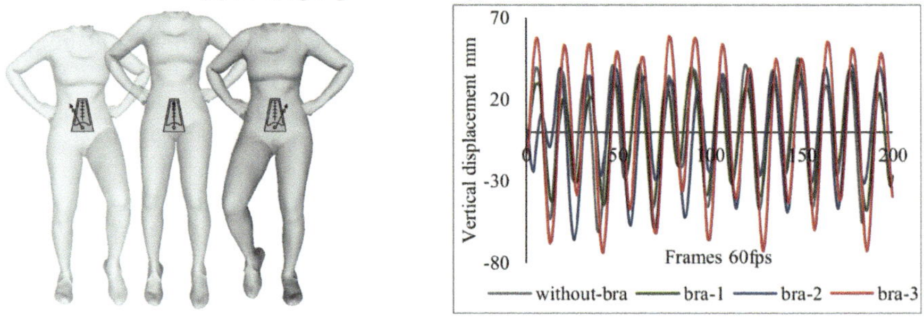

(b) Running on the spot with a metronom at 125 bpm

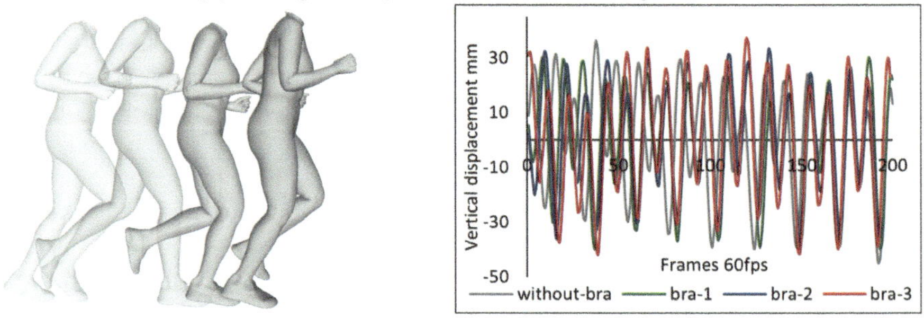

(c) Running on the treadmill at 7.5 $km\,h^{-1}$

Fig. 6.5 Body movement curve comparison in various motions

6.3 Materials

The 4D Scanner software by IBV (Sect. 5.3.1) processes the captured dynamic movements into homologous meshes. The export formats are *.obj, *.png and *.mtl files. The treadmill position is analogue to Sect. 6.2.1.1.

Two test subjects are scanned. Section 5.3.2 describes the two test subjects. Tested bras are described in Sect. 5.3.3.

6.4 Methods

The first step of analysing breast-bra interactions in motion is the vibration analysis (Sect. 6.4.2). Influences of treadmill velocities on the body and the breast movement is determined. Methods to retrieve comparable body movement curve segments to perform a breast deformation comparison are presented in Sect. 6.4.3. In the second part developments of a method to classify bra support is presented (Sect. 6.4.4). The novel method enables to compare dynamic behaviour of breast-bra interactions.

6.4.1 Scan Procedure

Scan plans of test subject-A and test subject-B are explained by Tables 6.1 and 6.2, respectively. All dynamic movements are performed on a treadmill. The treadmill is positioned in the middle of the scan area rotated by 45° (Fig. 6.3). The 4D scans on the treadmill are exported as a homologous mesh as an *.obj file. An example of such homologous mesh in a running motion can be seen Fig. 6.6.

Table 6.1 Scan plan of test subject-A for the movement analysis

Dynamic movements on a treadmill		**Walking**	**Running**				**Colours**
Treadmill velocity km h^{-1}		5	5	7.5	10		
Scan project (SP)		SP 1	SP 1	SP 1	SP 1	SP 2	
	Without-bra	A	A	A	A	A	
	Bra-1		M		M	M	
	Bra-2					M	
Bra variation						S	
	Bra-3	M	M	M	M	M	
		S	S	S	S	S	
	Bra-5		M	M	M	M	

Table 6.2 Scan plan of test subject-B for the movement analysis

Dynamic movement: Treadmill running **Treadmill velocity** km h^{-1}		**5**	**7**	**Colours**
	Without-bra	B	B	
	Bra-1	L	L	
Bra variation	Bra-2	L	L	
	Bra-3	L	L	
	Bra-4	75D	75D	
	Bra-6	85D	85D	

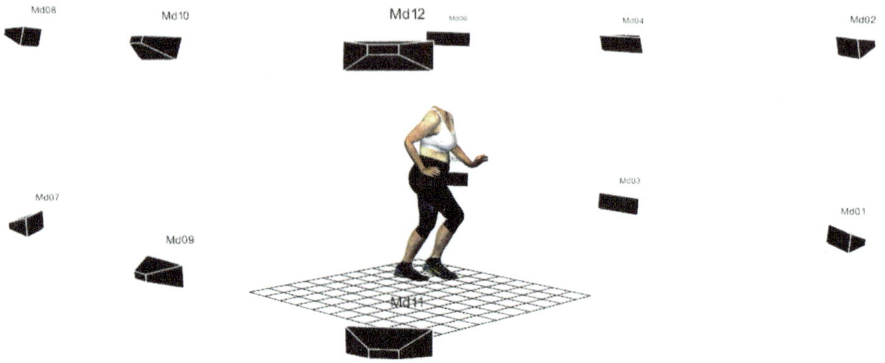

Fig. 6.6 Homologous mesh exported of test subject-B running on the treadmill

On test subject-A dynamic scans are performed at a treadmill velocity of 5 km h^{-1} in a walking motion and at 5 km h^{-1}, 7.5 km h^{-1} and at 10 km h^{-1} resulting in running motions. At 10 km h^{-1} was the maximum velocity test subject-A felt comfortable to run. The bra variations: without-bra, bra-3-M-A, bra-3-S and bra-5 were tested. To increase comparability, all scans are performed on the same day.

A second test run is conducted 5 months later for validation of the comparability method. The second test run parameters are

- test subject: A
- treadmill settings: velocity at 10 km h^{-1}, 45° rotation relative to symmetry axis of scan area
- bra variations: without-bra, bra-1, bra-2-S, bra-2-M, bra-3-S, bra-3-M-A and bra-5-M
- scanner settings: 60 fps for 400 frames

60 fps during the second test run is chosen to confirm the Nyquist-Shannon sampling theorem by comparing the scan data quality against the first test run at 30 fps (Sect. 6.2.1.2).

Dynamic scans of test subject-B on a treadmill at 5 km h^{-1} and 7 km h^{-1} are performed (Table 6.2). 5 km h^{-1} and 7 km h^{-1} provoke running motions of test subject-B. 7 km h^{-1} is the highest comfortable velocity for test subject-B. As test subject-B is older, fewer scans are performed in a single test run as for test subject-A. Motions where test subject-B wears bra variations: without a bra, bra-1–bra-4 and bra-6, are captured. The scans settings are 30 fps for 200 frames.

6.4.2 *Vibration Analysis*

The vibration analysis determines how changes in treadmill velocities affect the body and breast movement during running on a treadmill. The movement analysis module of BodyTex4D is applied (Sect. 4.7.1).

Scan datasets of test subject-A of the first test run are analysed in Table 6.1. Scan datasets in Table 6.2 of test subject-B is analysed. The vertical displacement of the body and breast is investigated.

Curves describing the body's movements are automatically obtained from scan datasets. The inputs for BodyTex4D's filter module are key-vertex-IDs and frames exploiting the homologous mesh capability of Move4D's scanner as presented in Sect. 4.7.1. The inputs indicate the filter method by mesh data (Sect. 4.4). The key-vertex-ID is 13905, representing the upper red marker point in Fig. 6.7a.

Curves describing breast movements are obtained by placing a coloured marker in bust point area (right marker point in Fig. 6.7a). The inputs for BodyTex4D's filter module are texture and frames exploiting the homologous mesh capability of Move4D's scanner as presented in Sect. 4.7.1. The inputs indicate filter method by mesh data and texture (Sect. 4.5.3). To filter the breast movement, the vertex-ID closest to the bust point of frame one per dynamic movement is selected via texture data of the marker bust point (Fig. 6.7b). By leveraging homologous meshes, the movement curves of vertical breast displacement are constructed by filtering the remaining scan dataset frames based on vertex ID, using BodyTex4D.

Results are movement curves of the vertical displacement of the body and breast while test subjects run at various treadmill velocities with different bra variations. The data is prepared to improve comparability and evaluability of vertical breast displacements for further analysis by following steps:

1. Breast and body movement curves are calculated.
2. Mean removal is performed on the breast and body movement curves (Fig. 6.8).

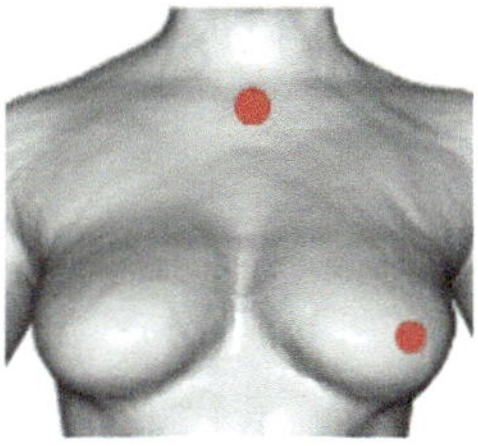

(**a**) Upper red marker point: body movements, lower red marker point: breast movement

(**b**) Selection of the breast vertex-ID by the identification method mesh and texture data

Fig. 6.7 Marker points for the movement analysis

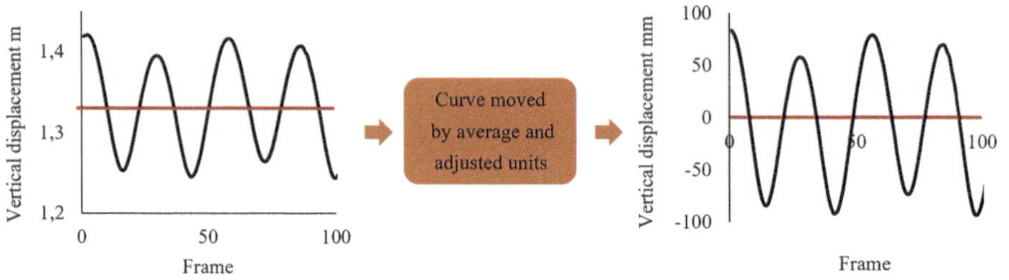

Fig. 6.8 Modification of extracted vertical movement data

3. Similar to cross-correlation, time-lag is reduced by phase-shifting the curves through setting the same t for the first maximum of each curve

Curves of each dynamic movement (scan run) are analysed based on following aspects.

1. similarity of frequency and amplitudes at one treadmill velocity
2. influence of various treadmill velocities on the frequency and amplitudes

This method predominantly demonstrates the use of the 4D scan data framework in the application of the breast-bra analysis. Single representative movement comparisons showcase the effects that can be determined.

6.4.3 *Analysis of Breast Deformation in Motions*

The second part of analysing the breast-bra interaction presents a method to compare the deformation of breasts in motion, focusing on the influence of different bra variations. Comparable frames of 4D scans are selected in Sect. 6.4.3.1. Selected frames are used to compare deformations of breasts in motion (Sect. 6.4.3.2).

6.4.3.1 Selection of Comparable Frames

To select comparable frames, two of three comparability criteria of Sect. 5.4.1 and Fig. 5.16 are applied. The comparability criteria are:

1. force similarity: applied (Fig. 5.16c)
2. body pose: applied (Fig. 5.16a)
3. breathing state: neglected (Fig. 5.16b)

The movement curves extracted in prior Sect. 6.4.2 are used to identify frames where force similarities occur. Force similarity denotes similar forces acting in the same regions of the body when comparing different frames. Similar movement curves lead to similar forces on the body because of comparable accelerations. Therefore, two

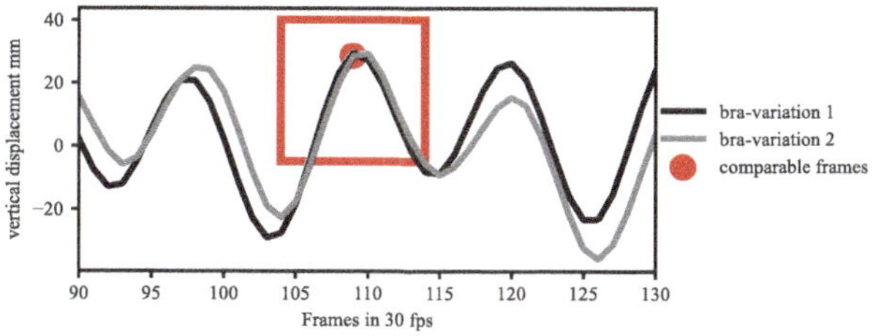

Fig. 6.9 Phase-shifted body movement curves for identifying similar curve sections

(a) Not similar body poses **(b)** Similar body poses

Fig. 6.10 Body pose comparison of selected frames in motion

body movement curves of two bra variations are phase-shifted to find similar curve sections (Fig. 6.9).

After aligning the movement curves by force similarity, investigation of pose similarity follows. While running, the most common pose difference is by opposing arm and leg positioning (Fig. 6.10a). The body movement curve can be the same, but the body is mirrored. In Fig. 6.10b, two poses fulfilling the body pose comparability criteria are depicted.

The similarity breathing state criteria Sect. 5.2.2 is neglected in this analysis because the effects of force similarity criteria and body pose criteria predominate the breathing state.

The data that sufficiently fulfil the comparability criteria are used to perform analyses of breast deformation (Sect. 6.4.3.2). The data preparation and breast deformation analysis is performed for test subject-B, investigating bra variations: without-bra, with bra-2, bra-4 and with bra-6. This analysis could be performed in the same way for other bra combinations or for test subject-A. However, the experiment is for showcasing the method as basis for future large-scale investigations.

6.4.3.2 Analysis of Breast Deformations

Selected comparable frames are used for a horizontal and vertical cross-section comparison. The cross-section reference points are near the bust points sliced horizontal and vertical.

6.4.4 Analysis of Amplification Factors

Prior section leads to the conclusion of running movements being superimposed harmonic signals. To analyse superimposed harmonic signals a Fast Fourier Transformation (FFT) is a profound and established method. From FFT an amplification factor analysis is derived. The amplification factor analysis is based on the developed method presented in Sect. 4.8.1. The method is capable of classifying bra support based on previously prepared body and breast movement datasets. According to the manufacturer and general understanding highly supportive sports bra should reduce relative breast movement compared to the body. A scientific based evaluation of bra support does not exist as of today. The section at hand provides a scientific based bra support evaluation method and making the method accessible by adding the method to BodyTex4D.

The amplification factor analysis is needed because previously discussed methods of synchronising movements by treadmill, metronome or data preparation are delivering insufficient results to instantiate a bra support metric (Sect. 6.2.2). Clear classification and comparison of bras variations is still impossible. The body movement (input) causes an breast movement (output). To simplify the analyses, a single dimension is evaluated: vertical movement component. The amplification factor analysis materialises to vertical movement component of the body as input, vertical movement component of the breast as output and the parameter for evaluating breast-bra interaction is the resulting amplification factor (Fig. 6.11). The movement curves the method is performed upon are extracted as described in Sect. 6.4.2.

The higher the amplification factor, the higher the relative breast movement resulting potentially in increased back pain, breast pain and discomfort during and after movements. The amplification factor is used to compare the bra support. The lower the amplification factor, the higher the bra support resulting in higher comfort.

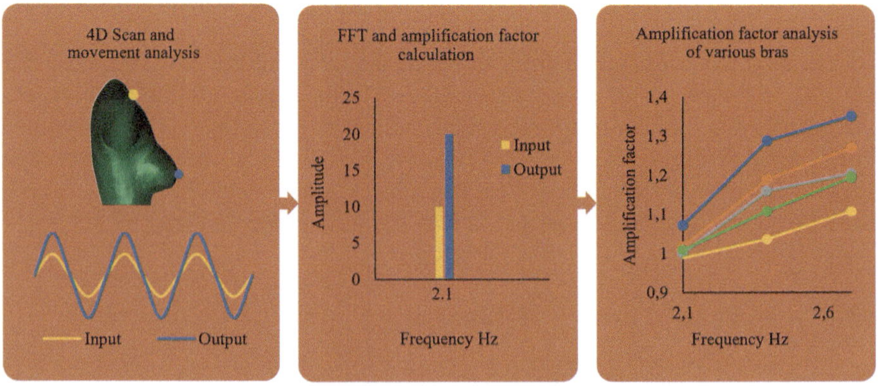

Fig. 6.11 Overview of the method: amplification factor analysis

For test subject-A, the amplification factor analysis is performed for all dynamic movements displayed in Table 6.1. Within the dynamic movements, running at $10\,km\,h^{-1}$ has been repeated with the same bra after five months. The repeatability of this method can be analysed and evaluated. For test subject-B, the amplification factor analysis is performed for dynamic movement scans described in Table 6.2.

6.5 Results

Based on 4D scan data and utilising BodyTex4D, breast deformation and bra support in running motions are investigated. The results of the developed metric to classify bra support by breast-bra interaction are presented in this section.

Firstly, the oscillation curves of breast and body movements at specific treadmill velocities are analysed (Sect. 6.5.1). Secondly, breast deformation due to the breast-bra interaction while running is investigated (Sect. 6.5.2). Lastly, the amplification factor analysis method to classify bras by support properties is presented (Sect. 6.5.3). Figure 6.12 displays the exported homologous meshes running on a treadmill. As shown, the treadmill is due to the positioning excluded from the scan data.

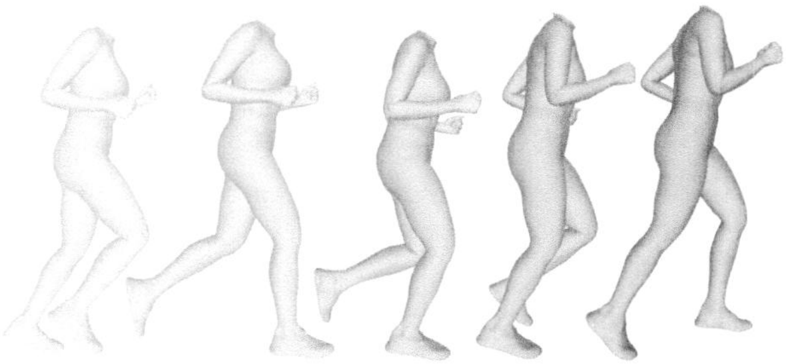

(a) Running frames of test-subject-A

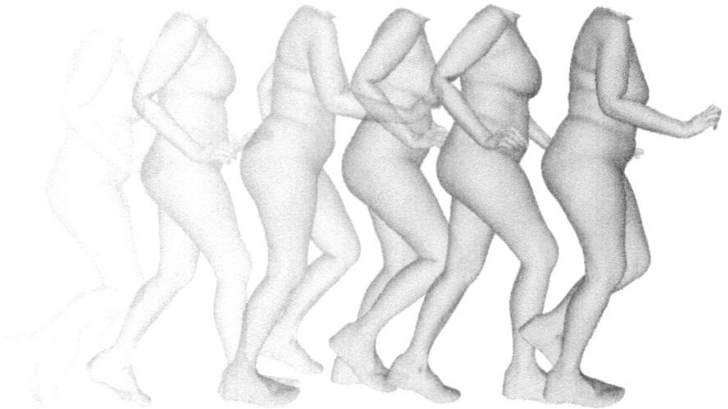

(b) Running frames of test-subject-B

Fig. 6.12 Homologous meshes of the test subjects running on a treadmill

6.5.1 Vibration Analysis

Results of the vibration analysis for test subject-A of the walking and running movements captured by the 4D scanner are depicted in figure 6.13. The analysis focuses on the body movements and the repeatability of capturing running motions.

Figure 6.13a and b shows the body movement curves for running motions at $5\,\text{km}\,\text{h}^{-1}$ and $10\,\text{km}\,\text{h}^{-1}$ with bra variations bra-1, bra-2, bra-3 and without-bra. It can be seen that the amplitudes and frequencies at $10\,\text{km}\,\text{h}^{-1}$ are higher compared to $5\,\text{km}\,\text{h}^{-1}$. Body movement curves at $10\,\text{km}\,\text{h}^{-1}$ are more similar compared to more dispersed curves representing running at $5\,\text{km}\,\text{h}^{-1}$. Lower velocities leads to more noisy data compared to data of running at a higher velocities. Reason being that at lower velocities, test subjects experience more flexibility in their style of running. Higher running velocities lead to higher effort pushing the body closer the limit. Locomotion at the limits of the body's capacity tends to become more uniform, enhancing the repeatability of the running motion.

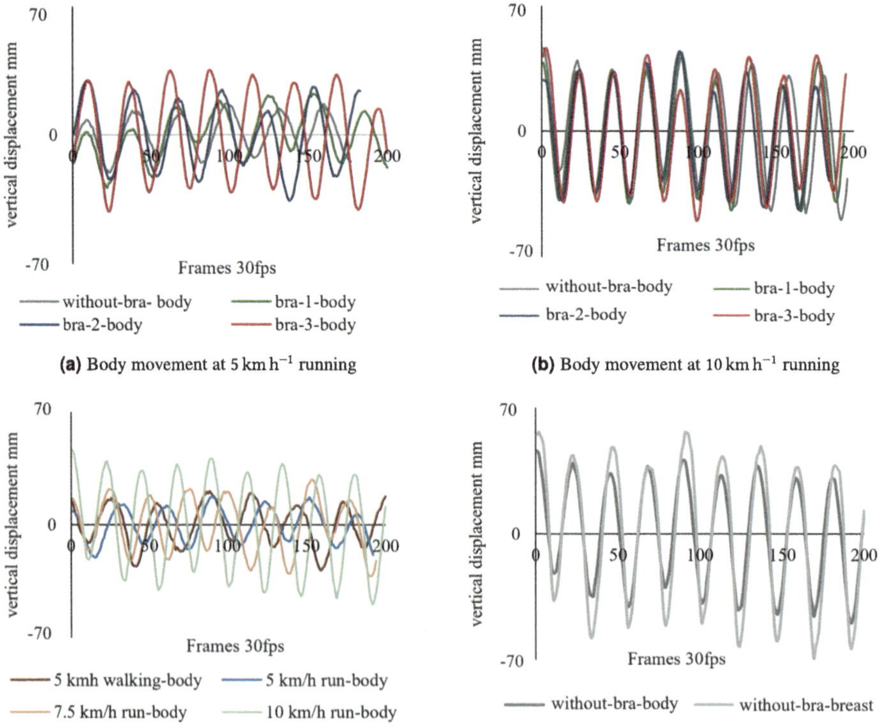

(a) Body movement at $5\,\text{km}\,\text{h}^{-1}$ running (b) Body movement at $10\,\text{km}\,\text{h}^{-1}$ running

(c) Body movement at various treadmill speeds of bra-1 (d) Body and breast movement at $10\,\text{km}\,\text{h}^{-1}$ without a bra

Fig. 6.13 Vibration analysis of test subject-A

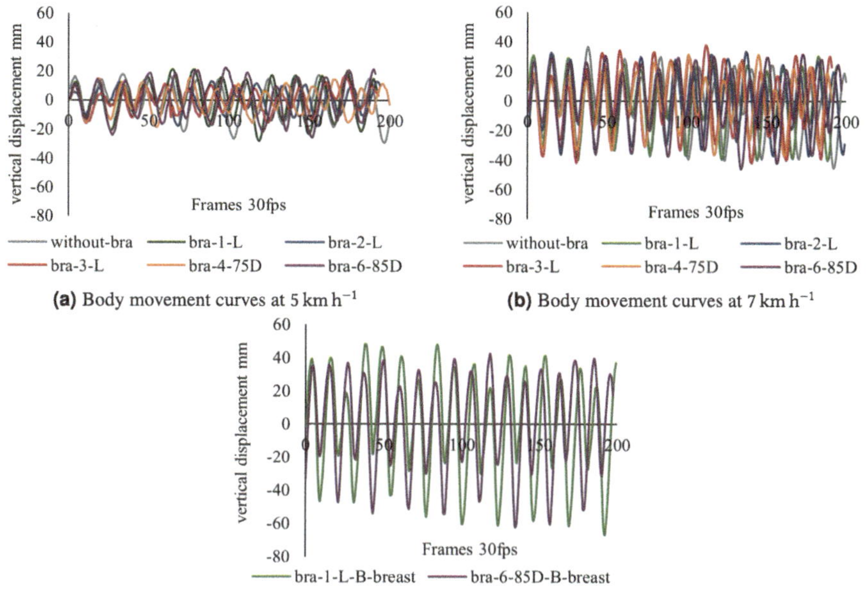

(a) Body movement curves at 5 km h^{-1}

(b) Body movement curves at 7 km h^{-1}

(c) Breast movement curves at 7 km h^{-1}

Fig. 6.14 Body and breast movement analysis at two treadmill speeds of test subject-B

Similar results could be determined when comparing the body movements of four different walking or running motions wearing bra-1-M-A (Fig. 6.13c). 5 km h^{-1} walking led to the lowest amplitudes and frequencies. The amplitudes and frequencies increase for running at 5 km h^{-1}. Running at 7.5 km h^{-1} and 10 km h^{-1} leads to the highest frequencies and amplitudes. Further analyses regarding the running frequencies are made in Sect. 6.5.3.

Figure 6.13d shows movements of body and breast of test subject-A running without bra at 10 km h^{-1}. The breast movement has more pronounced amplitudes and is slightly phase shifted compared to the body movement. The phase shift is caused by the higher mass of test subject-A's breasts, causing higher inertia. The results confirm, that treadmill running is not resulting in same movements and a method to make those movements comparable is needed.

Findings for test subject-B shown in Fig. 6.14 confirm results produced for test subject-A. Higher treadmill velocity increases body movement amplitudes, frequencies and results in similar movement curves. Body movement amplitudes of test subject-B are close to half the value of the amplitudes of test subject-A. The difference in amplitudes might be caused by differences in age and/or weight. The younger test subject, test subject-A, had more energy and confidence running on the treadmill. Nevertheless, the amplitudes and frequencies are more similar at a higher running speed, the breast movement curves are remain challenging to compare. In Fig. 6.14c, the breast movement of bra-1 and bra-6 of test subject-B at 7 km h^{-1} is displayed.

Curves bra-1 and bra-6 are similar, and a classification regarding differences in the bra support cannot be performed on breast movement curves. The amplification factor analysis solves this challenge and results can be seen in Sect. 6.5.3.

6.5.2 *Analysis of Breast Deformations in Motions*

The following section discusses results of the breast deformation in motion. Firstly, comparable frames are selected as described by Sect. 6.5.2.1. Secondly, results of the breast deformation analysis is presented in Sect. 6.5.2.2.

6.5.2.1 Selection of Comparable Body States

Comparability criteria force similarity is applied. Figure 6.15a and c show curves of test subject-B's vertical movement component of the body while running. Phase-shifting of the curves pronounces similar movement regions.

In Fig. 6.15a, the movement curve of bra-6-85D-B (magenta curve) has been phase-shifted by 14 time steps (frames). The body movement curves of without-bra-B and bra-6-85D-B are the same from frame 76 to frame 84. The accelerations and forces on the body are the same for this curve sections. Within this similar curve sections, two comparable frames are selected: Frame one at the maximum of the curve at frame 79 and frame two at the point of inflection at frame 83 from movement curve bra-6-85D-B. The selected frames are marked with a red dot. The force similarity parameter is fulfilled for these two movement states as the breast experiences equal accelerations.

Investigating the body pose similarity in Fig. 6.15b, both body poses are similar for frame 79 of the bra variation without-bra and bra-6-85D-B. The right leg is stretched out to the front, and the upper body pose is similar. In frame 83, the body poses are similar in the same fashion but the frames describe the movement 0.13 s after frame 79.

For the bra-variations bra-2-L-B and bra-4-75D-B the force similarity analysis results are presented in Fig. 6.15c. With a phase shift of 6 frames between movements curves of bra-2-L-B and bra-4-75D-B, a similar movement curve section is determined. The force similarity is fulfilled for this selected frame. As shown in Fig. 6.15d the body pose criteria is fulfilled.

The selected frames are defined as comparable for a breast deformation analysis. The frames form the basis for further investigation of influences of motions and bras on breast deformations.

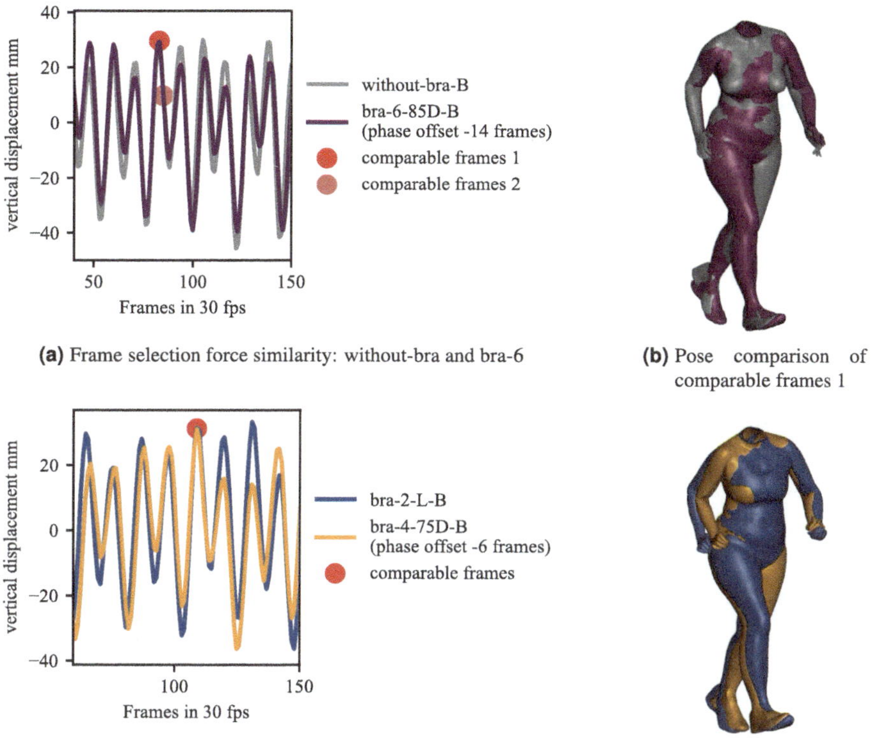

(a) Frame selection force similarity: without-bra and bra-6

(b) Pose comparison of comparable frames 1

(c) Frame selection force similarity: bra-2 and bra-4

(d) Pose comparison of comparable frames

Fig. 6.15 Selection of frames with force similarities of the body movement curve of test subject-B running at 7 km h^{-1}

6.5.2.2 Analysis of Breast Deformations

The breast deformation analysis of two breast-bra interactions while in motion are exhibited in Figs. 6.16 and 6.17.

Grey colouring indicates the breast shape without-bra-B. Magenta colouring indicates the breast shape of test subject-B wearing bra-6-85D-B (Fig. 6.16). Figure 6.16b and c display the shape deformation at comparable frame 79 (local curve maximum, equal accelerations, similar forces). Figure 6.16e and f display the breast shape deformation of without-bra-B and bra-6-85D-B at frame 83 of the movement, point of inflection. The comparison of thorax cross-sections reveals notable differences of breast deformations in motion of without-bra-B and bra-6-85D-B. Analysing bra-6-85-D-B, the breast is in all cross-sections closer to the body. The assumption is, having the breast closer to the body shifts the breast's centre of mass closer to the body's centre of mass reducing the leverage between the centre of masses resulting in reduction of forces acting on the breasts. The relative movement between breast

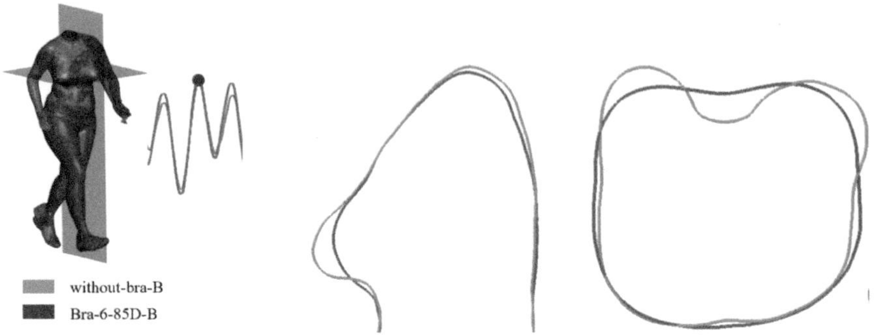

(a) Slice positions at movement state 1 **(b)** vertical cross-section at movement state 1 **(c)** horizontal cross-section at movement state 1

(d) Slice positions at movement state 2 **(e)** vertical cross-section at movement state 2 **(f)** horizontal cross-section at movement state 2

Fig. 6.16 Cross-section analysis of test subject-B running at 7 km h^{-1} without bra and with bra-6

and body lowers and decreases pain and discomfort. Between the two movement states, it can be seen that the breast without a bra is still moving upwards while the body is already moving down. The higher the relative displacement of breast to body, the higher the discomfort. Compared to the bra-6-85D-B cross-section, the bra and breast shapes are deformed minimally. Additionally, in Fig. 6.16f the side swing of the right breast of the bra variation without-bra-B can be seen.

With this breast deformation analysis not only the breast-bra interaction can be visualised, but also the understanding of how the breast deforms without a bra in motion is increased.

Figure 6.17 displays a cross-section comparison of the bra variations bra-2-L-B and bra-4-75D-B. Comparing the shapes of superimposed cross-section portrayed in Fig. 6.17b, different upper breast deformations are observed. Bra-2-L-B (blue) is moving upwards, while the breast shape in bra-4-75D-B is closer to the body and stiffer. This indicates lower breast movement occurs for bra-4-75D-B. A further investigation of the breast movement with the bra variations is discussed in section 6.5.3. Differences occur in the horizontal cross-section comparison are minor (Fig. 6.17c).

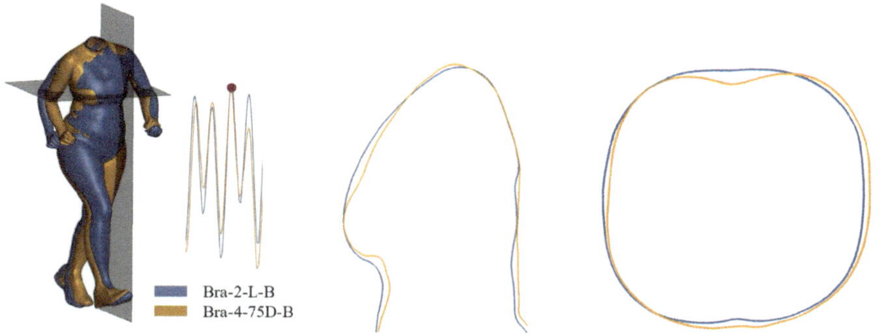

(a) Slice positions and moving state: maximum (b) Vertical cross-section at movement maximum (c) horizontal cross-section at movement maximum

Fig. 6.17 Cross-section analysis of test subject-B running at 7 $\mathrm{km\,h^{-1}}$ with bra-2 and with bra-4

In summary, the 4D scanner captures high quality data for deformation behaviour analyses of deformable body parts. Applying cross-section analysis, the deformable-body-textile interaction can be visualised during movements. Presented methods to prepare data by selecting comparable body movement states increases the evaluability, enabling deformation analyses. Nevertheless, it should be noted that this is only a snapshot of breast deformation conditions. To make precise statements about how each bra influences breast deformation during motion, further analyses are necessary. This can be achieved by quantifying curve similarity metrics, such as cross-correlation, applied to the obtained data. Quantification of all assumptions made are out of scope of this work as the focus lays on developing an overarching methodology framework from data collection to automatic classification of tightly fitted clothing properties and functionality evaluation. To build up a complete methodology, amplification factor analysis as exemplary metric is introduces by this chapter. The results are discussed in following section.

6.5.3 *Analysis of Amplification Factors*

The following section presents a quantification method for bra support; amplification factor analysis. Previous sections show the procedure to acquire comparable 4D scan data of running motions. In the context of this study, the amplification factor serves as an index quantifying the extent to which body movements excite or amplify breast motion. The higher the amplification factor, the large the relative movement between breast and body, the lower the support of given bra. In turn, high support of a bra is defined by the magnitude of reduced breast movement compared to other bra variations worn by the same test subject.

Main harmonic components of each body movement is extracted by FFT and can be seen in Table 6.3. Fundamental observation is the causality of treadmill velocity

Table 6.3 Main frequencies of body movement curves while treadmill running at various velocities determined by the FFT analysis

Bra variation	Test subject	Scan project	Size	Main frequencies Hz				
				Walking	Running			
				5	5	7.5	7	10
Without-bra	A	1		1.95	2.1	2.4		2.7
		2						2.7
	B				2.4		2.7	
Bra-1	A	1	M		2.1			2.7
		2	M					
	B		L		2.4		2.7	
Bra-2	A	2	M					2.7
			S					2.7
	B		L		2.55		2.7	
Bra-3	A	1	M	2.1	2.25	2.55		2.7
			S	2.1	2.25	2.55		2.7
		2	M					2.7
			S					2.7
	B		L		2.4		2.7	
Bra-4	B		75D		2.4		2.7	
Bra-5	A	1	M		2.1	2.4		2.7
		2	M					
Bra-6	B		85D		2.4		2.7	

to magnitude of the main harmonic component of given test run. The higher the treadmill velocity, the higher the main harmonic component. Running at 7 km h^{-1} for test subject-B and running at 10 km h^{-1} for test subject-A leads to the same main harmonic component frequency (main frequency) of 2.7 Hz. Causality is established, but no trivial correlation can be derived at this stage. The main influence on the frequency seems to be the test subject. The main frequency for test subject-A fall into the range of 2.4 Hz to 2.55 Hz for a velocity of 7.5 km h^{-1}. Lower main frequency range of 2.1 Hz to 2.4 Hz is observed for running at 5 km h^{-1}. It can be seen, that the main frequency differ within the same treadmill speed especially for slower running velocities. This results reinforces the observation in Sect. 6.5.1 that higher treadmill speeds lead to a more even running frequency. Analysing the body movement before a bra comparisons is necessary for reliable results.

For test subject-A, two amplification factor analyses are conducted. The first amplification method investigates the breast-bra interaction at different treadmill velocities wearing following bra variation: without-bra-A, bra-1-M-A, bra-3-S-A, bra-3-M-A and bra-5-M-A (figure 6.18). Figure 6.18 shows for all bra variations,

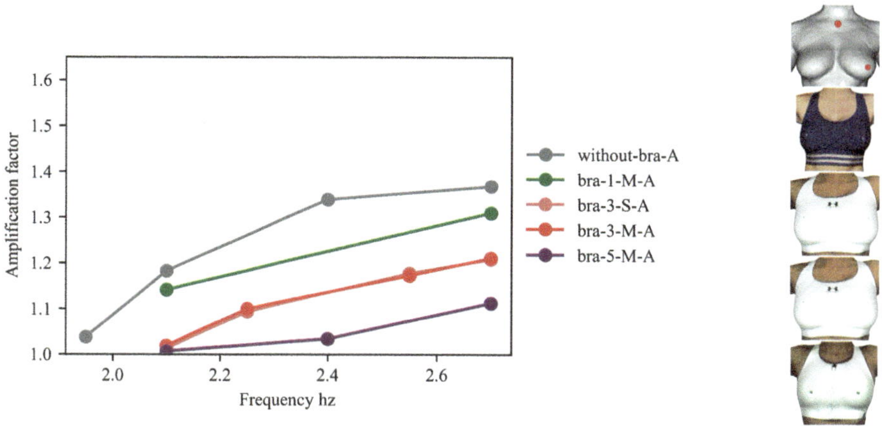

Fig. 6.18 Results of the amplification factor analysis of without-bra-A, bra-1-M-A, bra-3 in sizes M and S and bra-5-Ma using 4D scan data from scan project one for test subject-A

the higher the main frequency, the higher the amplification factor. Higher running velocity results in higher body accelerations. Higher body accelerations causes higher relative movements between breast and body. In conclusion, the faster the test subject is capable of running, the higher the bra support must be for comfort.

The highest amplification factor for test subject-A is measured without bra. The result is expected as the breast movement is not restricted in movement. Every bra leads to a lower amplification factor compared to running without bra. Following statement is proven, bras decrease breast movements in the vertical direction. The classification of the bras from lowest support to highest support is bra-1-M-A, bra-3-S/M-A and bra-5-M-A. Bra-1-M-A front and back parts are composed of textiles with low tensile strength. Compared to Bra-1-M-A, Bra-3-S-A is a medium supportive sports bra with higher textile strength and stiffer seams compared to bra-1-M-A. Bra-5-M-A is a high-compressive sports bra. Bra-1 is closer to the support of wearing no bra than bra-3-M-A. With decreasing running velocity the differences in bra support become minor. Test runs resulting in the frequency of 2.1 Hz show minor differences of amplification factors of bra-3-S-A and bra-5-M-A. Bra support diverges at higher frequencies (2.7 Hz). Bra-5-M-A is designed for high-intensity sport activities with stiffer materials and a tighter, compressive fit and thus results in the lowest amplification factors. Bras are designed application-specific. Sports bras are suitable for high-intensity activities exerting higher pressure but reducing breast movements. During low-intensity activities, the use of sports bras may be counterproductive, as the discomfort associated with bra pressure outweighs the impact of breast movement on perceived comfort. In turn, bras with light support exert less pressure but offer enough support for low-intensity activities. What seems to be common knowledge is quantified by this work. Confirming the expected support properties, shows the reliability and validity of the amplification factor analysis.

Unexpectedly, the difference in size of bra-3-S-A and Bra-3-M-A has no effect on the amplification factor. Smaller bras are assumed to decrease breast movement due to the tighter fit. It is hypothesised that tightness is not the single-strongest influence on support properties of bras. The overall fit must be considered. The underbust band moves upwards the chest slightly onto the breast reducing support for the smaller tested bra-3-S-A. This investigation leads to the assumption that the bra fit has a high influence on the breast-bra interaction. Higher breast-bra pressures acting from the bra on the breast does not cause higher bra support directly. The observation improves understanding of bra support and subsequently bra development.

Based on the broadened knowledge on bras newly developed bras are now testable and classifiable relative to existing bras based on the amplification factor analysis.

In the second amplification factor analysis of test subject-A the bra variations without-bra, bra-2-M/S-A, bra-3-M/S-A and bra-5-M-A are anaylsed at a treadmill velocity of 10 km h^{-1} to test the methodology (figure 6.19). The 4D scans are performed 5 months after the scan project 1 to cross-check effects and confounding factors. The effect of the running velocity is validated sufficiently and is kept constant at 10 km h^{-1}.

Comparing the amplification factor without bra of test subject-A in scan project 1 (Fig. 6.18 grey line) and project two 2 (Fig. 6.19 grey bar) at 2.7 Hz, the amplification factor of scan project 1 with the value of 1.37 is lower than scan project 2 with 1.4. Difference could be due to lower breast volumes or composition (e.g. water retention) of test subject-A. Within this time, the breast size may have changed due to body weight changes or because test subject-A is in a different phase of the menstrual cycle among many more possible influences which need to be explored in further studies. It follows, the amplification factors from scan project 1 are not comparable directly to scan project 2, but the bras are classifiable by support among one another. As in the analysis of scan project 1 bra-5-M-A is the bra with the highest support, followed by bra-3-S-A. As bra-2-M-A and bra-3-S-A have similar amplification factors with

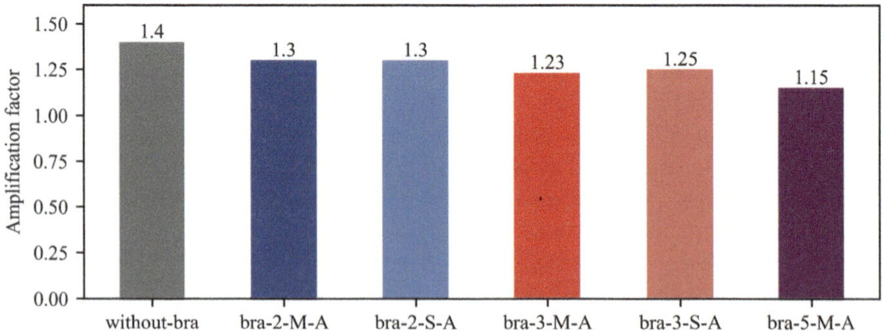

Fig. 6.19 Results of the amplification factor analysis of without-bra-A, bra-2 and bra-3 in sizes M and S and bra-5-M-A using 4D scan data from scan project two of test subject-A (running at 10 km h^{-1} and 2.7 Hz)

a difference of 0.05-0.07, higher support of bra-2 compared to bra-1 tested in scan project 1 is assumed.

Analogue to the amplification factor analysis of bra-3 in scan project 1, the size difference of bra-3-S-A and bra-3-M-A did not affect the bra support in scan project 2. The same effect could be seen for bra-2. The smaller size of bra-2-S-A did not increased the bra support. In contrast to bra-3, the fit of both bra-2-M and bra-2-S did not change during movement. The bra-band did not slip on top of the breast, reducing the support properties of bra-2. No bra support difference can still be seen, because bra-2 consists of low-tensile strength, elastic, one-layered materials. No seams to stiffen up the front and back parts exist as it is the case for bra-3. The seams are assumed to have a major influence on the bra support. The lower bra support fits to the results of the breast deformation analysis in a static pose (Sect. 5.5.2.1). In the T-pose analysis, the difference in the breast deformation between bra-2-S-A and bra-2-M-A was smaller than the difference between bra-3-S-A and bra-3-M-A (size difference S versus M). Indicating that higher-strength materials and seams do not compensate for the smaller cut pattern as bra-3 has higher-strength materials and seams while bra-2 does not. Overall, determining the same support classification for scan project 1 and scan project 2 proves the usability, validity and accuracy of the amplification factor analysis.

Figure 6.20 shows the amplification factor analysis results for test subject-B. Similarities in the results are observed to the amplification analysis of test subject-A of scan project 1 (Fig. 6.18). The higher the frequency, the higher the amplification factor. The order of the bras by amplification factor, in turn, bra support aligns with the results of test subject-A. Without-bra exhibits the smallest support, followed by bra-1 to bra-4. Bra-6 shows the highest support. Compared to test subject-A, the amplification factors are higher. At a running frequency of 2.7 Hz, the amplification factor of test subject-B without a bra is 1.615. At the same frequency, the application factor without-bra of test subject-A is between 1.35 and 1.4. Presumably, the differences are caused by larger breasts of test subject-B, and softer skin softer due to the higher age of test subject-B. Similar to test subject-A, the differences between the amplification factors of the bras are increasing with higher running frequency.

6.6 Conclusion and Discussion

This chapter presents various methods for analysing the breast-bra interaction in motion. Knowledge about the body's deformations and the body-textile interaction in motion is necessary for improving the fit, functionality and comfort of tightly fitted clothing. Running motions of test subjects are inconsistent. Methods are developed to transform 4D scan data of running motions into a comparable state utilising BodyTex4D. Two test subjects are 4D scanned with the Move4D scanner while running on a treadmill. Treadmill velocities from 5 km h^{-1} to 10 km h^{-1} in combination with multiple bra variations are captured. The following key findings are established:

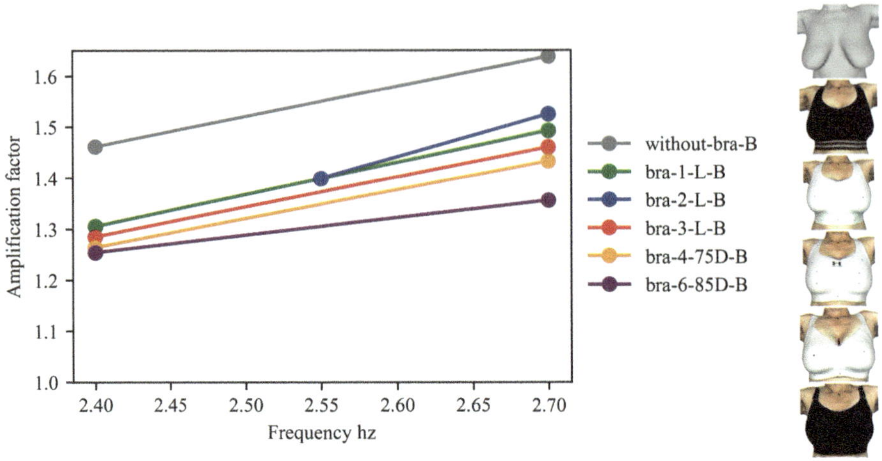

Fig. 6.20 Results of the amplification factor analysis of bra-1-M-A to bra-4 and bra-6-85D-B for test subject-B

1. **4D scan data and the BodyTex4D improve the analyses of breast-bra interaction in motion**: Using 4D scan data, the breast deformation and movement with and without a bra can be captured. The BodyTex4D filters and presents movement curves for analyses automatically. Application-specific methods as the amplification factor analysis could be implemented easily. BodyTex4D is valid and applicable to a variety of applications.
2. **Running motions performed on a treadmill does not result in reproducible motions**: Test runs of the same test subject at the same treadmill velocity with the same bra on the same day results in different frequencies and amplitudes for each test run. To improve the repeatability, longer capturing duration could be performed. As test subjects feel more comfortable running on a treadmill, repeatability could increase. However, a humans can never run the same way. Furthermore, pushing test setup further to reach repeatability, for example, by conducting experiments with professional athletes could lead to results not applicable to the broad population, contradicting the target of this study to implement a framework for more individualised clothing development; thus, simple movement comparisons are insufficient.
3. **The breast deformation with and without a bra can be analysed using 4D scan data and the BodyTex4D in motion**: Similar motion curves of treadmill running are chosen by compatibility criteria: force similarity and body pose similarity. The prepared data is processed to compare breast deformation for six different bras and without bra. Comparisons are performed based on vertical and longitudinal cross-sections of the test subjects and breasts. Most significant finding of the cross-section analysis is: cut pattern and material difference influencing the breast deformation in motion can be captured. This analysis improves the knowledge about breast deformation and the interaction with different bras in motion. Cut

patterns can be redesigned using this information, and newly developed bras can be tested. Till now, the selection of similar curve segments is manual, an improvement of the automatic curve selection, for example, by calculating cross-correlation depending on time-leg has to be made in following research work.

4. **A new method to classify bras has been determined—the amplification factor analysis for breast movements**: The amplification factor analysis is successfully used to classify bras by support properties and integrated into BodyTex4D. Correlations between the amplification factor and running velocities are verified. The amplification factor is the quotient of the results of a FFT at the main harmonic frequency of the body over the main harmonic frequency of the breast. The main harmonics are determined by FFT. The main harmonics frequency show an increasing tendency for higher running velocities. Differences in bra supports are observed and following significant observations are made:

   - Higher amplification factors are calculated for the test subject with larger and older breast
   - Smaller bra size does not cause higher bra support, as commonly assumed
   - Bra designs are application specific:
     - Bras for low intensity activities are unsuitable for high intensity activities as the bra compresses the breast insufficiently
     - Bras for high intensity activities are unsuitable for low intensity activities as the bra causes discomfort by high pressure zones

   The amplification factor analyses classified the bra support repeatedly while confirming the manufactures statements about the support categories low to high. It must be noted that the amplification factor is based on a linear amplification behaviour. Linearity is an approximation and works for movements similar to running. The method can be applied to other movements, such as jumping, but the results cannot be transferred from running to jumping. This method is unsuitable for non-linear movements, as this is a non-suspended system. The method should be applied to more test subjects and breast sizes in further investigations. This study investigated is focused on vertical movements of the body and breast. The next step is to include the upper-body rotation in the analysis. The upper body rotation can be calculated using the movement analysis script. Vertex-IDs on the shoulders can be tracked, and the rotation can be calculated.

References

Figure of a woman running on a treadmill. https://www.amazon.co.uk/Laufhome-Treadmills-Treadmill-Bluetooth-Protection/dp/B098PYD9FZ. (visited on 07/23/2024)

Hanley B, Mohan AK (2014) Changes in gait during constant pace treadmill running. J Strength Cond Res 28(5):1219–1225. https://doi.org/10.1519/JSC.0b013e3182a38796

Lorentzen D, Lawson L (1987) Selected sports bras: a biomechanical analysis of breast motion while jogging. Phys Sports Med 15(5):128–139. ISSN: 0091-3847. https://doi.org/10.1080/00913847.1987.11709355

Liu S et al (2023) Predicting the effect of bra pad specifications on breast deformation during jumping using a finite element method. Int J Cloth Sci Technol 35(5):779–798. ISSN: 0955-6222. https://doi.org/10.1108/IJCST-02-2023-0009

Salvatella G, Redo-Sanchez A (2010) Terahertz spectroscopy in medical and nutritional applications. https://api.semanticscholar.org/CorpusID:73650361

Chapter 7
Part II: Conclusion, Discussion and Outlook

Abstract This chapter summarises the findings from Chaps. 4–6. Research gaps that have been addressed are presented and discussed in relation to existing solutions. Additionally, the transferability of the developed methods to other application areas and their economic feasibility are considered. Overall, the results demonstrate the high usability of 4D scan data and BodyTex4D in clothing development and the analysis of body deformations. The effectiveness of the developed methods in analyzing breast-bra interaction is evaluated, and the key findings are summarized.

Keywords 4D scanning · Dynamic movement analysis · Automated scan data analysis · Body deformations · Breast-bra

7.1 Conclusion

Chapter 4 **presents an framework for analysing 4D scan data (BodyTex4D).** BodyTex4D automates data preparation by filtering frames from 4D scan datasets based on input parameters such as key-vertex-ID. BodyTex4D automates 4D scan data analyses by calculating and visualising motion curves, cross-sections, lengths, mean removal, FFT and amplification factors. The implementation of BodyTex4D in Python is described. The availability and accuracy of homologous meshes is exploited by BodyTex4D. Homologous meshes prove to be a powerful data structure to analyse motions and deformations flexible, time-efficient and from many points of view. Developed methods integrated into BodyTex4D has the advantage of being applicable to analyse a wide range of applications. In this work, three basic analysis methods have been developed, implemented and applied: The movement analysis (Sect. 4.7.1), length analyses (Sect. 4.7.2) and shape analysis (Sect. 4.7.3). More methods are presented as examples and templates for extensions of BodyTex4D. BodyTex4D is modular, whereas methods can be used individually or combined. In Chaps. 5 and 6, BodyTex4D is applied and verified in its usability to investigate breast-bra interaction in a static T-pose and in motion. Figure 7.1 shows an overview of the applied advanced methods. In purple, methods for analysing the breast-bra interaction are marked.

A.-M. Schmidt, *Novel Methodologies for Digital Investigations of Clothing-Body-Interactions*, Lecture Notes in Bioengineering,
https://doi.org/10.1007/978-3-031-91794-3_7

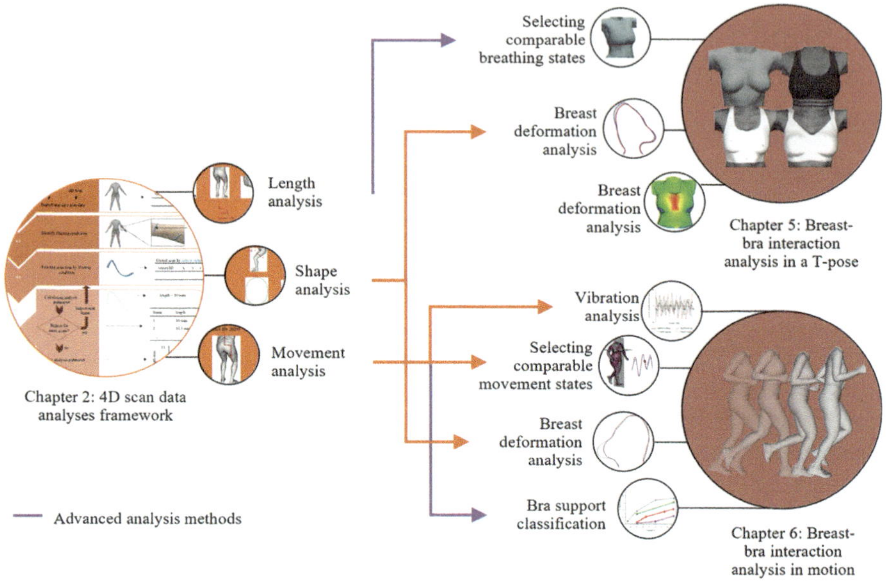

Fig. 7.1 Chapter conclusion of Part II

Chapter 5 **improves the breast-bra interaction analysis of static T-poses using 4D scan data by selecting comparable breathing states and breast deformation comparisons**. Experiments with two test subjects wearing different bra sizes and six different bra models are conducted. Firstly, the influence of the underbust circumference difference while breathing is observed and investigated. An extended-length analysis to determine comparable breathing states is developed and integrated into BodyTex4D. Comparable breathing states are achieved by taking frames from different 4D scan datasets where the test subject has the same underbust circumference. Using comparable frames, 3D mesh comparisons and 2D cross-section analyses are performed. Differences in breast deformation caused by the tested bra models and without bra are significant. The effect of bra sizes (S and M of the same bra model), cut patterns and materials are analysed. Accurate comparison of the bras is performed, enabled by the improved comparability of the frames. The presented breast deformation analyses can be used to compare newly developed bras with existing bras or to identify dependencies between breast deformation and bra efficiently for a faster and better bra development process in the future.

Chapter 6 **improves the breast-bra interaction analysis in motion using 4D scan data by a breast deformation comparison and a classification of the support of bras**. The motion chosen for analysis is running on a treadmill. Similar curve sections are selected of test runs to compare. Comparability criteria for selecting comparable frames are defined and applied. The comparability criteria used are force similarity and body pose similarity. Force similarity is defined by the state of acceleration of the breast. Body pose similarity is self-explanatory. Based on the comparable frames a breast deformation analysis is performed. Described methods lead to the classification metric by calculating amplification factors depending on running velocity and are integrated in BodyTex4D. The classification metric is developed based on an amplification factor analysis (Sect. 4.8.1). The amplification factor analysis method is applied to the two test subjects used in Chap. 5 running wearing different bra models at treadmill velocities in a range of 5–10 km h^{-1}. The bra models are classified according to their support. The higher the running frequency, the higher the difference in the amplification factors between the bra models. It is observed a smaller bra size does not automatically lead to a higher bra support, counter-intuitively. The fit can have a higher influence than the cut pattern part reduction of the smaller sizes. The amplification factor from the test subject having larger breasts and being older is higher due to the higher mass of the breasts and the softer connective tissue due to the age. The possibility of a breast deformation analysis is new in clothing development, enabled by the 4D scan technology. Availability of developed methods increase the understanding of the breast-bra interaction in motion and renders bra properties classifiable for the first time. Newly developed bras can be tested and compared to other existing bras.

In summary, the developed scan data analysis framework BodyTex4D showed its functionality, usability and validity in different steps of analysing the breast-bra interaction. BodyTex4D is modular to be applied to a wide variety of analyses and be easily extended for specific applications. The automatising becomes trivial and manual input is low. Additionally, the breast-bra interaction could be analysed in static poses and motion. Differences caused by bra cut patterns and materials are investigated in static poses and running motions. Utilising 4D scan data opens up new development possibilities in clothing development.

7.2 Discussion

A detailed discussion of each chapter can be found below. This comprehensive section focuses on the overall contribution to the field of research. Section 7.2.1 discusses the addressed research gaps and compares key findings to existing technologies. Section 7.2.2 discusses the transfer of developed technologies and methods to other application. Section 7.2.3 discusses the results from an economical stand point and estimates the amount of work still needed for implementation in the industry.

7.2.1 Addressed Research Gaps

Developing the 4D scan data analysis framework BodyTex4D and applying BodyTex4D to the breast-bra interaction addresses the research gap in the following areas:

1. **Using 4D scan data for the analysis of tightly fitted clothing-body interaction**: Leveraging 4D scanner technology, closes the data acquiring research gap. 4D scanner are made for capturing 3D surfaces of subjects in motion. Subsequently, 4D scanner present themselves as outstanding measurement technology for analysing body deformations and clothing-textile interactions in motion. The exposure time for each frame is significantly shorter compared to that of a conventional 3D scanner. Therefore, the noisy is lower as the subject has less time to move. The 4D scanner software of Move4D ships with their product an automatic conversion of the raw data into homologous meshes, enabling the user to track any point on the body easily. Chapters 4–6 presented movement, length and deformation analyses. The body movements could also be analysed using acceleration sensors, such as the breast movements while wearing different bras. The disadvantage compared to 4D scanning arise from the number of sensors. Sensor placement influences the resulting data and is not trivial. Results cannot be put in the context of the whole body movement, and no visual validation of the results is possible. As the 4D scanner captures the whole body, it is possible to analyse the motion of any body part after the capturing process. Subsequently, the region of interest can be chosen in the evaluation phase and is not a pre-set of the test setup. This gives a great advantage in the flexibility of the analysis. The advantage of acceleration sensors is that the sensors produce more accurate data for movement analyses of a single point compared to 4D scan data. It is assured that the sensors stay at the same place in the body, which is not the case with the homologous mesh. As the homologous mesh is fitted newly for each frame, the vertex-IDs will not always be at the exact same place. Further disadvantage of the 4D scanner is the price and the amount of data. The price of a 4D scanner compared to a 3D scanner is currently approx. 16 times higher. Nevertheless, as new information about individualised body deformation can be captured and the comparability of movements is increased, the 4D scanner opens up new possibilities for tightly fitted clothing development.
2. **Analysing automatically 4D scan data**: Currently, 4D scan data can be analysed, for example, with Move4D's software. This software is commercial and is licenced with their scanners. Analysing the 4D scanner data with the scanner software has many disadvantages. Firstly, only lines and cross-sections pre-defined on the internal avatar can be used for analyses. An individual selection of vertex-IDs or geometries intersecting the mesh is not possible. Secondly, no Application Programming Interfaces (APIs) are exposed rendering the software non-extensible for external users. Automating the analysis process is only possible via exporting raw data, homologous meshes, pre-defined lines or pre-defined cross-sections are available for further processing, for example, using Python. Thirdly, as the limited analyses can only be performed within Move4D's software, the number of people performing scan data analyses is limited by commercial licences. Pei et al. (2021a) misses an automatic 4D scan data analyses tool for their dynamic breast deformation investigations. This limitation is eliminated by BodyTex4D. The 4D scan data analysis framework, BodyTex4D, is developed to eliminate this research prohibitor. Various methods to identify and filter 4D scan datasets

are discussed, and analysis methods are built on top. The regions of interest can be chosen flexibly by different means: single vertex-ID, multiple vertex-IDs and texture. The analyses tool box is extensible easily. The result is the possibility of analyses of any region of the body. Body deformations due to movements can be visualised. As Python is an open-source, easy to learn, mighty, widely used by researchers and professional software developers all over the world programming language, many free packages are available. The data analyses are not limited to the scanner software licences.

3. **Improving the comparability of static T-poses through breathing state analysis**: Pei et al. (2021b) stated limitations in the comparability of body poses in static bra comparison. Most 3D scan comparisons of clothing are made without considering the current state of breathing of the test subject. Especially in bra analyses, the breathing state has significant influence on the fit and the evaluation results as the upper body shape changes significantly due to breathing. If a random breathing state is taken, comparison is nearly impossible and definitely leads to inaccurate results. Comparability limitation are eliminated using comparability criteria such as the breathing state analysis performed on the basis of 4D scan data.
4. **Analysing the breast deformation in motion with and without wearing bra**: Analysing deformation behaviour in motion is rare and innovative. Understanding breast deformation improves the design process of future bra models. To the authors knowledge, Pei et al. (2021a) is the only study exploring breast deformations in motion using a 4D scanner. The authors of Pei et al. (2021a) state limited comparability. The limited comparability originates from the method of data preparation. The authors choose a minimum, maximum and the corresponding point of inflection for each motion and based the comparison on it. Comparison criteria like force similarity through a curve similarity were not taken into account. Applying comparison criteria proven by this work improves the outcome of 4D data preparation, resulting in a truly comparable selection of frames.
5. **Classifying bra support**: Currently, breast movements running on a treadmill have mainly been compared directly without analysing the body movements, only by sensors. Running on treadmills does not automatically lead to comparable running motion despite constant running velocity. The newly developed method to calculate the amplification factor for various bra models makes movements comparable and can classify bra models by support which is shown repeatedly.

Overall, significant contributions could be made in the field of tightly fitted clothing-body interaction. Developed methods improve interaction analyses. Applying the developed methods to the breast-bra interaction shows their usability and improves understanding of tightly fitted clothing-body interaction. Further studies with more test subjects need to be conducted to refine the methodology and improve properties of bras to increase the ratio of women who feel comfortable while wearing bras.

7.2.2 Transferability

This section presents and discusses the transferability of the developed 4D scan data analysis methods. BodyTex4D is modular and not specialised for breast-bra interaction analysis. BodyTex4D is applicable to other applications. In Chap. 4, examples have been shown, for example, the use case of investigating circumference changes while sitting down. Further example is the deformation of the buttock while leaning backwards can be analysed using movement and shape analysis to improve car seats. Body angles in leaning backwards motions can be calculated, and the buttock deformation can be measured at similar states. In Fig. 7.2, an example is depicted. A female test subject is sitting and bending forward while being scanned (Fig. 7.2b). Using the extended 4D scan data framework, the angle between upper body and vertical plane is calculated. The frame most similar to the male's sitting angle (Fig. 7.2a) is selected, and the buttock deformations are compared. This example shows that movement and shape analyses can be applied to other body parts, test subjects and applications easily.

It is also possible to analyse the leg shape deformation behaviour in various sitting positions. Jeans can be designed based on this information to make them more comfortable in sitting positions. An example is displayed in Fig. 7.3. A male test subject sits in various sitting positions on a chair while being scanned with the Move4D scanner. Shape analysis (Sect. 4.7.3) is performed using vertex-ID 30281 as origin and vertex-IDs 30350 and 30294 for the normal vector definition.

Methods such as selecting comparable breathing or motion frames can also be applied to compressive sport shirts and trousers targeting activities such as running or cycling.

The bra classification method is applicable to measurement data of acceleration sensors. The only input parameters are body and breast movement curves.

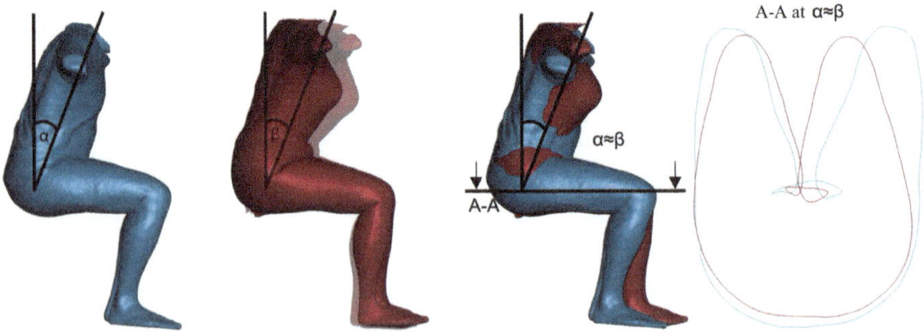

(a) Sitting male **(b)** Sitting female **(c)** Same upper body angle **(d)** Cross-section results

Fig. 7.2 Selection of comparable sitting positions using the automated 4D scan data analysis framework BodyTex4D

(a) Sitting positions **(b)** Shape analysis

Fig. 7.3 Automated shape analyses of the leg deformation in three sitting positions

For scan data without homologous mesh property, retrospectively, homologous meshes are fitted by using the open-source data-driven skinned multi-person linear body model (SMPL) Loper et al. (2015) human body model. Testing SMPL lead to the assumption of missing the key-vertex-ID feature compared to Move4D's software and meshes are less dense (Schmidt et al. 2024). Still, SMPL is a helpful tool for preparing scan data to be used for analysis requiring homologous meshes.

Overall, the following properties of BodyTex4D support transferability:

- Easy extension of functionality
- Combining flexibly methods for analyses
- Methods are applicable to different areas of the body
- Methods are applicable to different test subjects
- Application-specific adaption of methods is possible through open-source access

7.2.3 *Economic Value*

The economic value is assumed to have great potential due to the generalisation, problem abstraction, innovation and open-sourcing of BodyTex4D.

The basis for the improved experimental analysis of body-clothing interaction is the availability of a 4D scanner. 4D scanners with 12 cameras are very expensive. However, the methods could also be applied to a 4D scanner with fewer cameras, reducing the costs.

Furthermore, the automated 4D scan analyses enables to understand the body in motion better and, for example, to derive dynamic pattern designs fitted to individual's needs. Individualisation results in better-fitting clothing and give companies a competitive advantage. Individualisation opens up new markets for the textile industry while having the opportunity of increasing sustainability. Sustainability is achieved through utilising the methodology of this work for digital cloth fittings reducing

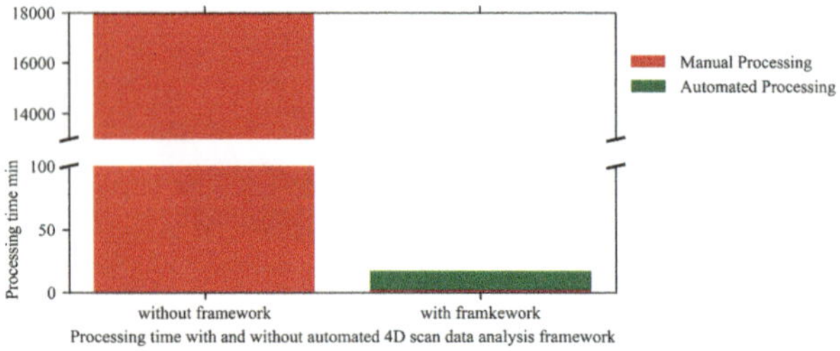

Fig. 7.4 Estimated evaluation time for comparing manual 4D scan data analysis with the 4D scan data framework over 3600 frames

costs and resources for transport. By analysing various tight garments interacting with the body, influencing factors such as material parameters, cut patterns and the test subject's physique can be recorded, and development based upon. Clothing with improved functionality can be developed more rapidly.

Using BodyTex4D is highly time-efficient in data evaluation. For the breathing state analyses in Chap. 5, more than 3600 3D object have been processed in the underbust calculation within less than 17 min. The manual work has been 2 min to set the import and check the results for reliability. Compared to a manual underbust circumference analysis and frame selection (approx. 5 min per frame), as it is commonly practised currently, analysis duration of scan data has been cut down immensely to less than 1 % (Fig. 7.4). Additional algorithms contained in BodyTex4D are not profiled which leads to the assumption of a couple orders of magnitude optimisation potential.

7.3 Outlook

A user-friendly interface would benefit the usability of BodyTex4D greatly. As the focus has been on implementing functionalities, no Graphical User Interface (GUI) has been created, and basic programming knowledge is needed. Improving the software architecture and adding a User Interface (UI) will reduce the programming knowledge needed further.

Steps presented in the development, such as the 3D point cloud comparison not performed in Python, could be transferred to Python utilising open-source libraries. To improve bra development, more scans should be performed with a wider variety of test subjects of different breast sizes, fitness levels, life styles and ages. More bra models need to be analysed. The influence of seam strengthening and seam positions on breast deformation and bra support is open for research applying BodyTex4D.

Overall, BodyTex4D can be used in the tightly fitted clothing development to test new prototypes' functionality or to broaden knowledge of individualised clothing in motion. Based on this knowledge, individualised cut patterns or parameterised template cut pattern can be developed (Fig. 7.5). Clothing that can be improved in its development includes bras, sports tights, compression socks, shape-wear and many more. Additionally experimental evaluated data can be used to set up and validate digital body-clothing models. These steps are shown in the following part III.

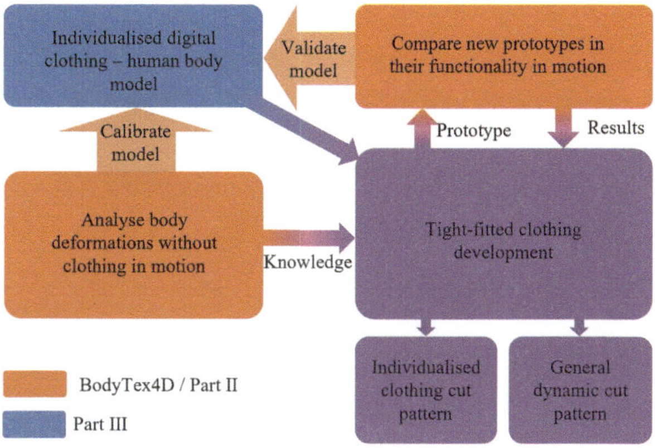

Fig. 7.5 Usage of BodyTex4D and the methods developed in Part II in clothing development.

References

Loper M et al (2015) SMPL: a skinned multi-person linear model. ACM Trans Graph 34(6):1–16. ISSN: 0730-0301. https://doi.org/10.1145/2816795.2818013

Pei J et al (2021a) An exploratory study of bust measurements during running using 4D scanning technology. Int J Fashion Des Technol Educ 14(3):302–313. ISSN: 1754-3266. https://doi.org/10.1080/17543266.2021.1938699

Pei J et al (2021b) The detection of the upper boundary of breasts using 4D scanning technology. Int J Fash Des Technol Educ 14(1):1–11. ISSN: 1754-3266. https://doi.org/10.1080/17543266.2020.1829097

Schmidt AM et al. (2024) Evaluation of automated skeleton fitting to 4D human body scan data using Open-Source SMPL- and OSSO models. 3D Body Tech J, vol. 1. https://doi.org/10.15221/24.40

Part III
Numerical Methodologies to Analyse Clothing-Body Interactions Using FEM and 4D Scan Data

Chapter 8
Introduction and Objectives

Abstract This chapter introduces Part 3, which focuses on the digital analysis of clothing-body interaction using the finite element method (FEM) and 4D scan data.

Keywords FEM · Soft avatar · Individualised · Digital clothing development · Human Body Model (HBM) · Clothing · 4D scanning

State-of-the-art of developing tightly fitted clothing is designing new collection manual in iterative work. The time and cost consumption is immense, and the results are often unsatisfactory in terms of individual fit, comfort and functionality. Some developments are based on digital tooling using clothing simulation software such as CLO3D (2024) or VStitcher (2024). Digital avatars are used by the software products, but they are mostly rigid. Non-deformable, rigid avatars are not beneficial for tightly fitted clothing, which deforms the body while worn. Clothing software products like CLO3D, VStitcher and Style3D have released deformable avatars in the past months. Besides the fact that they were only released a short time ago, these avatars are still not beneficial in developing tightly fitted clothing. The deformations are unrealistic (Fig. 8.1) and the pressure between body and clothing cannot be calculated. Trials of the software leads to the assumption that the "soft avatars" are only shell or thicked solid models (Fig. 2.20b) with elastic material properties. Body stiffness cannot be individualised for different body areas. Additionally, these software are primarily made for design purposes. Importing experimentally tested material data for textiles or body stiffness is difficult, which decreases the simulation results. Clothing-body interaction is essential for the resulting body deformations and pressures, especially for tightly fitted clothing such as compression socks or bras. Individualised, deformable human body-clothing models are needed for accurate and improved clothing development.

Finite Element models (FE models) calculate accurate stresses, strains and pressures. Experimentally measured material data can be used to calibrate FE models. Using FE models in the tightly fitted clothing development is promising. However, fundamental research methods are missing to generate individualised, deformable human body models of real test subjects and clothing models. Methods to combine human and clothing models are missing. After the combination of human and clothing FE models, the model validation is essential for an accurate FE model.

A.-M. Schmidt, *Novel Methodologies for Digital Investigations of Clothing-Body-Interactions*, Lecture Notes in Bioengineering,
https://doi.org/10.1007/978-3-031-91794-3_8

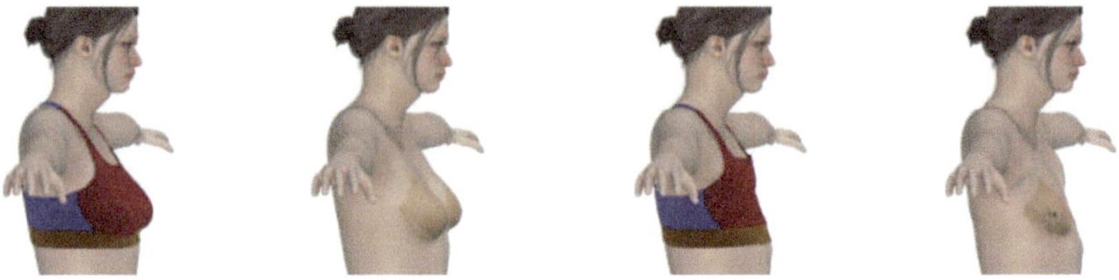

(a) bra-2-M on a rigid avatar **(b)** bra-2-M on a soft avatar

Fig. 8.1 CLO3D soft avatar deformation by a sports bra

Modelling real humans leads to the challenge that the experiments are not exactly repeatable, and the experimental data does not fit the FE model exactly. A simulation gap will always persist. 4D scan technology is an innovative approach to calibrate and model human body-clothing FE models. Captured 4D scan data promises to reduce the simulation gap by improved comparability (Part II). Additional, FE models

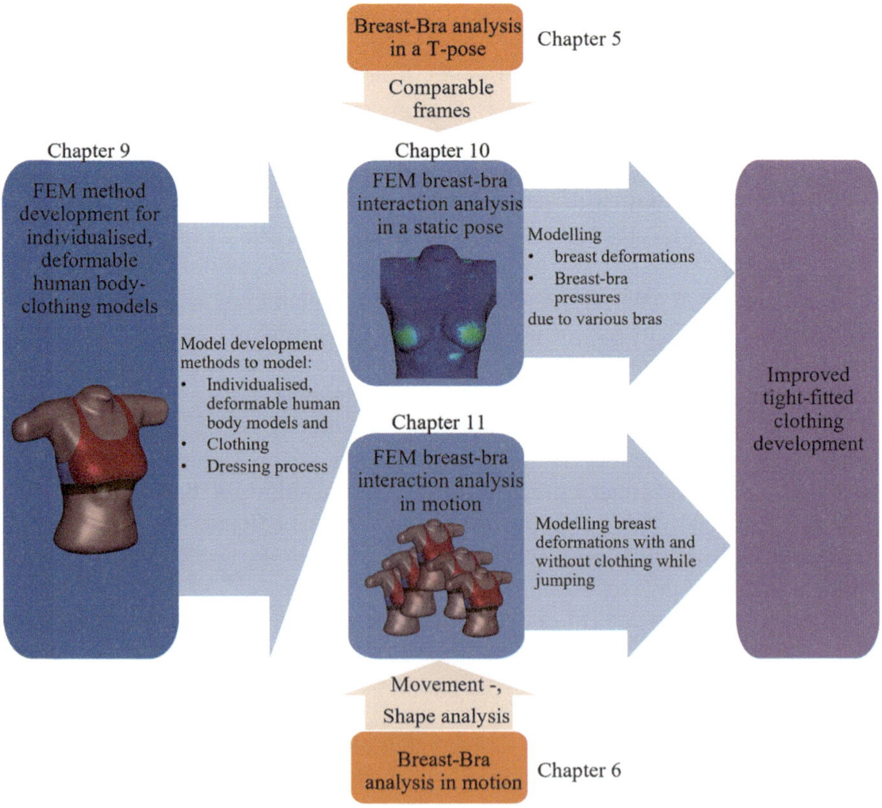

Fig. 8.2 Overview of Part III

analysing the clothing-body interaction in motion are verifiable though the body deformation behaviour captured by 4D data which is impossible with 3D scanner or sensor data.

This part presents general methods to generate individualised, deformable human body-clothing FE models (based on 4D scan data). Developed methods are evaluated and discussed in Chap. 9. Developed methods are applied to model the breast-bra in a static pose in Chap. 10. Chapter 11 presents methods for dynamic breast-bra FE modelling. Figure 8.2 shows an overview of the chapters of this part. Model generation and validation is based on the results of part II.

References

CLO3D (2024) CLO Virtual Fashion. Version 7.3.0. https://www.clo3d.com/en/
VStitcher (2024) Browzwear Solutions Pte Ltd. https://browzwear.com/

Chapter 9
Methodology Development for Deformable Human Body-Clothing Finite Element Models

Abstract This chapter presents general modelling methods for developing an individualized, deformable human model that interacts with a close-fitting garment. Various modelling approaches are introduced, newly developed, and evaluated in terms of their usability for clothing development. The outcome is a general guideline for building a human-clothing model efficiently. Additionally, the integration of 4D scan data for model validation is explored in detail. Based on these methods, most tightly fitted clothing can be modelled.

Keywords FEM · Soft avatar · Individualised · Digital clothing development · Human-Body-Model (HBM) · Clothing · 4D scanning · Breast-bra

9.1 Introduction and Objectives

This chapter develops methods to generate an individualised, deformable human-body-clothing FE model.

The interaction of a tightly fitted clothing item and the wearer is highly individual. At the same time, this interaction and individual clothing fit define how well the functionality and comfort of the clothing are. To improve the development of tightly fitted clothing, it is beneficial to model these individual clothing-body interactions digitally before production. In the digital model, body deformations, pressure distributions, clothing strains and more can be analysed in static poses and in motion. A common tool for accurate analyses of deformations, pressure and strains is the FEM. To achieve meaningful results with the FEM, accurately calibrating the mechanical properties of the body and clothing material is essential.

The literature review shows no generalised methods to develop individualised, deformable human body-clothing FE models (ID-HB-Cloth-FEM). Methods for material model calibration of the body's soft tissue are absent from the clothing industry today. Part I discusses how most clothing modelled with FEM is not based on original cut patterns (Sect. 2.4.3.1). Often, the geometry of tightly fitted clothing is generated by an offset of the body's surface, which is inaccurate because it does not render how the actual garment would interact with the body. The approach does not

A.-M. Schmidt, *Novel Methodologies for Digital Investigations of Clothing-Body-Interactions*, Lecture Notes in Bioengineering,
https://doi.org/10.1007/978-3-031-91794-3_9

have to deal with the challenge of having an undersized clothing item. Clothing item models smaller than the body, as tightly fitted clothing items are, cannot be merged with the model of the body because of intersections. Consequently, the procedure of dressing must be modelled. Very few approaches are found in literature solving the dressing problem. However, comparisons of dressing approaches have not been compared by the available literature.

This chapter develops new methods to generate ID-HB-Cloth-FEM. Thereby, the following research gaps are addressed:

- Comparing, evaluating and developing methods to generate an individualised, deformable human body-clothing Finite Element Model (ID-HB-Cloth-FEM).
- Developing new material calibration methods to model bio-mechanical properties of humans soft tissue.
- Developing clothing FE models based on the original cut pattern.
- Developing fast and generic dressing processes for undersized clothing items to be applied to FEM.

Figure 9.1 depicts the structure of this chapter. In Sect. 9.2, methods to generate individualised human body models are displayed. Various geometry development, meshing and material calibration methods are described, compared and evaluated. In Sect. 9.3, geometry development methods focused on modelling realistic clothing items are presented. Meshing and material calibration methods are explained and evaluated. Section 9.4 describes various methods to merge a human body model and an undersized clothing item, modelling the dressing procedure. Finally, in Sect. 9.5 methods for validating developed models are presented. Most examples are based on the breast-bra modelling. Nevertheless, the presented methods are applicable to other clothing items and body parts. The comparison and evaluation of the following methods are based on the usability in the clothing development.

9.2 Method Development for Individualised, Deformable Human Body Finite Element Models

An overview of generating an individualised, deformable human FE model (ID-HB-FEM) is portrayed in Fig. 9.2. The procedure is based on the general modelling steps of FEM described in Sect. 2.3. In Fig. 9.2, orange boxes mark approaches using 4D scan data for the modelling steps: geometry development, meshing and material calibration.

9.2.1 Geometry Development

A 3D or 4D scanner can capture human body surfaces. Alternatively, measurements can be taken from a test subject to create a virtual avatar using software such as

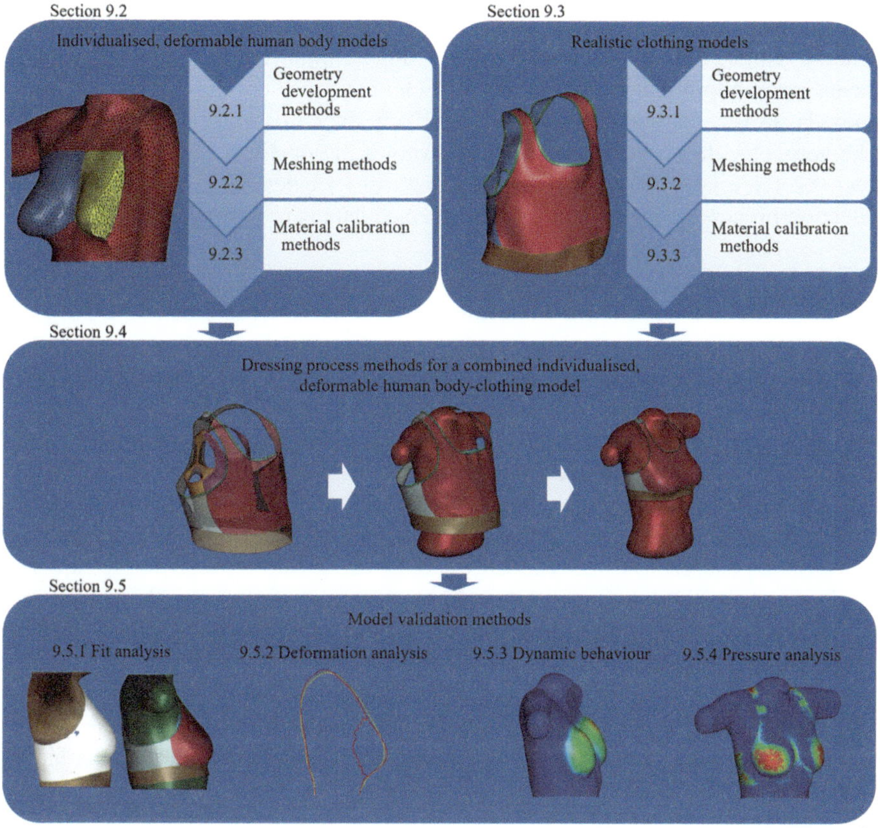

Fig. 9.1 Overview of Chap. 9

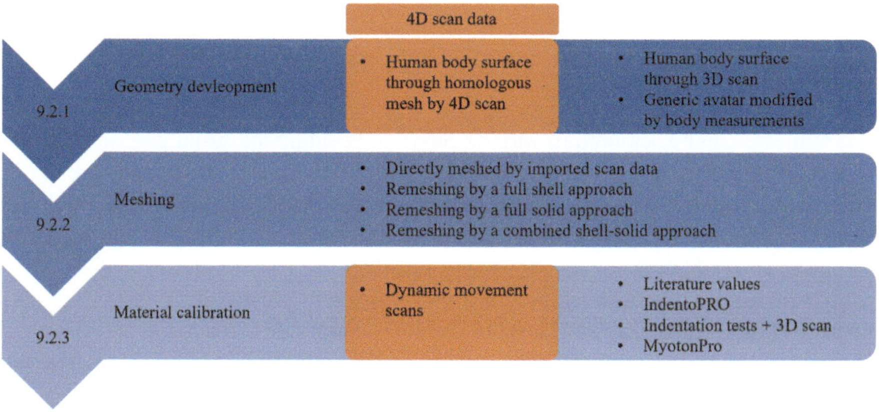

Fig. 9.2 Integration approaches of experimental and 4D scan data (orange) into a deformable human body Finite Element model

MakeHuman (2024). Individualisation is strongly limited for pre-modelled, standard avatars. Avatars are often symmetrical and limited numbers of pre-defined measurements are configurable which are not sufficient to represent the details of a human body. However, this method does not need any expenses on scanner technologies. The geometry development technology using a 3D or 4D scanner is presented in Sect. 9.2.1.1. Firstly, methods for surface capturing are described. Secondly, the pose selection process based on the captured surface data is presented.

9.2.1.1 Surface Capturing Technology by 3D or 4D Scan Data

Individual's body surfaces can be captured using a 3D or 4D scanner. Generally, capturing a single frame by 4D scan is faster than by 3D scan. The exposure time of 4D scanners falls in the order of 1 ms. The exposure time of 3D body scanners depends on the technology applied. Handheld 3D scanners and similar 3D scan approaches capture the body section-wise. For example, the test subject needs to stand on a rotating plate. Consequently, reconstructing the 3D object from multiple exposures or longer exposures. Multiple exposures are by design slower than single exposures of 4D scanners. 3D scanning duration are in the range of 10–120s. Within that time period, the body pose of the scanned test subject likely changes due to breathing or wiggling. Subsequently, exposure durations of 3D scanners are inherently more inaccurate compared to 4D scanners. The 4D scanner model Move4D by IBV is, additionally, capable of generating homologous meshes, which results in neglectable manual post-processing work. Mostly, scan data of handhold 3D scanners have to be post-processed manually. Homologous meshes for 3D scan data are achieved by applying third-party algorithms.

In summary, the 4D scan technology presents itself as a superior tool for generating the geometry of an individualised human body model.

9.2.1.2 Pose Selection

Test subjects can be scanned in various poses to build up an initial geometry model. This work is based on the T-pose. Nearly the complete body surface can be captured while a test subject is in a T-pose (Fig. 9.3). To avoid deformation of the body, it is important that the scanned body is not pre-deformed by clothing in the areas that will be examined later.

A-Pose are less suitable as they are more prone to mesh defects due to the arm position. The armpits are out of sight of the 4D scanner due to the lower arm positions. The breast is not lifted. Subsequently, more manual mesh post-processing is required (Fig. 9.4).

In summary, the pose selection should fit the application and cover the maximum amount of body surface to limit the manual work of handling mesh defects.

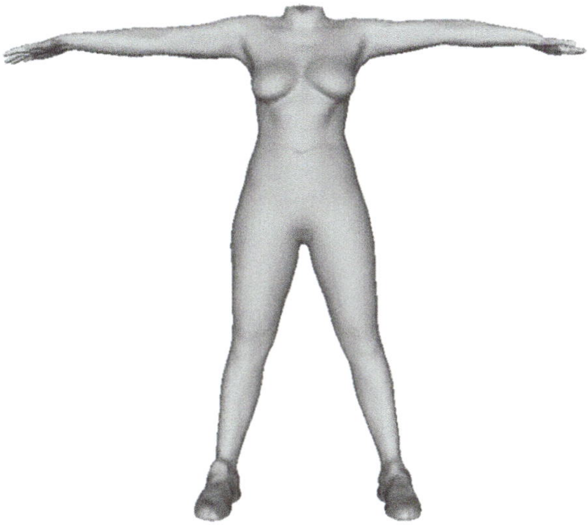

Fig. 9.3 Pose selection (T-pose) for the geometry development of an individualised, deformable human body Finite Element model

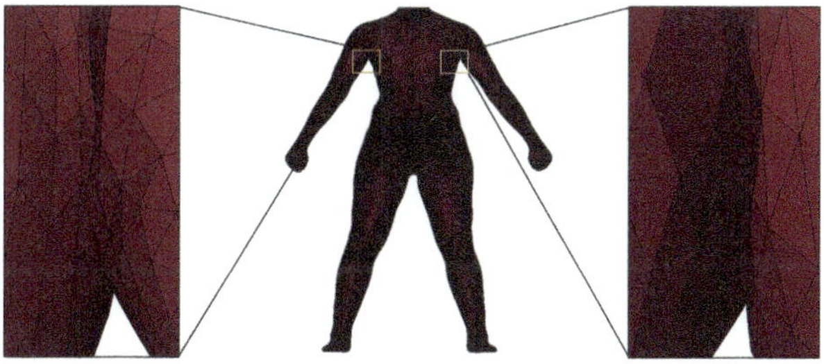

Fig. 9.4 Mesh intersection issues in the A-pose

9.2.2 Meshing Approaches

Three meshing approaches are discussed in the following (Fig. 9.5). Discussed meshing approached are full shell (Fig. 9.5a), full solid (Fig. 9.5b) and combined solid-shell (Fig. 9.5c). The combined solid-shell approach is developed by this research work. The full-solid and full-shell approach exists in the literature (Sect. 2.4.2.2). Literature describing full-solid and full-shell meshing approaches in detail is not publicly available. A comparison between the combined solid-shell approach and existing approaches is performed. The meshing approaches are compared based on modelling breast behaviour. The three FE models are designed to consist of deformable and rigid regions. The breast is modelled to be deformable while the rest of the body is rigid. This segmentation process is typically performed manually, which results in the highest model accuracy. The shell approaches are improved by the author by an automated breast selection to reduce manual labour.

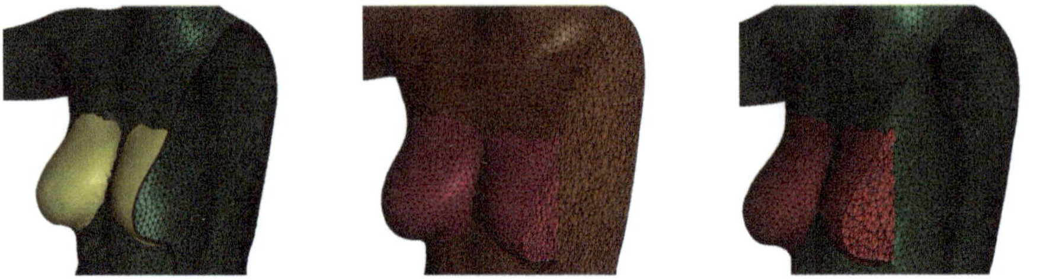

(a) Full shell model (b) Full solid model (c) Combined shell - solid model

Fig. 9.5 Three meshing approaches for deformable human Finite Element models

9.2.2.1 Meshing Deformable Human Body Finite Element Models Fully by Shell Elements

The meshing approach by full-shell models is the simplest approach. If needed, the mesh exported from the 3D/4D scanner can be remeshed using software tools (e.g. LS-PrePost, MeshLab, ArtecStudio or MeshMixer). The avatar geometry is segmented into breast and body, and shell thickness is defined.

The full-shell approach is refined by this section to decrease the model setup time and generate an automatic FE model. A template-avatar, homologous-mesh-based body part segmentation approach is developed. It follows a concise overview. Details of the algorithm are explained in the next paragraph.

Regions of interest are selected on the template avatar. Figure 9.6 visualises the example of segmenting the right and left breast. As template-avatar test-subject-A is chosen (Fig. 9.6 upper left). The algorithm is tested by segmenting the breasts of test-subject-B based on the marked template avatar of test-subject-A. The output of the algorithm are three FE parts: left and right breasts separately, and the body (Fig. 9.6 right). The template-based part selection approach increases the automation in the model setup immensely as each region of interest has to be selected only once. Setting up higher numbers of similar FE models for many individuals becomes reasonable. The manual model setup time is less than 1 min as only the input files for the Python script: template-k-file-path and *.obj-file-path (Listing 9.1) have to be defined.

The template-based body part segmentation algorithm is based on the requirement of having homologous meshes, consequently, consistent vertex-ID and face ID data is necessary. In this case, homologous mesh data is retrieved from the 4D Scanner by Move4D. The script is developed to extend the FE software LS-PrePost, which has a human-readable input file format called k-file. For other FE software, the output of the format could be changed. In future work, an intermediate representation can be developed from which arbitrary file formats can be generated.

Firstly, one homologous 3D mesh of a human body has to be imported as an FE k-file. The import process of *.obj-files needs to keep the order of vertex-IDs and face-IDs as the order manifests the homologous property. The self-implemented reader of obj-files is used. Common mesh processing Python libraries dissolve the homologous property of meshes by reordering the data sequence inconsistently over

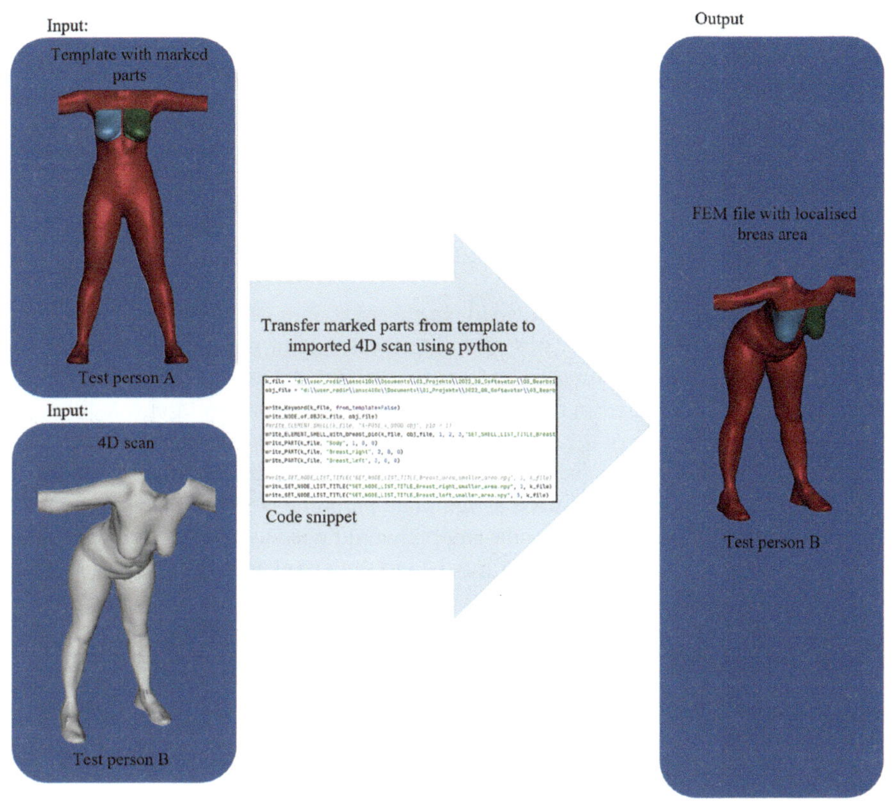

Fig. 9.6 Automated template-based segmentation of 3D scan data of a human body using a developed python script

multiple loading cycles. Subsequently, the self-implemented obj-read function of BodyTex4D is used in function obj_2_node_k_file to import the vertex-IDs and coordinates consistently as homologous mesh (Listing 9.1). Import of face information is analogue (Listing C.1).

Listing 9.1 Template-based body part segmentation code snippet to import *.obj to k-file - nodes

```
def obj_2_node_k_file(obj_file_path, k_file_path):
    vertex_coordinates =
    get_vertex_coordinates_from_obj(obj_file_path)

    with open(k_file_path, "w") as file:
        file.write("*NODE\n")
        file.writelines(
        (f"{i},{coordinate[0]:.8f},{coordinate[1]:.8
    f},{coordinate[2]:.8f},0,0\n" for i, coordinate
    in enumerate(vertex_coordinates, 1))
        )
```

The k-file is opened in LS-PrePost to generate the template avatar. Arbitrary regions can be defined and saved. Additionally, shell-sets are created part-wise using *SET_SHELL_LIST_TITLE. This k-file is defined as the template for further human model scans, which should undergo similar segmentation.

To apply the segmentation to another human body scan, the following input parameters are needed (Fig. 9.6):

- k-file of template avatar and
- *.obj-file of the individual human body to be segmented into the same parts

The *.obj-file is imported as described. In the next step, the face-IDs are filtered from the template-avatar-file. Corresponding faces IDs of the mesh of interest are picked, from which the segmented body part is composed as new *SHELL_SET. The code-snippet is located in the appendix (Listing C.2). Finally, a new k-file from the imported as *.obj-file of the mesh to be segmented by the template-avatar method is saved and can be used for further studies. Other constant process parameters such as boundary conditions can be used as template parameters to be selected once and applied to multiple meshes. The code-snippet would need to be extended to include more parameter keywords to transfer the data to the template k-file.

9.2.2.2 Meshing Deformable Human Body Finite Element Models Fully by Solid Elements

The full-solid meshing approach uses solid-volumetric elements applied to the scanned surface. For example, meshing can be performed using triangular meshes in LS-PrePost (DYNAmore GmbH 2023) or in Gmsh (Geuzaine and Remacle 2009). The prerequisite is a high-density surface mesh. After meshing, the deformable regions of the human body model are defined (Fig. 9.7). These deformable regions can be defined individually.

This approach is less automated than the full-shell template approach as the meshing is performed manually. The vertex-IDs vary due to volumetric meshing, subsequently, the template-based segmentation code-snippet is incompatible. Nevertheless, this approach requires only a minor manual setup time. In the presented example in Fig. 9.7 the manual setup time has been 5 min.

9.2.2.3 Meshing Deformable Human Body Finite Element Models by Solid and Shell Elements

The newly developed combined solid-shell breast-body model generates a high-density solid mesh in the region of interest. Solid elements are used as they accurately model deformations and pressures of volumetric geometries. Regions of less interest are meshed as shell. Shell elements require less computational resources causing decreased computation duration. Figure 9.8 exhibits an example of volumetric meshed breasts with a body meshed by shell elements.

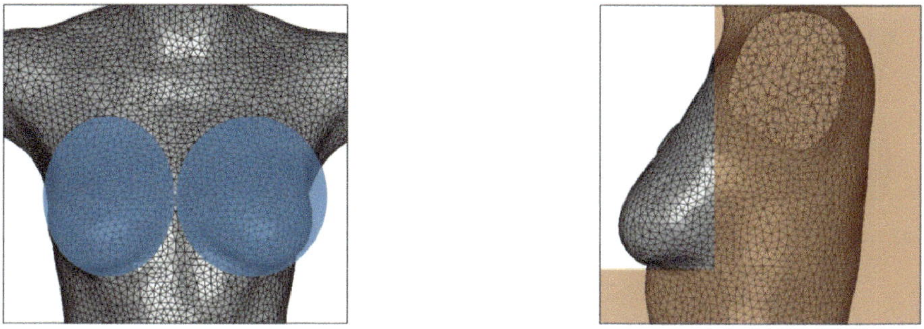

(a) Part selection **(b)** Part deselection

Fig. 9.7 Full solid meshing approach for human body Finite Element models

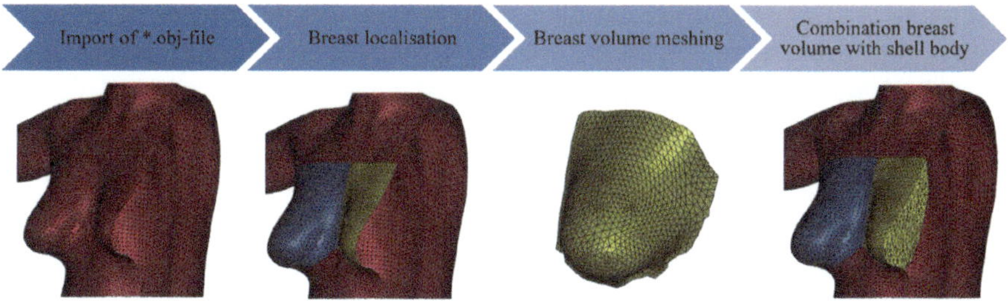

Fig. 9.8 Combined solid-shell meshing approach for human body Finite Element models

Firstly, the region to be deformed is divided into segments. Segmentation can be accomplished manually or using the template-based segmentation approach described in Sect. 9.2.2.1. The segmentation results in separated parts: regions of interest and peripheral regions (Fig. 9.9a). Regions of interest are meshed by solid elements. Prerequisite for solid meshing is a closed geometry. Gaps in the geometry caused by the segmentation step are closed by connecting all free edges (Fig. 9.9b). The resulting surfaces are meshed (Fig. 9.9c). The closed geometry is meshed by tetrahedrons (Fig. 9.9d) and, merged again with the shell body part (Fig. 9.8). This method requires more manual setup time (in the example in Fig. 9.8 15 min) compared to the shell and solid meshing approaches. The meshing can be performed in LS-PrePost or Gmsh.

9.2.2.4 Comparing Meshing Approaches for the Modelling of Individualised, Deformable Human Body Finite Element Models (ID-HB-FEM)

In Sects. 9.2.2.1–9.2.2.3 the development and implementation of three meshing approaches to model human body models are explained and in the following com-

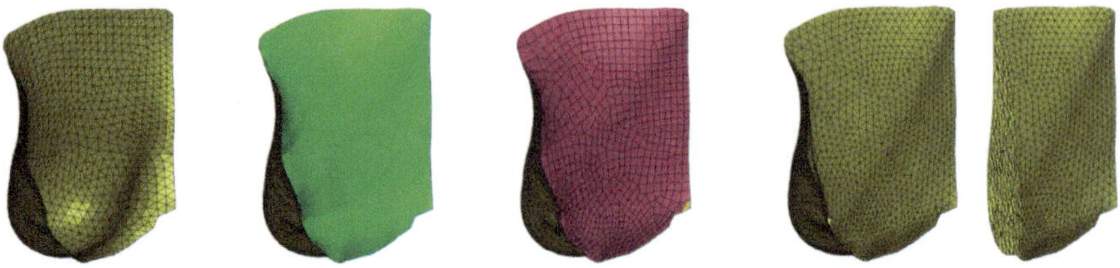

(a) Breast surface (shell elements) (b) Breast surface closed by new surface (c) New surface meshed (d) Breast with solid elements

Fig. 9.9 Generating a volume mesh from a surface mesh

pared and evaluated. The focus of the evaluation is the improvement of clothing development. The evaluation is conducted with the example of a bra model. Other clothing-body interaction examples such as a compression sock-leg or sleeve-arm interaction could have been chosen too. Meshing approaches are evaluated along the following evaluation criteria:

- manual setup time
- realistic deformation behaviour
- computational stability
- computation duration

The boundary conditions, material data and deformable-modelled areas are the same. The dressing method used is described in Sect. 9.4.3. The breast region is modelled by an elastic material model, and the body by a rigid model. The full-shell model has been modelled in two different shell thicknesses of 2mm and 4mm. After modelling the breasts the dressing procedure is used to place the bra on top of the breast model, causing the breasts to deform. In Fig. 9.10, results of the meshing approach are shown. The meshing approaches are in the following evaluated:

- **Manual setup time**: describes how much manual effort measured in time is required for setting up the numerical FE model. A shorter setup time is beneficial as this leads to higher productivity. The manual setup time for the full-shell approach is generally the lowest, and the combined shell-solid approach demands the highest manual effort. The full-shell approach is completely automated and the manual setup times are less than 1 min, when the template avatar has been defined. The combined shell-solid approach needs manual meshing for the separated volumetric regions, which is currently made manual and needs a manual setup time of approximately 15 min. The full-solid model demands minor manual work comparatively. Meshing the complete model is fast, and the segmentation process of regions of interest is done quickly. Nevertheless, the segmentation process is still manual (manual setup time: 5min).
- **Realistic body deformation behaviour triggered by tightly fitted clothing**: The modelling results must be realistic enough to see the influences of cut patterns and materials on the human body, close to the effect the real clothing invokes on the

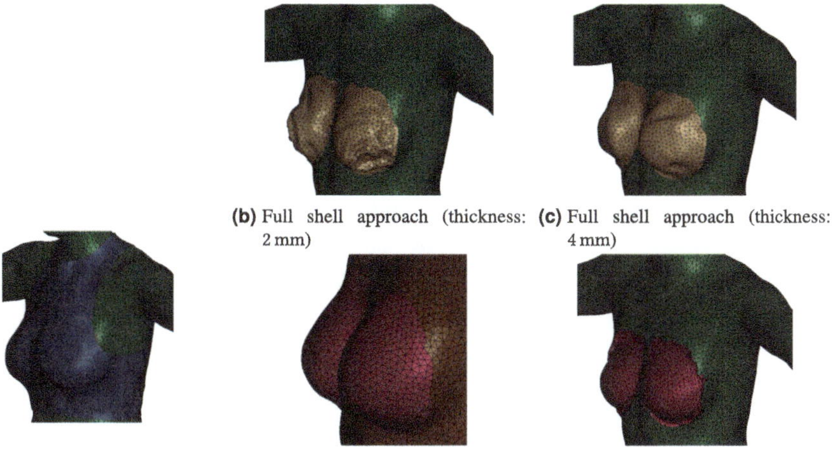

(a) Clothing - ID-HB-FEM interaction model

(b) Full shell approach (thickness: 2 mm)

(c) Full shell approach (thickness: 4 mm)

(d) Full solid approach

(e) Solid-shell approach

Fig. 9.10 Comparing meshing approaches by modelling body deformations through tightly fitted clothing (published by author in Schmidt and Kyosev 2023)

real body. The closer the effect to reality, the higher the usability of models for the clothing industry becomes. Examples of the full-solid approach (Fig. 9.10d) show the most realistic deformation behaviour. The full-shell approach (Fig. 9.10b and c) produces unrealistic deformations. The shell collapses due to the model being hollow. The hollow shape is unable to distribute applied pressure evenly or transfer it to lower model layers. Thicker shells lead to a higher resistance against the pressure of the tightly fitted clothing item, but the deformations are unrealistic. In conclusion, the full-shell meshing approach is not suitable for the clothing-body interaction simulation. The solid-shell approach (Fig. 9.10e) shows unrealistic deformation and is unusable at its current state. Improved contact and boundary conditions could increase the realistic deformation behaviour. Restricting the backside of the deformable breast region would lead to the same result as the full-solid approach. Another promising boundary condition is a sliding surface on the inside of the breast. This could result in larger, more realistic breast movements than at the full-solid approach. However, the contact area increases by adding a sliding plane behind the breast which leads to contact instabilities and longer computational times.

- **Computational stability**: indicates how stable the numerical model is. The element behaviour, contacts and boundary conditions influence the stability of numerical models. High stability leads to less manual effort and faster results because less adjustments and reruns are required. Consequently, higher model stability increases the usability of clothing development on an industrial scale. The full-shell and full-solid approaches showed higher computational stabilities than the combined shell-solid approach. The shell-solid approach is more prone to errors due to the boundary conditions between the solid breast model and the shell body model.

Additionally, contact losses are observed at the breast edges when using fixed node-to-surface contact conditions. The solid-shell approach leads to higher instability, and further research into the boundary and contact conditions is required.

- **Computational time**: describes the calculation time demanded by the FE model. Shorter computational time is beneficial. The computational durations of all three models are similar. The shell approach proved to be the fastest (5min), followed by the solid approach (20min) and, the solid-shell approach (40min). These computational times are based on non-calibrated material models and display tendencies of the computational duration. The higher computational duration of the shell-solid approach is provoked by more complex boundary conditions.

In Fig. 9.11, result of the evaluation are illustrated. The thickness of the beams indicates the degree of fulfilment of the given criteria. Green colouring indicates the evaluation criterion to be advantageous for clothing development. The red colour indicates the less advantageous impact of the evaluation criteria.

In conclusion, the full-solid (Sect. 9.2.2.2) is the most promising approach to model the deformable-body-clothing interaction. It has the highest usability in clothing development due to a lower manual setup time, a highly realistic deformation behaviour, and stability during the calculation. Additionally, compared to the more complex solid-shell approach, it needs less knowledge about contact and boundary condition of FEM. Even though the full-solid approach is slower than a full-shell approach, the computational duration is still within a reasonable order of magnitude. The FE model decreases the development time compared to manual prototyping by a huge margin. Using supercomputers could lead additionally to lower the computational time.

9.2.3 Material Calibration

The next step in setting up an FE model of a human body model with deformable regions is identifying the individual material properties. Most FE models in the literature use soft-tissue properties based on literature data. Two main methods for material calibration are determined. One method is based on indentation tests (Sect. 9.2.3.1) or based on dynamic behaviours (Sect. 9.2.3.2). Several methods to measure the soft-tissue properties under indentation or breast behaviour mechanisms are tested in this chapter. A promising measurement device for indentation measurements of soft tissue is IndentoPRO (ThermoTrigger 2024). The method to calibrate the soft-tissue properties with the IndentoPRO has not been used before in FEM or clothing development. The same is true for the material calibration by 4D scan data. This chapter evaluates the technologies for their usability in clothing development. The best-evaluated testing approach is described in more detail.

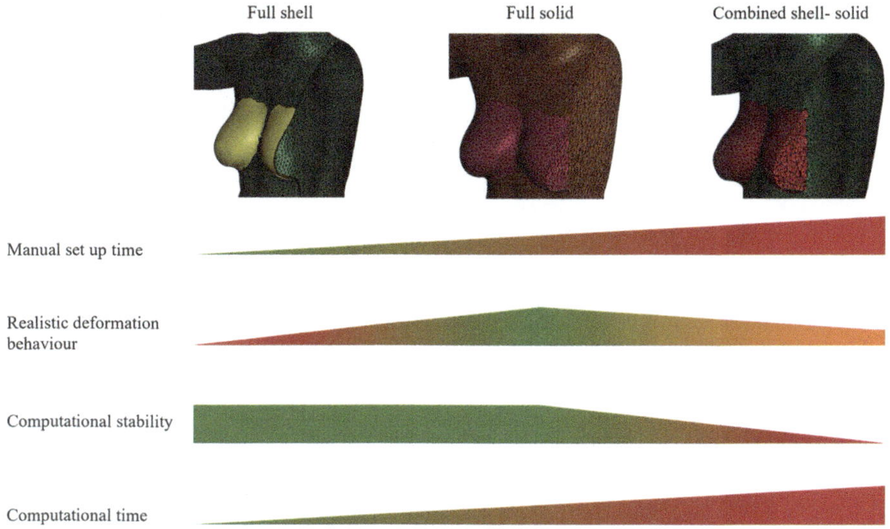

Fig. 9.11 Evaluation of three meshing approaches to model deformable human body Finite Element models (green colour: beneficial in clothing development, red colour: less beneficial in clothing development)

9.2.3.1 Material Calibration by Indentation Tests

The indentation test is an in-vivo testing mechanism which moves a stamp into the soft tissue, measuring the force and displacement of the stamp. For example, for amputees, self-build indentation test setups are described in Tonuk and Barbara Silver-Thorn (2003). "The human touch" (Pai et al 2018) measures reaction forces and the deformation optically. Nevertheless, neither the indentation test mechanism is available for purchase nor the testing device presented in Pai et al (2018). Two new approaches and an approach inspired by "the human touch" are evaluated. The first method is reusing an indentation test machine for the tensile test of textiles. Tensile test machines are standard equipment in the textile testing lab (Fig. 9.12a). The second indentation test uses the IndentoPRO device (Fig. 9.12b). The IndentoPRO is a handheld measuring device that measures the force and distance of a stamp, which is pushed into the soft tissue. The third tested approach was inspired by "The human touch". Loads are pushed into the soft tissue while the deformation is captured by 3D scan (Fig. 9.12c).

9.2.3.2 Material Calibration by Dynamic Breast Behaviour

Two approaches for material calibration of dynamic breast behaviour models are evaluated. The first method is based on the measuring device MyotonPro (MyotonPro 2024). MytonPro sends pressure impulses to the skin to measure the mechanical

(a) Tensile test machine (b) IndentoPRO (c) Load cell and 3D scan

Fig. 9.12 Material calibration methods by indentation test mechanisms

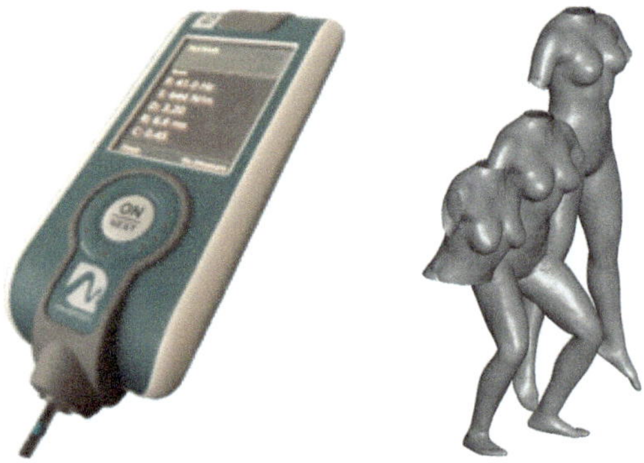

(a) MyotonPro (b) 4D scan data

Fig. 9.13 Material calibration methods by dynamic behaviour

behaviour of soft tissue. The second method is developed by this research work and is based on the 4D scanner to calibrate the material's behaviour (Fig. 9.13b). A movement is captured by the 4D scanner and the soft-tissue deformation is analysed. Based on these deformation data, the FE model is calibrated. The second method proves to be especially useful for calibrating breast's tissue behaviour in motion. Further details are presented in Sect. 9.2.3.5.

9.2.3.3 Comparison of Soft Tissue Measuring Devices and Methods

The soft-tissue properties differ due to genetics, age, weight, and fat-muscle distribution amongst other factors. Five methods have been tested to measure the soft-tissue properties of an individual. Three methods are based on the indentation test and two on the dynamic breast behaviour. The measuring mechanisms are evaluated based on the following criteria. A summary of the comparison of the measuring methods can be found in Table 9.1. A "++" indicates a high fulfilment of the evaluation criteria and a "-" a low fulfilment. Green indicates high suitability for clothing development and red a low one.

Table 9.1 Comparison of soft-tissue measuring devices

Evaluation criteria	Measuring methods/ devices				
	3D Scanner	MytonPro	TensileTest machine	IndentoPro	by 4D scan
Flexibility	++	++	-	++	++
Handling	-	-	++	+	++
Post-processing time	–	++	++	++	++
Data usability	o	-	++	++	o
Accuracy	-	+	++	+	+
Costs	+	o	-	+	-

- **Positioning freedom-The flexibility to measure various body parts**: Except for the tensile test machine, all presented methods are flexible in positioning and measuring different body parts.
- **Handling-The level of experience required to perform the measurements**: The method tensile test machine, IndentoPRO and 4D scan are the most simple to handle. 3D scanning is more challenging due to its process duration of approximately one minute. Inaccuracies can arise from the subject's movements and load cell instability, especially when using handheld devices. Load cells are force sensors that, when used in combination with 3D scanners, allow precise measurement of the relationship between object deformations and the forces acting upon them. However, this approach has drawbacks: load cells often cover the regions experiencing the greatest deformation, limiting visibility. Additionally, the need to accurately position the load cells complicates the scanning process. The mechanism of the "The human touch" device presented in Pai et al (2018) promises better handling. Nevertheless, as this device is not available, it is excluded from the discussion. The MyotonPro showed difficult handling, especially for upper-body measurements. Upper body measurements are easily interrupted by the breathing and body movements of the test subject due to the small tolerance range of the measuring device.
- **Post-processing time-The time required to extract soft-tissue properties**. The processing time of all measuring devices is minor (less than 2min), except from 3D scanning (approx. 15min). Curves of indentation tests or single skin damping values (MytonPro) are directly exported. The 3D scan-based method requires manual work for scan data processing. 4D scan data can be directly processed as homologous meshes, exported and analysed by BodyTex4D.
- **Data usability-The usability of the extracted soft-tissue properties in the field of deformable-body-clothing interaction**: The acquired data by IndentoPRO and 4D scanner is observed to be the most sufficient in the context of clothing development. IndentPRO presses a stamp up to 15mm into the soft tissue. IndentPRO is able to directly measure the force-displacement curve, capturing non-linear material behaviour of soft-tissue trivially. 4D scan material calibration is based on oscillating motions of soft tissue. The data usability by 4D scan is high as the whole body is measured simultaneously. Material property measurements by 4D

scan are inferred compared to direct force-displacement curves of the IndentoPRO. Likewise to IndentoPRO, the tensile test method an indentation test, analogue, the data usability is high. The data usability of the MyotonPro is evaluated as poor, as mainly the skin properties are extracted by MyotonPro. Measurements by 3D scanner showed high potential in addition to the force-displacement curves, the deformation behaviour is included in the material calibration process. However, due to the current missing 3D information in the areas of largest deformation, this measuring method is still now not high in its usability.
- **Accuracy: The accuracy of the measured data**. All measuring devices are accurate except the 3D scan method. The deformation region is covered by the load cell and, consequently, the deformation in the region of interest can not be captured accurately. Furthermore, the exposure time is too long causing a reduction in data accuracy. The IndentoPRO is not as accurate as the tensile test machine and slightly dependent on the operator because the test device must be placed on the human body manually. Measurements are influenced by the angle between the translation direction of the stamp and the surface of the skin. Varying angles due to inconsistent placement by the operator cause measurement deviations. Keeping the angle constant leads to repeatable results, rendering IndentoPRO applicable.
- **Costs**: IndentoPRO is at a cost of approx. 500 € is the most affordable measuring device compared in this chapter. Followed by the MytonPro with approx. 2000 €, 3D Scanner with approx. four to five figure € amount, the 4D scanner (more than approx. low to mid six-figure € amount) and the tensile test machine come with a higher undisclosed price tag.

Overall, the IndentoPRO and 4D scan methods are determined to be the most usable measuring method for clothing development. The IndentoPRO is flexible in positioning, fast in processing, highly useable in ist measured data and low in costs. The material calibration by 4D scan has the advantage of including body deformations and the dynamic tissue behaviour in a material calibration, which is especially for dynamic FE models beneficial. Both methods are in the following further described in their functionality and implementation into a FE model.

9.2.3.4 Material Calibration Method using the IndentoPRO and Parameters from Literature

Calibration of a deformable body part using the IndentoPRO is new in individualising human body FE models for clothing development. Figure 9.14 shows the required steps.

After the experimental testing of the material behaviour of certain body parts, the same test is remodelled by FE software. Therefore, a stamp of the same size as the IndentoPRO is remodelled and pushed into the same region corresponding to the physical experiment. If the experimental and numerical force-distance curves align, the calibration of material properties is done. Material parameters such as soft-tissue densities or Poisson ratios are taken from the literature (Sect. 2.4.2.3).

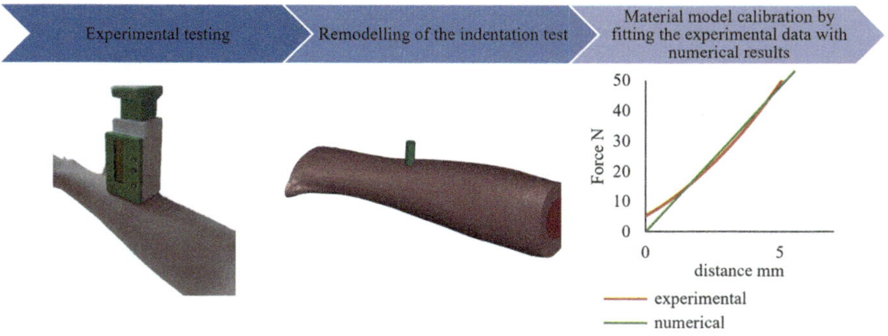

Fig. 9.14 Material calibration by IndentoPRO

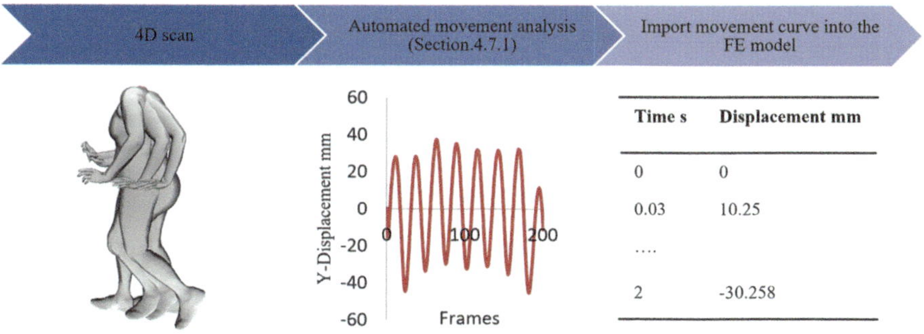

Fig. 9.15 Material calibration by 4D scan data

9.2.3.5 Material Calibration Method using Dynamic 4D Scan Data and Parameters from Literature

Material calibration based on dynamic 4D data extracts material properties from movements and resulting deformation. This method is specifically usable for breast modelling as the breast has a relative movement to the body and neither in-vivo nor ex-vivo methods discussed in Part I lead to individual material properties supporting FEM sufficiently. The following five steps are required to calibrate the breast behaviour using 4D data. Firstly, the breast movement is captured by a 4D Scanner (Fig. 9.15). Secondly, the body movement is exported using the movement analysis presented in Sect. 4.7.1. Thirdly, extracted movement curves of the body are used as a boundary condition of the FE model. Fourthly, the simulated breast deformation and breast movement can be compared with the experimental 4D scan data. Fifthly, the breast's mechanical property parameters are adjusted. The five-step process is iteratively repeated until convergence is achieved between the numerical and experimental breast behaviour. Breast density and position ratio are based on parameters from the literature.

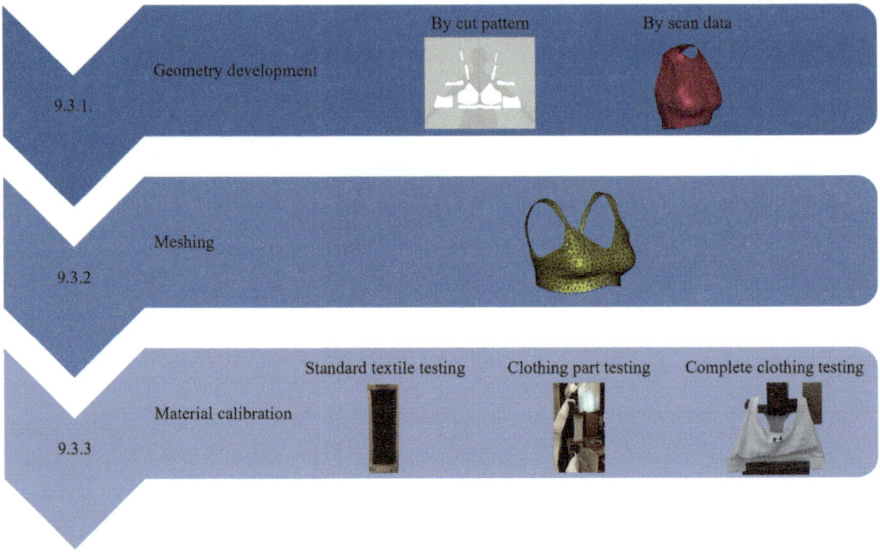

Fig. 9.16 Method development to generate a clothing Finite Element models

9.3 Method Development for Clothing Finite Element Models

This section describes and develops methods for generating **cloth**ing **FE m**odels (**CLOTH-FEM**). In Fig. 9.16 the modelling steps and structure of the section at hand are visualised. This section evaluates various methods for geometry development and material calibration, focusing on usability for digital clothing development.

9.3.1 Geometry Development

The development process of geometries of clothing items is divided into cut pattern design and scan data-based approaches (Fig. 9.17). The geometry development by cut patterns consists of digitising the cut pattern based on the original cut patterns or by a reconstruction of the cut patterns when the original cut patterns are not available. The scan data approach is divided into scanning the clothing item while worn (Fig. 9.17b) and scanning the clothing item not worn (Fig. 9.17c). The approaches are described and evaluated in the following.

(a) Based on cut patterns: original or reconstruction

(b1) Clothing worn **(b2)** Clothing unworn

(b) Based on scan data

Fig. 9.17 Geometry development of a clothing item

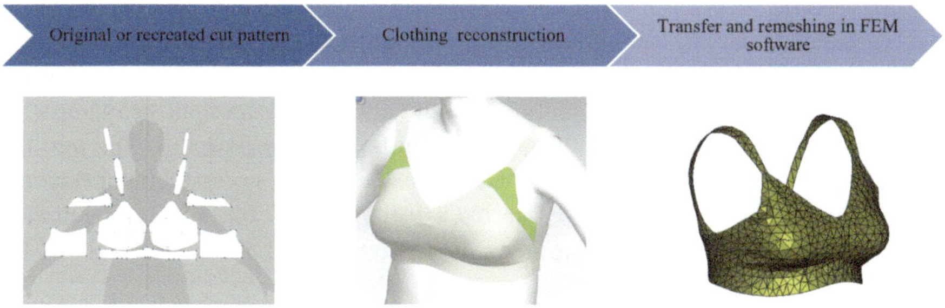

Fig. 9.18 Process flow to create a clothing geometry based on cut pattern data

9.3.1.1 Development of the Clothing Item Geometries by Cut Pattern

If information of the cut patterns is available, commercially available clothing software such as CLO3D (CLO Virtual Fashion 2024) or VStitcher (Browzwear Solutions Pte Ltd 2024) offers digitization functionality (Fig. 9.18). 3D geometries are derived from the cut pattern by listed software products. During the process of generating the 3D geometry of the cut pattern, it has to be ensured that no initial deformation of the clothing occurs within the software.

In the context of research, original cut patterns of purchased garment products are usually unavailable. The alternative is to unstitch the garment's seams to retrieve the cut patterns. Unstitching is declared as a fallback as it requires time-consuming manual labour compared to directly using digital representations of cut patterns. Digital cut patterns are usually a side product of clothing development, which renders unstitching a minor issue in clothing development. Nevertheless, it is recommended to unstitch random samples after production to validate the assembly algorithms of listed software products. Digital 3D representations of garments can be exported as *.obj-files. *.obj-files render the model compatible with BodyTex4D. *.obj-files can be directly imported into human-readable FE-files using the developed function `obj_2_node_k_file` and `obj_2_face_k_file` presented in Listing 9.1 and C.1. Cut pattern-based modelling of garments 3D geometry biggest advantage

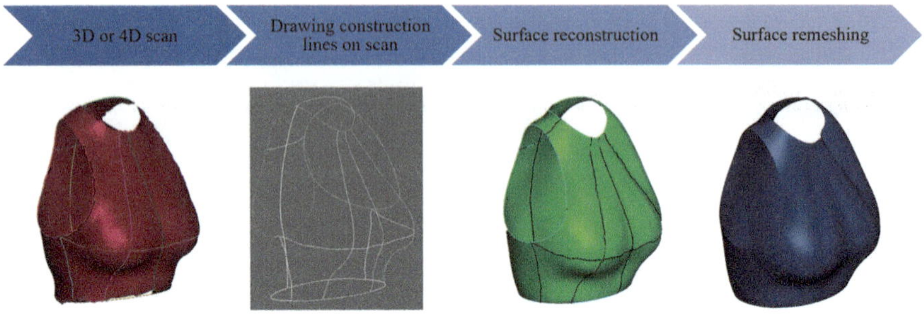

Fig. 9.19 Process flow to create a clothing geometry based on scan data

is the generality for build up a FE model. Cut patterns are interchangeable easily, leading to high usability in clothing development. Furthermore, the procedure has an inherent validation option by recreating the cut pattern from unstitching the finished product. Current downsides are the cost of commercial licensed software, and that tightly fitted garments tend to get stretched in the dressing process causing differences to the result of the assembly of the garment from cut patterns. Lastly, unstitching destroys the clothing.

9.3.1.2 Development of the Clothing Item Geometry by Scan Data

Scan data has great potential to improve the development of clothing item geometries. Clothing items are scanned while worn and unworn, for example, on a hanger. Followed by manual export of scan data of clothing geometries.

Scan data of peripheral regions (e.g. test subjects) not related to the clothing must be removed. Construction lines represent the abstract shape of the clothing and are defined for example in the open-source software LS-PrePost. The function Create N Side Surface –> Simple fit, inserts surfaces between the construction lines, resulting in the green reassembled clothing item (Fig. 9.19). As an alternative to reconstructing the garment, pre-processing could also be done by smoothing the edges, e.g. with ArtecStudio (commercial software) or MeshLab (open-source software).

In conclusion, the manual effort (approx. 60min for one bra) of garment reconstruction based on scan data is high. The manual steps are too specific to be automated easily. Machine learning might be an approach but the required training data is not available as of today. Nevertheless, the approach can be used to analyse the garment's current status and functionality. Attention must be paid to the pre-stretched state of the garment in the worn state. Scan data in the non-worn state reflects the current initial geometry of the garment. Capturing garments non-worn is difficult as regions of the surface are covered by many folds and wrinkles, especially with the increasing amount of textile per clothing item. The biggest advantage lies in capturing the influence of the production process on the garment (e.g. moulding) which is not possible from construction by cut pattern.

9.3.1.3 Comparison of Approaches to Develop Geometries for Clothing Finite Element Models

Presented approaches to create clothing geometries vary in starting conditions. The initial requirements for the evaluation of the geometrical functionality for clothing development are:

1. Availability of the original cut pattern
2. Possibility to damage clothing

The geometry development approaches are evaluated based on criteria estimating the ease of use in clothing development. In the following, evaluation criteria are defined and evaluated. The evaluation levels are set to low (-), medium (o), high (+), and very high (++). The evaluation and comparisons are based on experience.

- **Availability of cut pattern**: The usability for clothing development is high if the cut pattern is available digitally. Allowing to modify and optimise cut patterns in small, quick steps iteratively. The optimised cut pattern can be tested, leading to a reduction in physical prototypes. Cut patterns are available as a side product of the manufactures current design processes or by reconstruction based on unstitched clothing items. The initial process of unstitching the seams and digitising the cut pattern is time-consuming. Unstitched cut patterns enable to reconstruct of any clothing item geometry in 3D and it opens up the possibility of analysing the influences of production on the textile which renders it highly usable for clothing development. The accuracy is medium as it is a reconstructed cut pattern, errors always occur. Using 3D or 4D scan data of worn or non-worn clothing does not lead to original cut patterns, the usability is ranked as low.
- **Processing time**: Describes the duration data has to be pre-processed manually to be applicable for FEM. A shorter duration is beneficial for the clothing development process. Pre-processing based on the original, digital cut pattern is low, as clothing software products such as CLO3D or VStitcher assemble 2D cut patterns in real-time into 3D geometries. The time to create an initial clothing item depends on the complexity of the clothing item. For a simple bra, the time is 7min. A change in parameters of an available digital cut pattern takes less than 2min. Reconstructing cut patterns from existing clothing takes medium manual time. For a simple bra, it takes approx. 20min to open the seams and 7min to digitalize the cut pattern. Generating 3D geometries based on 4D scan data of worn or non-worn clothing items leads to high manual work (approx. 60min for a simple bra), as the scan data have to be processed manually.
- **Geometrical accuracy**: describes the geometrical similarity between the digital clothing item and the real item. A higher accuracy indicates a more similar geometry. Having a high similarity between digital and produced clothing items

Table 9.2 Comparison of geometry development approaches for clothing Finite Element models

Textile geometry development	Starting requirements		Evaluation criteria		
	Original cut Pattern availability	Clothing destruction Needed	Cut pattern Availability	Processing Time	Accuracy
Reconstructed cut pattern	No	Yes	Yes	o	o
Original cut pattern	Yes	No	Yes	-	o
Clothing worn	No	No	-	++	-
Clothing unworn	No	No	o	++	+

improves the accuracy of the clothing-body interaction analysis and, consequently, improves clothing development by increasing the reliability of the digital process. The geometry of scanned non-worn clothing items shows the highest accuracy in terms of similarity to the original clothing geometry. Geometry changes due to the production process by moulding, seams, thermal treatments, etc., are captured. Influences of production are not respected by cut pattern approaches. Scan data of worn clothing items deforms the item rendering it not usable. Nevertheless, the scan of a worn clothing item can be used to validate the digital dressing procedure.

In conclusion, cutting patterns could be investigated as the most promising approach in digital clothing development. The high usability and little manual processing time are the biggest advantages. The accuracy depends on the cut patterns and production processes. Cut pattern-based approaches are the recommended pre-processing step for setting up FE models. Clothing by scan data can be used to validate the model by comparing the clothing FE model to the physical model. Table 9.2 summarises the evaluation of the geometry development approaches. Green indicates a high usability for the clothing development process.

9.3.2 *Meshing and Part Selection*

Boundary conditions for meshing clothing or textiles are fibre properties (micro-scale), yarn properties (meso-scale) or textile properties (macro-scale). Due to modelling on the fibre level, the micro-scale causes the highest computational costs. Macro-scale is the most coarse approach with the lowest computational costs. This chapter focuses on the meshing approach based on the macro-scale. Due to the application of modelling the deformable body-clothing interaction, the focus is on analysing the effect of various cut patterns and materials during the interaction, not the fibre or yarn displacement within the clothing.

After importing the clothing geometry by *.obj-file, the geometry is meshed using the open-source software LS-PrePost. The element size influences the computational time; smaller elements lead to more nodes and lead to higher computational costs.

On the contrary, smaller element sizes lead to better contacts and more accurate simulation of deformation behaviours.

Shell elements are applied to mesh the 3D clothing geometry, as the thickness compared to the surface area of clothing items is small. Generally, membranes could also be used for modelling clothing. However, membranes have been investigated in preliminary studies to not be as stable as shell elements in simulations of the interaction of tightly fitted clothing and the human body.

Depending on the clothing structure, triangle or quadratic elements are suitable. Triangular shapes model complex surface structures more accurately, while quadratic elements lead to more stability in computation.

After meshing, the mesh is divided into parts with different material properties. Depending on the level of simplification and relevance, the seams' areas can also be modelled separately. Each part is assigned a calibrated material model.

9.3.3 Material Calibration

Three different approaches for material calibration are analysed and evaluated. Figure 9.20 displays the general workflow to calibrate clothing items. The main approach to calibrate material models in FE analysis is to recreate the material testing setup within the FE environment. The material model is chosen and constant material parameters are determined and set. All other material parameters are iteratively adjusted until a sufficient alignment between experimental and numerical material testing is reached. The material behaviour regarding deformation and forces are included as calibration parameters.

The approaches for testing clothing material "standard textile testing", "tensile test of clothing parts", and "tensile test of complete clothing items" to calibrate FE material are described in the following sections. The procedure of selecting sections of interest in the characteristic material property curves is presented. Finally, the approaches are compared.

9.3.3.1 Standard Textile Testing

Standard textile material tests include tensile-, bending- and shear testing. The tests are remodelled in FE environments (Fig. 9.21). Standard textile testing leads to a material database upon which arbitrary material combinations can be tested within the simulation environment. Concerning digitised clothing development, standard textile testing should be preferred. Standard textile testing requires all materials to be available in pre-production conditions to exclude disturbance of properties due to production processes. Consequently, the disadvantage follows to be a slight mismatch between the FE model and the real clothing item due to influences of the production process on the textile such as stresses.

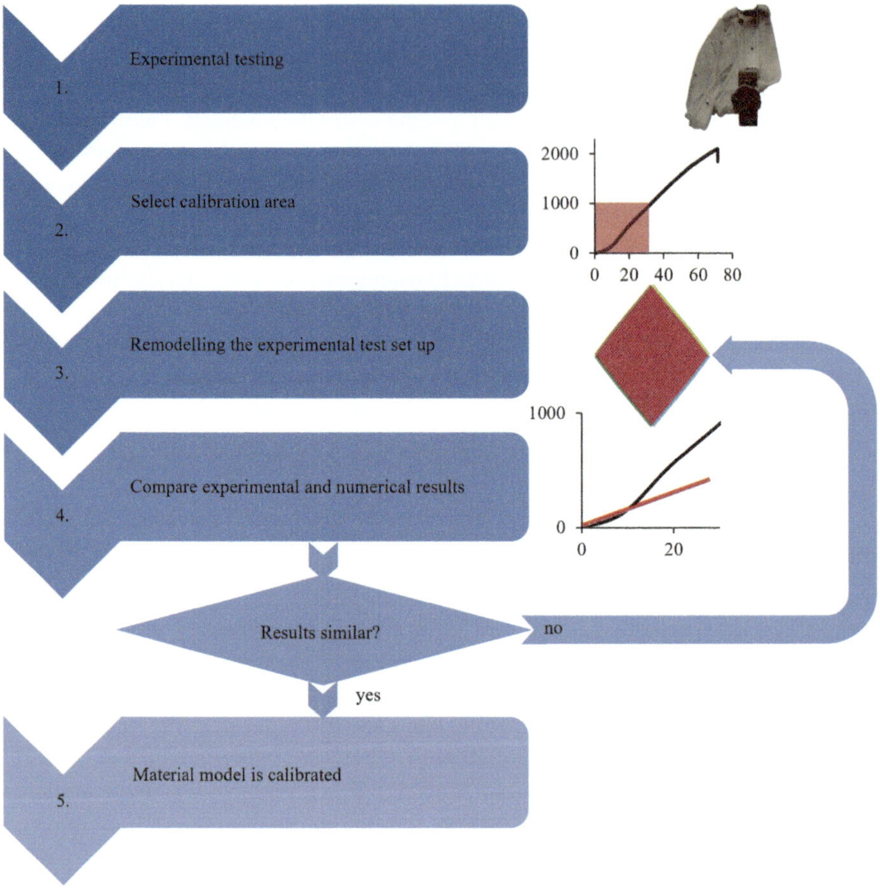

Fig. 9.20 Workflow for a material model calibration

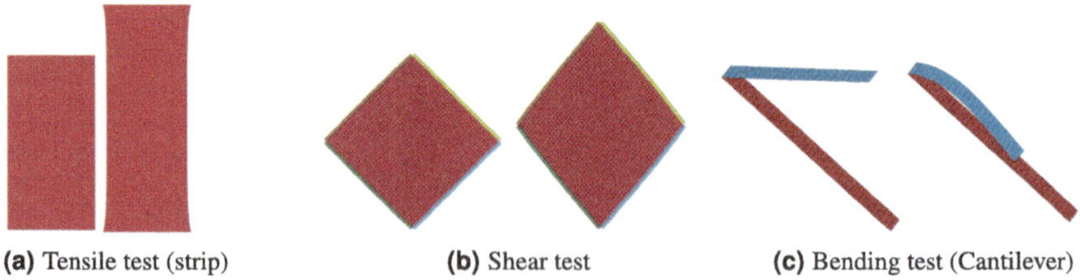

Fig. 9.21 Material calibration approach by standard textile material testing

(a) Tensile test bra strap (b) Tensile test bra cup (c) Tensile test bra seam

Fig. 9.22 Material calibration approach by clothing part testing

9.3.3.2 Tensile Test of Clothing Parts

A tensile test of clothing parts is performed on clothing items after production. Therefore, a product of the clothing production process must be available and is destroyed. Various parts of the clothing item with different material properties are tested. For example, for testing bras, tested parts are straps (Fig. 9.22a), cups (Fig. 9.22b) and seams (Fig. 9.22c). The test should be applied in two directions if the area of the textile allows it. Testing two directions retrieves information about the anisotropic nature of textile materials. Parts consisting of multiple textile layers lead to valuable insights into combined mechanical behaviours.

The strongest advantage of testing the mechanical properties of part of clothing items is that it includes the influence of the production process on the tested material properties. Especially, evaluating the behaviour and impact of seams and moulding. The disadvantage can be attributed to not all standard textile tests can be conducted. For example, shear or bending tests cannot be performed due to the limited material availability for a bra, but for a compressive shirt, it would be possible. The disadvantages are negligible for tightly fitted clothing items as the tensile loads are typically the major factor for the functionality of tightly fitted clothing items. "Tensile test of clothing parts" is particularly useful for garments where raw textiles are unavailable. Furthermore, calibrating models by this method leads to more accurate results as the effects of production and combinations of materials are considered.

9.3.3.3 Tensile Test of Complete Clothing Items

A tensile test of complete clothing items is a method to test the complete system in a ready-to-wear state. The entire clothing item is clamped into the tensile testing machine. Figure 9.23 presented and remodelled tensile test of entire bras in a FE simulation environment. The advantage is to acquire data on interactions between the clothing item's various materials, parts, and seams for calibrating the FE model.

Fig. 9.23 Material calibration approach by testing the complete clothing system

The disadvantage is primarily due to the inability to apply the determined material properties to cut pattern parts. This material test is particularly suitable for validating the performance of an entire clothing item; however, it is less effective for assessing local material properties.

9.3.3.4 Comparison of Material Calibration Approaches for Clothing

The presented and developed material calibration approaches are useful and valid for different purposes. The material calibration method based on standard textile testing requires a large amount of unprocessed textiles. The calibration methods based on the clothing parts or complete clothing require the post-production clothing item.

The highest usability and impact on digitised clothing development is the approach of standard textile testing. Various materials, material combinations, final fit of the clothing item, can be tested before producing the first prototype. The clothing item does not have to be produced physically, decreasing development iteration times and required resources.

The approach based on testing clothing parts is especially useful when setting up a simulation environment. This approach results in the most accurate description of the clothing behaviour, representing the material behaviour after the clothing production. Results of clothing part testing could enrich the database produced by standard textile testing.

The approach of testing the complete clothing item system is useful for validating the accuracy of models resulting from the standard textile tests. It can be validated, if the modelled material interactions of the different clothing parts based on standard textile tests reflect reality sufficiently.

In summary, all calibration approaches are useful in developing a clothing model. In the following, the main use cases for the developed approach can be seen:

- **Clothing development**: textile standard testing
- **Accurate calibration of clothing parts**: tensile test of clothing parts
- **Validation of full clothing item system**: material testing of complete clothing

The most accurate FE model is generated by combining all three material testing approaches. First, the mechanical behaviour of the raw textiles is tested and each clothing part is modelled. The material test of the clothing part validates the clothing part models. Finally, the digitally assembled clothing items containing various material properties and assumptions about seam and moulding properties can be validated by testing the complete system post-production. The goal of applying the three methods sequentially in future studies is to develop models that accurately extrapolate from standard textile tests to the performance of a full clothing item. This approach aims to reduce the reliance on costly and time-consuming physical prototyping and production test runs.

9.3.4 Selection of Characteristic Material Property Curve Sections and Material Models

Following experimental material testing, an important step is to select regions of interest in the experimental data by which the calibration of the FE model is performed. Three approaches are determined and discussed below (Fig. 9.24).

9.3.4.1 Complete Curve Fitting

The first selection approach is performing a complete curve fitting (Fig. 9.24a). This approach selects a material model and material parameter that best aligns with the complete measured material property curve. In general, characteristic material property curves are non-linear. Consequently, complete curve fitting requires non-linear material models. The advantage is that the material behaviour of the whole system can be captured. The disadvantage arises from the high complexity of creating and

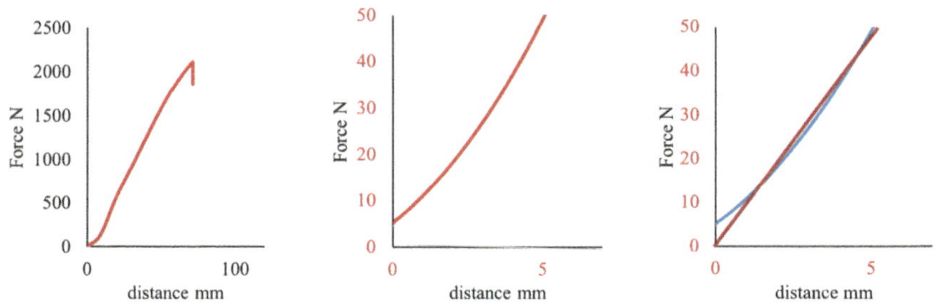

(a) Complete curve calibration **(b)** Complete curve calibration in selected calibration area **(c)** Linear calibration in selected calibration area

Fig. 9.24 Approaches to select curve sections for the material model calibration fitting non-linear material models. Additionally, the computational cost increases with more complex material models.

9.3.4.2 Curve Fitting for Closed Set of Experimental Data

This approach is based on selecting a dataset of interest for the material model calibration process. Pre-tests determine the range of strains and stresses the clothing item is put under in real life to define the borders of the dataset of interest. Fitting the curves focuses on achieving the highest possible accuracy for the selected set utilising the simplest possible material law. Therefore, the resulting model is valid for a smaller range of applications compared to complete curve fitting (Sect. 9.3.4.1). Compared to complete curve fitting, the calibration process is faster due to the focus on the closed dataset. Nevertheless, the parameter fitting and material model are often non-linear and take longer to optimise than linear material models.

9.3.4.3 Linearisation for Closed Set of Experimental Data

This approach further simplifies the material calibration process compared to the method presented in Sect. 9.3.4.2. Similar to "curve fitting for a closed set of experimental data", a set of interest of the experimental data is selected at first. Secondly, a linear-elastic material model is approximated to fit the measured material curve (Fig. 9.24c). Linear-elastic material models have low complexity, quickly calibrate and have low computational costs due to a low number of parameters describing the model. FE modelling is generally computationally costly with respect to elastic materials, deformations and contact conditions. Reducing computational costs is important to make FE modelling a feasible approach for the clothing development process.

9.3.4.4 Comparison of Curve Fitting Methods to Calibrate Material Models

Figure 9.25 concludes the comparison of the material model calibration approaches. Green indicates the positive effects of the criteria on clothing development. Red indicates the negative effects of the criteria on clothing development. It can be determined, that accuracy and cost are strongly dependent. The higher the accuracy, the higher the cost in terms of time and computation. The fastest computational and calibration times are achieved by linearisation. Whether the linearisation of the material model is appropriate and does not produce unrealistic results of the clothing item's behaviour must be decided on a case-by-case basis.

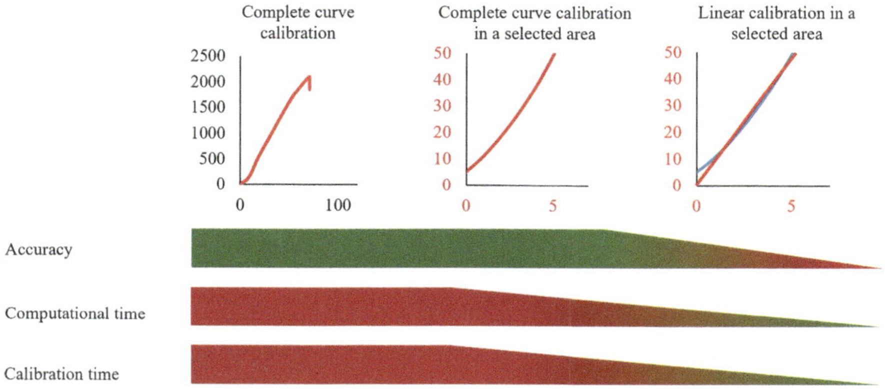

Fig. 9.25 Comparison of approaches to select curve sections for the material calibration (green: beneficial in the clothing development, red: less beneficial in clothing development)

9.4 Method Development of Dressing Processes for Tightly Fitted Clothing in Finite Element Modelling

After modelling the individual, deformable human body and clothing FE models are combined to investigate the deformable-body-clothing interaction. This combined model is further called **i**nividualised, **d**eformable **h**uman **b**ody–**cloth**ing **FE** **m**odel (ID-HB-CLOTH-FEM). The interaction of clothing and body primarily influences and defines the functionality of the clothing item. Tightly fitted clothing items are essentially undersized. Undersized clothing is geometrically smaller than the body. Undersized clothing items cause the following interactions with the human body:

- body and/or clothing deformations,
- pressure between body and clothing or
- movement restrictions of soft tissue

Modelling the effects of interactions enhances the clothing development process by establishing a digital knowledge base that captures the influence of cut patterns and textile materials on deformation, strain, pressure, and movement of both the body and the clothing item.

A central challenge in modelling and combining an unclothed body with undersized clothing results from the clothing item being smaller than the body. In Fig. 9.26, the issue is depicted for the case of breast-bra interaction analysis. To overcome the intersections and model the tissue-clothing interactions realistically, dressing the body model virtually is required. Different methods of the dressing process can be found in the literature (Sect. 2.20). In the following section, existing methods are implemented, and newly developed approaches to dressing processes of undersized clothing are presented and evaluated. Lastly, approaches for validating the combined model are presented.

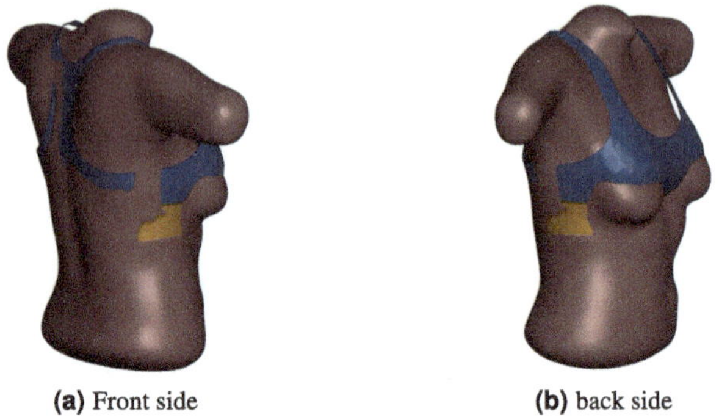

Fig. 9.26 Dressing challenge of undersized clothing: Clothing is smaller than the body

Two categories of approaches to dress the human body model in undersized clothing items are described. The first category is approached where the human body model is modified, and in the second category, the clothing item model is modified. Modifying clothing has the main advantage of being closer to reality as the clothing gets stretched to fit over the human body model. Shrinking and extending the body size to dress clothing is unrealistic. The dressing process has a major influence on the fit of the clothing item, therefore, the dressing process must be as realistic as possible. The amount of elements of the clothing model is generally less and modelled with shell or membrane elements, causing lower computational cost than a modification of solid elements of the human body model. In conclusion, dressing processes by modifying the clothing instead of the body are preferred. Methods based on body modifications are omitted. Three dressing methods presented in the following are geometrical expansion (Sect. 9.4.1), load expansion (Sect. 9.4.2) and thermal shrinkage (Sect. 9.4.3). Thermal shrinkage and load expansion methods are developed in this work. The method of geometrical expansion has already been developed to dress legs with compression stocking (Kaiser 2019). The methods are evaluated by

1. model setup time,
2. computational cost and
3. the quality of the results.

9.4.1 Dressing Method: Geometric Expansion

A geometrical expansion method for an undersized clothing item can be found in Kaiser (2019) for dressing legs with compression socks. Nevertheless, in the literature, the method is not applied to more complex body parts and clothing items such as breasts and bras. The method is modified to be applicable to the example of the breast-bra interaction. Geometric expansion falls in the preferable dressing method category of dressing by modifying the clothing item.

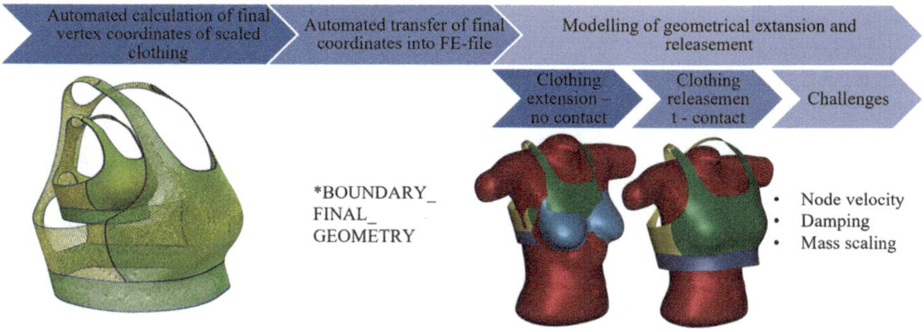

Fig. 9.27 Dressing method: Geometrical expansion

A Python function is implemented to automate the dressing process. Figure 9.27 shows the process flow of the geometrical expansion dressing method.

Firstly, the clothing item scale factor is determined. The clothing item needs to be scaled up to a size where it does not intersect with the body. The clothing item must be outside of the body surface completely. Secondly, a Python function is developed to calculate the final vertex coordinates of the scaled clothing item. The Python script has been developed to modify k-files from LS-PrePost. The inputs for the automated scaling process are:

- point cloud extracted from *.obj-file of the clothing item
- determined scale factor
- vertex-ID, which should be declared as origin or origin in coordinates

The clothing item is imported into the Python script as *.obj-file and dumped into a k-file using the function `obj_2_node_k_file` and `obj_2_face_k_file` presented in Sect. 9.2. An array is created with the original vertex-IDs and x-,y- and z-coordinates in the form of a point cloud. If the bra is already imported into LS-PrePost and is only available as k-file, the same array can be generated using the function shown in listing C.3. The point cloud is scaled by the function `scale_point_cloud` (Listing 9.2).

Listing 9.2 Scaling of a point cloud

```
import pyvista as py
import numpy as np

def scale_point_cloud(vertex_array: np.ndarray,
    scale_factor: float, origin_vertex_id: np.ndarray
     = 1) -> pv.PolyData:
     # Find the coordinates of the vertex declared as
     origin
     origin = vertex_array[vertex_array[:, 0] ==
    origin_vertex_id, 1:4][0]

     # Move vertex array to make vertex declared as
    origin overlap with world origin
```

```
    vertex_array_at_origin = vertex_array[:, 1:4] - origin
    # Scale the translated points
    scaled_vertex_array = vertex_array_at_origin * scale_factor
    # Reverse the translation to restore the fixed vertex
    scaled_vertex_array += origin
    # Create pv.PolyData from scaled point cloud
    scaled_point_cloud = pv.PolyData(scaled_point_cloud)
    scaled_point_cloud.point_data['Node IDs'] = vertex_array[:, 0]

    return scaled_point_cloud
```

The result of the scaling is dumped into the k-file of the ID-HB-FEM using the function *BOUNDARY_PRESCRIBED_FINAL_GEOMETRY. The function assigns a target position to each node, which should be reached after a certain time. The Python code-snippet can be seen in the appendix, Listing C.4.

Concatenating the function `scale_point_cloud` for determining the target of expansion and *BOUNDARY_PRESCRIBED_FINAL_GEOMETRY for performing the geometrical expansion automates the dressing process based on the scale factor. No contact conditions between the clothing item model and the human body model are set during this expansion. After the expansion is finalised, the boundary restriction is suspended, causing the induced strains to act on the bra. The bra shrinks until the force equilibrium between bra strain and body deformation occurs. Consideration of contact conditions between body and bra is activated for the shrink process. Triggering shrinkage by releasing the boundary condition to expand the bra presents several challenges, primarily due to the high accelerations induced by the strain within the material. A higher scale factor of the bra results in greater final strains, leading to a more rapid increase in absolute accelerations of the bra following the release of the boundary condition after the expansion. The high velocities cause instabilities. To lower velocities, the damping factor of the bra material can be increased. The damping factor has to be determined in several iterative steps. Additionally, mass damping influences the time step, contact and material behaviour. This method is recommended for clothing with lower scale factors, as it allows for a lower damping parameter, reducing instability and incorrect material behaviour. Nevertheless, this method contributes to the automation of FE-file generation and virtual dressing through geometrical expansion.

9.4.2 Dressing Method: Load Expansion

The developed dressing method by load expansion realises the clothing item expansion by applying loads onto each segment. The acting direction of the load is defined

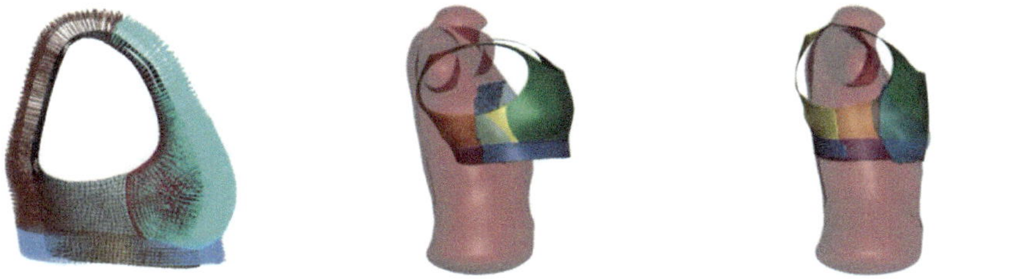

(a) Segment normals (b) Moving bra due to imbalance of segment forces (c) Constrains do not lower the imbalance of segment forces easily

Fig. 9.28 Dressing method: Segment load expansion

by the segment normals. Figure 9.28a shows each segment's normal vector. Loads are increased until the clothing item is expanded to an extend of no intersections with the human model's surface. Shrinkage is realised by decreasing continuously the load back to zero. During expansion, contact conditions between the clothing item model and the human model are deactivated and reactivated for the contraction.

The advantage of load expansion is the low implementation effort and low computational cost. To use load expansion, a load curve has to be defined. The challenge arises in defining the appropriate load conditions. Increasing the load too fast leads to instabilities. Furthermore, the distribution of surface areas greatly impacts clothing expansion. Clothing patterns with unevenly distributed cutting parts can lead to an imbalance of the acting forces. Areas of higher material density experience greater forces. In Fig. 9.28b, the result of imbalanced segment forces is visualised. The clothing tends to move forwardly out of the chest, because of the bigger surface of the frontal bra part. Applying boundary restrictions to the clothing item solves the problem. Due to the force imbalance almost always occurring, the load expansion method requires a high number of manual iteration steps currently (Fig. 9.28c). Load expansion is better suited for clothing items with more evenly distributed surface areas and, consequently, evenly distributed segment density (e.g. compression socks). Additionally, the material properties have a significant influence on the expansion. For example, the higher the e-module, the higher the required loads. Subsequently, the load curves are highly dependent on the material model. Currently, fitting load curves to the material property is time-consuming. Nevertheless, load expansion is a promising approach with the potential to be automated by calculating the load forces based on normal vectors of the segments and the material model.

9.4.3 Dressing Method: Thermal Shrinkage

The dressing method by thermal shrinking is newly developed within this work. The dressing process is visualised in Fig. 9.29. The first step of the thermal shrinkage method is to scale the clothing item to resolve any intersections with the human

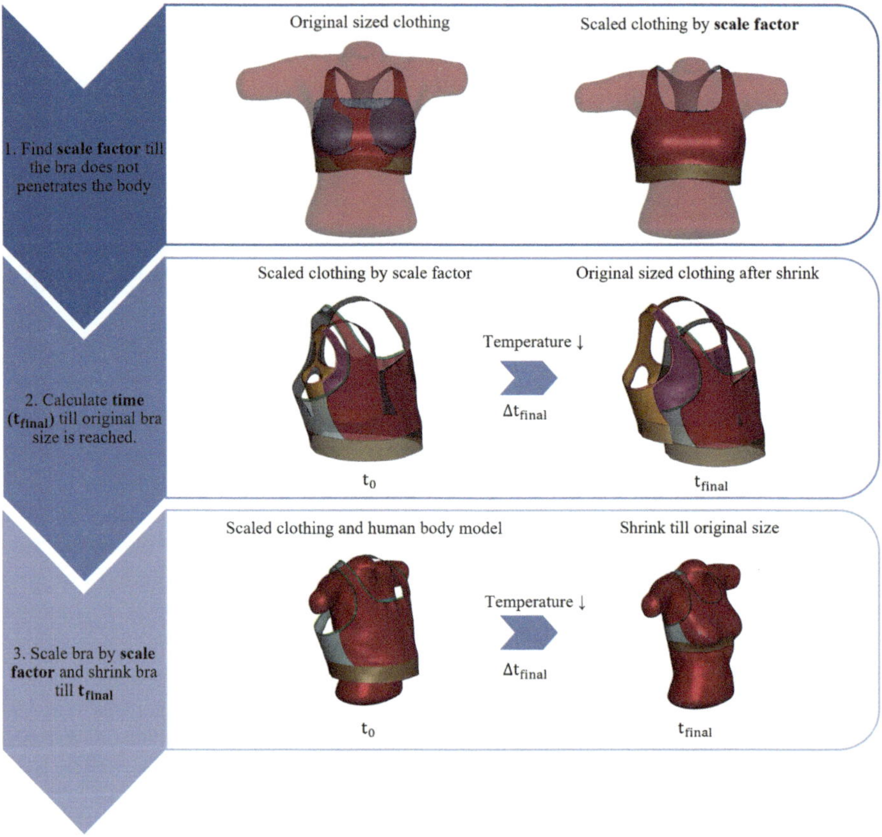

Fig. 9.29 Dressing method: Thermal shrinkage

body model. The scaling process can be performed by reusing the Python function presented in Listing 9.2 or by the FE software. Using the open-source software LS-PrePost, the scaling can be performed by Element Tool –> Transform –> Scale. Generally, clothing of similar sizes can be scaled by the same scale factor.

The second step of the thermal shrinkage method is to calculate the time Δt_{final} until the bra contracts back to its original size. The shrinkage is realised by a change in temperature. The influence of temperature can be added to the material model. Temperature change is introduced to the model, and the clothing shrinks depending on the heat transfer coefficient added to the material model. The challenge in calculating the shrinkage time results from the shrinkage per time step being dependent on the shrinkage of the previous time step forming a recursive sequence. Consequently, the change in length for each time step is different and is calculated on the previous time step. Eqs. 9.1–9.6 show the derivation for calculating the shrinkage time for a single length.

The initial length is $l_{initial}$, and after scaling the length by the scaling factor, the length changes to l_{scaled} (Formula 9.1).

$$l_{scaled} = scale\_factor * l_{initial} \tag{9.1}$$

In the first time step (Δt), the scaled length shrinks by $\Delta l_{\Delta t1}$. $\Delta l_{\Delta t1}$ can be calculated by the strain equations displayed in Formulas 9.2. Thereby, the strain is defined as ϵ, the heat transfer coefficient α and the temperature change in that time step is ΔT.

$$\epsilon_{t1} = \frac{\Delta l_{t1}}{l_{scaled}} = -\alpha * \Delta T \Leftrightarrow \Delta l_{t1} = -\alpha * \Delta T * l_{scaled} \tag{9.2}$$

The new length at time step t_1 is $l_{t1} = l_{scaled} - \Delta l_{t1}$. For the second time step, the new length can be calculated by Eq. (9.5)

$$\epsilon_{t2} = \frac{\Delta l_{t2}}{l_{t1}} = \frac{\Delta l_{t2}}{l_{scaled} - \Delta l_{t1}} = -\alpha * \Delta T \tag{9.3}$$

$$\Delta l_{t2} = -\alpha * \Delta T * (l_1 - \Delta l_1) \tag{9.4}$$

$$l_2 = l_{scaled} - \Delta l_{t1} - \Delta l_{t2} \tag{9.5}$$

For several time steps, the following equation can be defined (Eq. 9.6). Using the equations above, the needed time Δt_{final} and temperature difference ΔT_{final} till the current length $l_{curr(t)}$ equals the initial length l_{inital}. A Python function has been implemented to calculate the time until the bra reaches its original size. The Python function is part of the appendix, Listing D.1.

$$l_{curr(t)} = l_{scaled} - \sum_{1}^{i} \Delta l_i \tag{9.6}$$

The third step of the thermal shrinkage method is to merge the human model with the scaled clothing item model. Contact conditions between the human model surface and the clothing item surface are defined. The shrinking process causes the clothing item to interact with the body. The shrinkage until a force equilibrium is established between the textile strain and body deformation. The force equilibrium depends on the clothing item's material model and the material model of the human body model. The clothing-body interaction can be analysed after t_{final} is reached.

The advantage of the thermal shrinkage method is the low setup time and the computational costs are the lowest amongst the presented approaches. Thermal expansion does not work as the clothing item starts to wrinkle and forms large folds intersecting with the body. The effect is shown in Fig. 9.30.

Fig. 9.30 Wrinkling due to thermal expansion of a clothing item

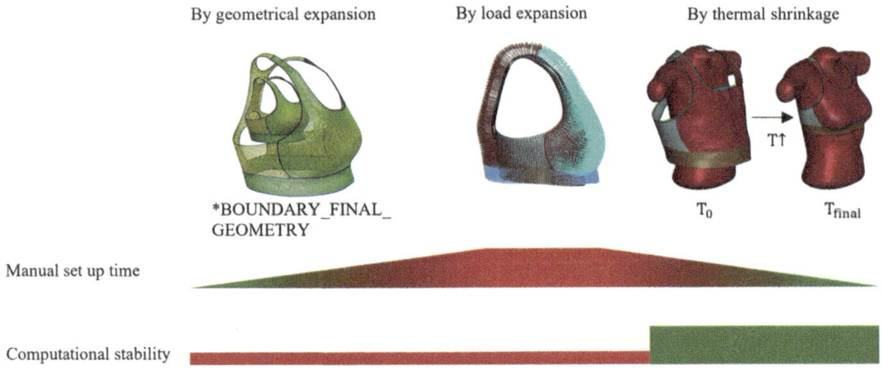

Fig. 9.31 Comparison of the dressing methods

9.4.4 Comparison of Dressing Methods

Three dressing methods involving clothing modification approaches have been implemented, improved, tested, and evaluated. Figure 9.31 provides an overview of the evaluation.

The dressing method by geometrical expansion is easy to use and quick due to the implemented Python functions, which automatically extract the geometrical target coordinates of the up-scaled clothing item, resulting in low manual effort. The process has been automated to the point of the scale factor determination. The disadvantage arises due to the high forces which are induced by the clothing expansion. The high forces cause high velocity during the contraction rendering the model unstable after the expansion process. Especially, when the clothing item has to be scaled up with a bigger scale factor, as is the case for bras, the tendency to instabilities increases.

The stability can be increased by introducing damping, but it must be ensured that the material and contact behaviour are not affected.

The dressing method by load expansion has a high manual setup time and lower computational stability comparatively. Especially when using non-symmetrical shapes and non-homogeneous material-distributed clothing items, the load distribution has to be adapted on a case-by-case basis. This adaptation has not been automated until now. An approach for automation would be calculating the load normal vectors and introducing relation to the material model element-wise. In the case of symmetrical clothing items with homogeneous material distribution (e.g. compression socks), the load expansion method is promising as it closely resembles the real process. Stability issues are analogous to the geometric expansion method for the same reasons.

The dressing method by thermal shrinkage is the most promising approach. The manual setup time is low as most steps are automated by Python functions. The computational cost is low, and stability is high. This dressing method is recommended for the fitting of tight clothing on a human body FE model.

9.5 Validation of Finite Element Models to Investigate the Deformable-Body-Clothing-Interaction

The deformable-body-clothing interaction is compared to experimental data to validate the FE model. After the validation, the FE model can be used in the development of new clothing items.

It is important to not fit the FE model too strictly to the experimental data. Every experimental measurement is a snapshot of the current state influenced by many confounding factors like influences of production or the state of the body of the test subject. A direct comparison of simulated and experimental data is not possible when working with humans, as humans cannot generate reproducible states. It is important to understand that the same condition cannot be recreated in the numerical model as in the experimental setup. The target is *similar* behaviours of the FE model and reality, not the exact fitting of the FE model to the experimental. Exact fitting would cause a specialised FE model to be the experiment at hand and not be generalised for the behaviour of interest. In the validation process, the behaviour of new cut patterns or material properties of clothing items on the body should be compared to calibrate the FE model. Similarly, the validation methods presented below aim to compare the effects of clothing on the body in both experimental and simulated conditions. Four different validation methods are in the following presented.

9.5.1 Fit Analysis

The fit analysis is determined as a validation method, especially for tightly fitted clothing items. The metric is to validate how similar different parts of the clothing item are placed on the body by the numerical model and the experiment. In Fig. 9.32a, an example can be seen. The fit comparison can be performed based on 2D images, 3D or 4D scan data. The fit analysis validation takes the following parameters into account:

- clothing parts placement on the body
- clothing deformations due to strains
- folds of clothing item

9.5.2 Body Deformation Analysis

The body deformation analysis is another validation process for comparing a clothing-body model to experimental data. This comparison is based on cross-sections (Fig. 9.32b) or 3D point cloud comparisons.

The best practice to capture body deformation caused by clothing is to perform a 4D scan. Various body postures are captured in rapid succession, and manual processing time is negligible. The body deformation is compared with the modelling results. The validation of the numerical model is successful if similar effects are observed. However, the aim of this analysis is **not** to achieve identical body deformation, as this is unrealistic and unattainable due to human variability. A method to improve the selection of comparable body poses is to use 4D scan data and BodyTex4D. Examples are given in the following chapters.

9.5.3 Dynamic Behaviour

The validation parameter by dynamic behaviour compares the effect of the deformable body-clothing interaction during dynamic movements. Using a 4D scanner, the interaction comparison is performed by comparing the movement of certain body parts or by performing a deformation analysis at specific movement points. The 4D scanner has the advantage of capturing body deformations during movements, which can also be compared. Alternatively, just a movement comparison can be performed based on acceleration sensor data.

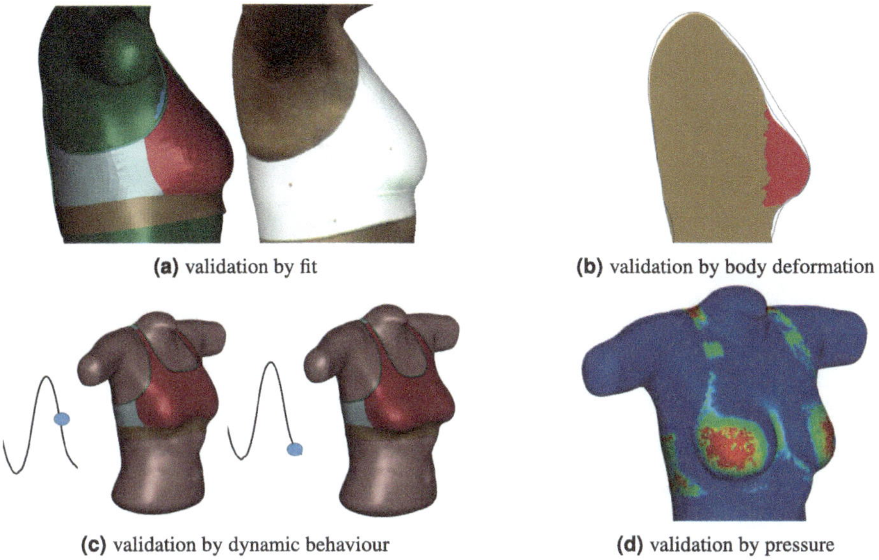

Fig. 9.32 Validation methods to validate a clothing-body Finite Element model

9.5.4 Pressure Analysis

Pressure analysis is a sensor-based validation method for numerical models (Fig. 9.32d). Pressure sensors are placed to measure the pressure at specific areas between the body and the clothing item. The validation is done by comparing the magnitude of the pressures of the experiment and the numerical model. Analogue to the previous discussed validation method, only a similarity of pressures is targeted. Pressures depend on the body pose, breathing state, the current clothing fit, the measuring process and muscle activity. A direct comparison of absolute values is impossible, as a person cannot reproduce the same position or movement consistently.

9.6 Conclusion and Discussion

This chapter presented methods for generating an individualised, deformable human body-clothing Finite Element Model (ID-HB-CLOTH-FEM). Methods from the literature are compared to newly developed approaches or improved by a higher degree of automation. The main modelling steps for the model generation are the development of an individual human body model (ID-HB-FEM) in Sect. 9.2, the development of a clothing model (CLOTH-FEM) in Sect. 9.3, the combination of both models by a dressing method in Sect. 9.4 and the validation of the combined model in Sect. 9.5.

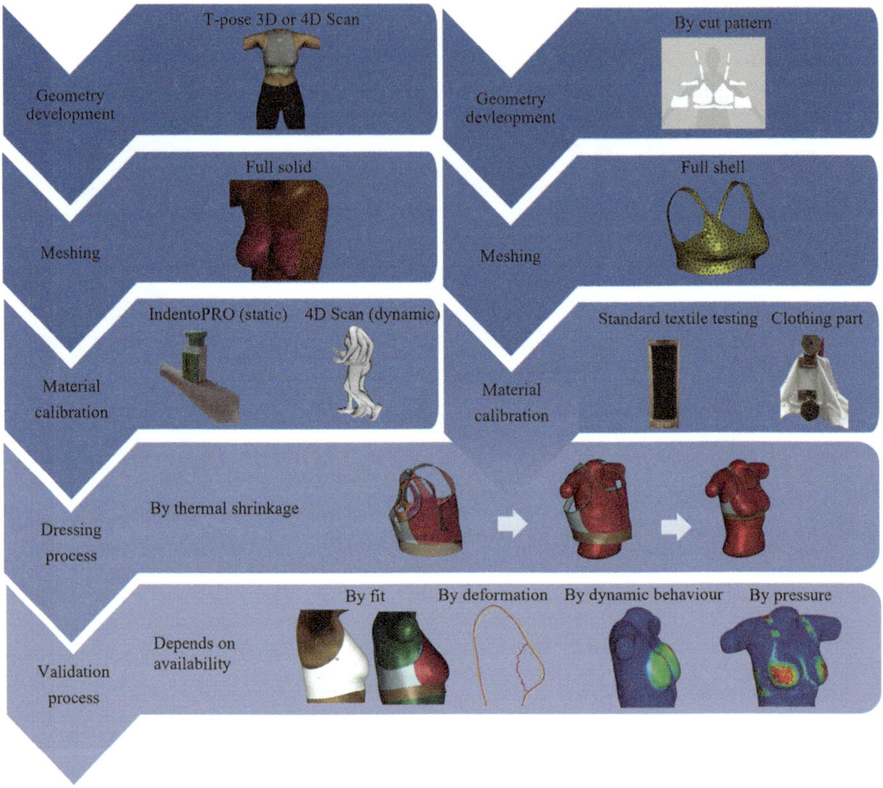

Fig. 9.33 Generalised method steps to generate a deformable human body–clothing Finite Element model

The generalised method steps, evaluated to be most suitable for the clothing industry, are displayed in Fig. 9.33. Each modelling step is described and discussed in the following:

1. **ID-HB-FEM: Geometry development by 4D scans**: The model is based on real human body data generated by 4D scans. Utilising 4D scan data has the advantage of having homologous meshes combined with fast and accurate capturing of the human surface. The 4D scanner has the advantage of being less prone to errors caused by the movements of the test subject during the scanning process due to low exposure durations. The 4D scan exposure duration is with 1ms much lower than a 3D scanning process, which can take up to 3min. 4D scanners are an order of magnitude more expensive than 3D scanners. 3D scanners are more suitable for transportation, rendering 3D scanners a valuable measurement tool next to 4D scanners. The T-pose is determined as the most sufficient pose for different applications as most of the body's surface is captured, and the resulting meshes have fewer defects.

2. **ID-HB-FEM: Meshing by a full-solid approach**: The full-solid meshing approach showed the most realistic deformation behaviour, is stable and needs minor manual setup times. The full-shell meshing approach causes the body to be hollow. The hollow space leads to unrealistic deformations. The newly developed shell-solid approach showed promising results as only the parts of interest are meshed in greater detail with solid elements. However, in its current state, the solid-shell approach needs further investigation into boundary conditions to produce realistic deformations. A template-based automatic body part selection algorithm is developed and implemented using Python. The algorithm selects the deformable areas automatically from homologous mesh data. Template-based, automated model setup is advantageous compared to the full-solid approach as the selection of a body part has to be performed once and can be transferred to all subsequent test subjects. Using the full-solid approach, the meshing has to be performed in LS-PrePost or Gmsh, and the part selection is manual. The manual selection takes approx. 2min. To lower the manual setup time in future, the avatar with a homologous shell mesh can be transformed into a homologous solid mesh. A template-based selection process can be implemented based on generated consistent mesh data.
3. **ID-HB-FEM: Material calibration by IndentoPRO or 4D scan data**: IndentoPRO is a portable indentation test device. IndentoPRO is applicable to any body area. IndentoPRO measures a force-displacement curve which can be used for in-vivo soft-tissue material model calibrations. IndentoPRO is used for the first time to calibrate an ID-HB-FEM within this work. The testing device is low in costs (approx. 500 €), can measure any body position, can display non-linearities, has short processing times, and is portable. Listed properties render IndentoPRO a valuable measurement tool in the context of digital clothing development. A movement can be captured using the 4D scanner and then remodelled in the ID-HB-FEM. ID-HB-FEM is calibrated based on the dynamic deformation behaviour of the deformable body parts of interest in motion to be as realistic as possible.
4. **CLOTH-FEM: Geometry development by cut patterns**: Currently, most FE models of clothing items in the context of textile-body interactions are not based on real cut patterns. The most common approach is to model the clothing item by an offset of the body's surface. Using real cut patterns offers the advantage of modifying and testing the patterns virtually. By utilising clothing software such as CLO3D or Vstitcher, the clothing item is stitched digitally, resulting in a 3D geometry which can be exported. Currently, the standard clothing software cannot create realistic ID-HB-FEMs. The deformation behaviour of the body parts and the level of individualisation of the deformable body parts is very limited. Additionally, this software is unable to accurately model seams, which have a substantial impact on the properties of clothing items and, consequently, the tightly fitted-clothing-body interaction.
5. **CLOTH-FEM: Meshing by the full-shell approach**: Generally, tightly fitted clothing is thin compared to its area. Shell elements can approximate the textile behaviour with low computational cost and have been determined to be more stable compared to membrane elements in the tight-clothing-body interaction.

6. **CLOTH-FEM: Material calibration by tensile tests**: Tensile tests are performed on raw, pre-production textiles. A material data bank can be created based on data generated by tensile tests of the raw textiles for further model development. If a clothing item is assembled virtually based on raw textile data, it should be validated by comparisons with the material properties of the post-production clothing item. During the production processes, the material properties change. The dataset of interest for calibration of the CLOTH-FEM is selected based on the application of the clothing item. The application defines the occurring loads for which the model is calibrated. Calibrating application specific saves time in the calibration process and can simplify the material model, which saves computational costs.
7. **ID-HB-CLOTH-FEM: Dressing method by thermal shrinkage**: Dressing by thermal shrinkage is mostly automated, has high stability, is application-related and can be applied to any clothing item. Two aspects are currently performed manually and should be automated in the future. Firstly, the alignment of the CLOTH-FEM relative to ID-HB-FEM. Secondly, the scale factor for expansion to resolve intersections between ID-HB-FEM and CLOTH-FEM.
8. **ID-HB-CLOTH-FEM: Validation by 4D scan data and pressure sensors**: The result of combining ID-HB-FEM and CLOTH-FEM by a virtual dressing process to an ID-HB-CLOTH-FEM can be validated by analysing deformation, fit, dynamic behaviour or pressure. Method(s) are chosen depending on the application and availability of measurement equipment. Combining multiple methods increases the accuracy of the validation and, consequently, the usability of the ID-HB-CLOTH-FEM in clothing development. Using 4D scan data and the BodyTex4D framework improves and automates the selection process of comparable 4D scan data. All validation methods except the pressure analysis can be performed based on 4D scan data.

References

Browzwear Solutions Pte Ltd.: VStitcher. 25 May 2024. https://browzwear.com/

CLO Virtual Fashion (2024) CLO3D. Version 7.3.0. 25 May 2024. https://www.clo3d.com/en/

DYNAmore GmbH, an Ansys Company: LS-PrePost. www.dynamore.de. (visited on 08/26/2023)

Geuzaine C, Remacle J-F (2009) Gmsh: a 3-D finite element mesh generator with built-in pre- and post-processing facilities. Int J Numeric Methods Eng 79(11):1309–1331. ISSN: 0029-5981. https://doi.org/10.1002/nme.2579

Kaiser C (2019) Vorgehensmodell zum Aufbau einer validierten Simulationsumgebung zur Bestimmung des durch Kompressionsstrumpfe induzierten Kontaktdrucks als Basis fur eine virtuelle Funktionsanalyse. Dissertation

MakeHuman. https://static.makehumancommunity.org/index.html. Accessed on 05-05-2024

MyotonPro. https://www.myoton.com/. (visited on 05/25/2024)

Pai DK et al (2018) The human touch. ACM Trans Graph 37(4):1–12. ISSN: 0730-0301. https://doi.org/10.1145/3197517.3201296

Schmidt A-M, Kyosev Y (2023) Finite element modelling of textile-soft material interaction using 3D/4D scan data. In: 14th European LS-DYNA conference 2023, Bade-Baden, Germany

ThermoTrigger: IndentoPRO Gewebe-Compliance-Messgeraet. 25 Jan 2024. https://www.thermotrigger.de/shop/indentopro-gewebe-compliance-messgeraet/

Tonuk E, Barbara Silver-Thorn M (2003) Nonlinear elastic material property estimation of lower extremity residual limb tissues. IEEE Trans Neural Syst Rehabil Eng 11(1):43–53. ISSN: 1534-4320. https://doi.org/10.1109/TNSRE.2003.810436. (A publication of the IEEE Engineering in Medicine and Biology Society)

Chapter 10
Analysing the Breast-Bra Interaction in a T-Pose Using Finite Element Models, 4D Scan Data and BodyTex4D

Abstract The methods developed in Chap. 9 are applied and validated in a static T-pose breast-bra interaction model. A customised human model of a female test subject with a deformable breast and skin is created. Four different bras are modelled and combined with the human model using a novel dressing method. Validation is performed through cross-sectional comparison, fit analysis, and pressure analysis. The first two validation steps utilise 4D scan data and BodyTex4D. The results show a high usability of the digital models, allowing differences between the four bra designs to be identified based on pattern and seam construction. The digital pressure analysis proves to be particularly advantageous compared to experimental methods. Overall, 4D scan data and BodyTex4D enhance the validation of FEM models, leading to improved accuracy in digital simulations.

Keywords FEM · Soft avatar · Individualised · Digital clothing development · Human-Body-Model (HBM) · Clothing · 4D scanning · Breast-bra

10.1 Introduction

This chapter presents how individualised, deformable human body models with deformable breasts and dressed in a bra are modelled (example of ID-HB-CLOTH-FEM) to analyse the breast-bra interaction in a static pose digitally.

Many women wear bras in their daily lives. Bras serve different purposes. Bras shape the breasts, reduce back pain or restrict breast movements. As breasts are highly individual in shape and soft-tissue properties, finding a bra that is comfortable and fulfils the wearer's purpose is challenging. Studies have shown that few women can easily find a well-fitting bra (Greenbaum et al. 2003). In Fig. 10.1, the main factors influencing the breast-bra interaction are displayed.

The interaction between the breast and the bra is influenced by both the bra's design and the characteristics of the breast. On the bra side, material properties and cut pattern play a crucial role. On the breast side, factors such as volume, shape, tissue composition, and overall body shape significantly affect this interaction. Additionally, body posture and movement alter how the breast and bra interact over time.

A.-M. Schmidt, *Novel Methodologies for Digital Investigations of Clothing-Body-Interactions*, Lecture Notes in Bioengineering,
https://doi.org/10.1007/978-3-031-91794-3_10

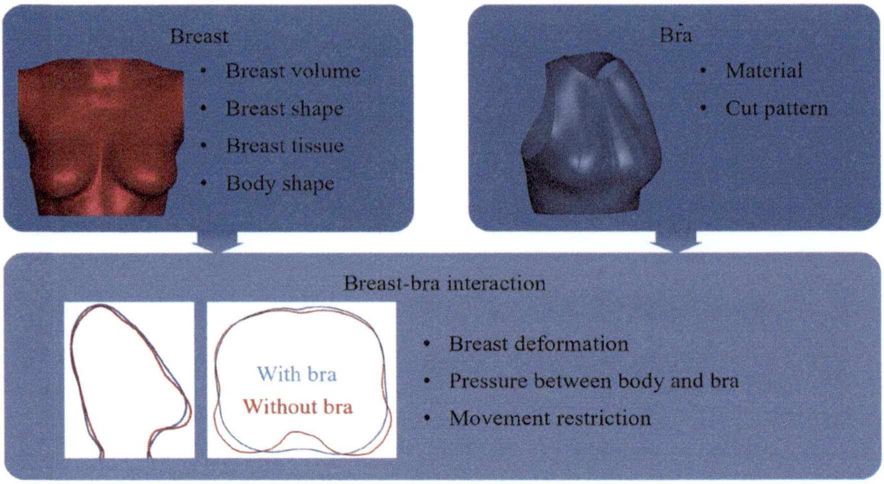

Fig. 10.1 Influencing parameter on the functionalities of the breast-bra interaction

This interaction can lead to breast deformation, localized pressure, and restricted movement. The specific performance goals of a braâŁ"whether focused on support, comfort, or mobilityâŁ"should be optimized based on its intended application. For example, sports bra development focuses on the maximal restriction of breast movement while minimal pressure peaks are exerted.

Currently, bras are developed mainly in iterative prototyping processes. Bra prototypes are produced, tested on a test subject, redesigned, and tested again after the production process. The current process is time consuming, material intensive, and labour costs are high. The production duration causes the test subject's body composition to change due to the period cycle, fat loss/gain or water intake in the soft tissue amongst other factors. Consequently, comparisons between the produced prototypes worn by the test subject are challenging. Even if the body composition of the test subject is constant, the difficulties of reproducing body states and comparable experimental data remain. In summary, the comparison of bra models and test subjects is complex. The challenges affect commercial bras but also individualised ones. Rigid manikins are used to reduce the source of error of not having reproducible bodies and body poses. However, the manikins are physical as well as the prototypes of the bras. Furthermore, most manikins are rigid and do not resemble breast properties of natural breasts. The amount of breast shapes is restricted, the breasts are symmetrical, and individualisation is not feasible. Subsequently, manikin-based bra development is inaccurate, unsuitable, and the root cause of the failure to provide women application specific, comfortable bras. A technology to make bras comparable, and decrease development effort while improving the degree of individualisation is needed. The work at hand presents technologies to base digital bra development.

The digital bra development promises to develop comfortable, individually fitted bras adapted to the wearer's specific purpose by using digital, individualised, realistic human body and clothing models.

Finite element models calculate accurately deformations, stresses and pressures. Most FE breast-bra models currently available are not generalised to form a basis for the digital development process. Most models are designed for one-time use, which makes them unattractive in clothing development. The breast material properties were mostly based on literature values and not individualised by data of real test subjects. Only the approach of determining the damping breast factor by an external breast lift is seen by Ruixin et al. (2020), Cai et al. (2018) and Yiqing (2016). Additionally, the literature does not present validation methods for selecting comparable body poses. In conclusion, the current breast-bra FE model's shortcomings are:

- A generalised description of model setups
- An individualised breast tissue material calibration based on indentation tests
- An accurate bra cut pattern FE model
- An interface between existing clothing software and FEM
- Bra seam modelling
- Accurate validation methods based on comparable body states

This chapter presents a Finite Element model to analyse the breast-bra interaction of four bras. Methods presented in Chap. 9 are used to generate the model. The interaction analysis focuses on the different influences of the bra models on breast deformation and pressure distribution in a static body pose. Influences of the bra material, such as cut pattern and size, are determined. 4D scan data are integrated into the FE model validation process. This chapter proves the usability of the generalised method presented in Chap. 9.

10.2 Methods

To develop the breast-bra FE model, firstly, the human body FE model (ID-HB-FEM) with deformable breasts is generated (Sect. 10.2.2). Secondly, four bra models (CLOTH-FEM) are generated (Sect. 10.2.3). The bra models differ by material and cut pattern. Thirdly, both models are combined and validated. In Fig. 10.2, an overview of the modelling steps with the used modelling approaches to create a breast-bra model is outlined.

10.2.1 Finite Element Environment

The numerical model is built with LS-PrePost (DYNAmore GmbH 2023). LS-PrePost is an open-source software tool. LS-PrePost loads a human-readable input

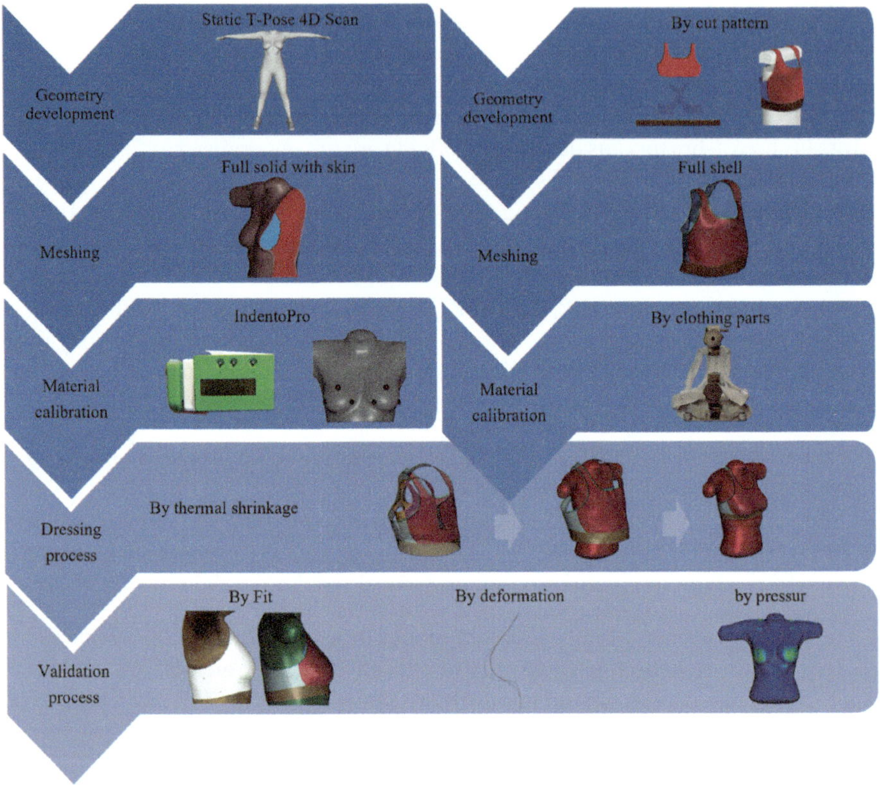

Fig. 10.2 Overview of the methods used to develop a breast-bra Finite Element model in a T-pose

file format containing all necessary information to run the model. A human-readable input file has the potential for easily implementing automated modifications. LS-PrePost software is designed to model non-linear material behaviour and large deformations. The units are set to mm, sec, ton, K (Listing 10.1).

Listing 10.1 Units used in the Finite Element models (Ls-PrePost k-file)

```
*CONTROL_UNITS
$#    length      time      mass      temp
          mm       sec       ton         K
```

10.2.2 *Development of an Individualised, Deformable Human Body Finite Element Model with Deformable Breasts*

The development of a human body FE model with deformable breasts includes the steps of geometry development (Sect. 10.2.2.1), meshing the geometry (Sect. 10.2.2.2) and material calibration of the meshed geometry (Sect. 10.2.2.3). Methods presented in Sect. 9.2 are applied. A detailed description of the methods containing advantages and disadvantages is presented in Sect. 9.2.

10.2.2.1 Geometry Development

The model is based on a 29-year-old female test subject with a cup size of 75 C. The test subject is the same as described in Sect. 5.3.2 named test-subject-A. The geometrical model is based on a static 4D Scan (Sect. 9.2.1.1). The Move4D scanner by IBV (Parrilla et al. 2019) is used. As the exposure time of the Move4D scanner is 1 ms, the scans are less prone to errors due to the movements of the test subject during the exposure compared to 3D scanners. T-pose is chosen (Fig. 10.3a) by the pose selection process (Sect. 9.2.1.2). The test subject's breast surface is fully captured in the T-pose. Less meshing defects occur compared to an A-Pose (Fig. 9.4). The T-pose is portrayed as seen in Fig. 10.3a. The test subject has an underbust circumference of 770 mm in the selected pose. The selection process is described in Chap. 5. This information is relevant for the validation process described in Sect. 10.2.5. The T-pose is processed as a homologous mesh (Sect. 2.1.3.2) and exported to an *.obj-file (Sect. 2.1.4.2). Meshing defects are observed for fingers and legs. Fingers and legs are not regions of interest and are cropped out, resulting mesh gaps are closed (Fig. 10.3b). The software Artec Studio 15 (2016) is used for mesh modifications as it is fast and easy to use. However, open-source software such as MeshLab (Cignoni et al. 2008) can also be used.

10.2.2.2 Meshing

The cropped human body scan data is imported from the *.obj-file into LS-PrePost. Meshing by full-solid elements (Sect. 9.2.2.2) is used for the human body. As in Sect. 9.2.2.4 described, solid elements have the advantage of low manual setup time, good calculation stability and model deformations realistically. Solid meshing is performed in LS-PrePost by the tetrahedron mesher. The tetrahedron mesh element size of 5 mm is applied as a compromise between deformation accuracy and computational cost.

In preliminary tests, three different mesh sizes are tested while performing the material calibration test described in Sect. 10.2.2.3. 7 mm is too coarse to model breast deformation by an indentation test realistically (Fig. 10.3a). 3 mm increased

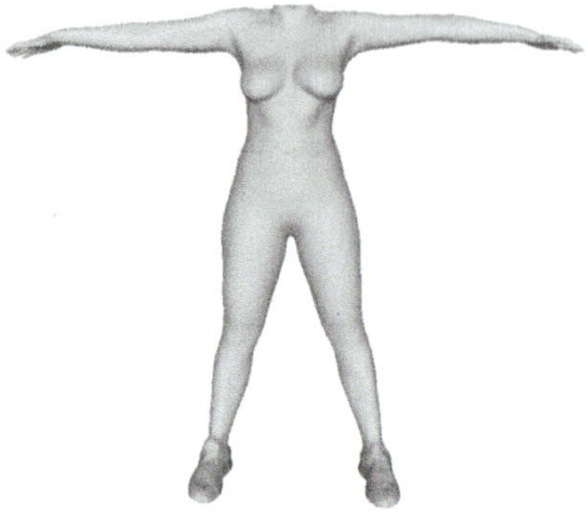

(a) T-Pose captured by 4D Scanner

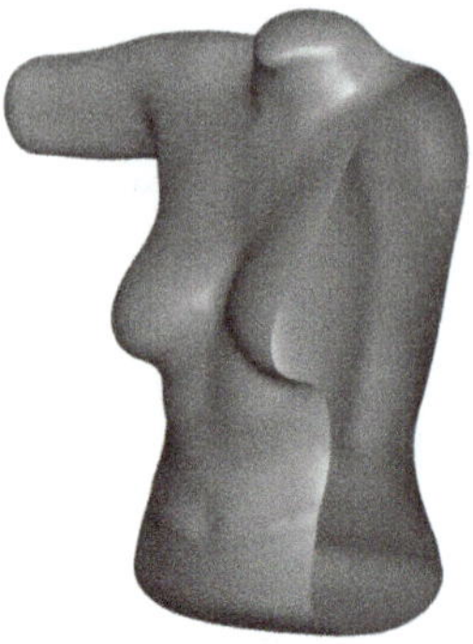

(b) Modified T-Pose mesh

Fig. 10.3 Selection and modification of a T-pose scan to model the deformable human FE model

the computational time by a factor of 12 to 6 h compared to an element size of 5 mm resulting in computational time of 20 min. 7 mm elements take 7 min to compute, but the deformation behaviour is not sufficient.

The body model is separated in three parts: breast volume, skin and the rest of the body. The skin is defined by selecting the most outer element layer of the body part by the LS-PrePost function `Element Tools -> Move or Copy` an option `3DSurface on`. Skin thickness results in the element size of 5mm. The breast is defined by the LS-PrePost function `Element Tools -> Move or Copy`. The breast is modelled as a homogeneous part as a more detailed breast model requires individual MRI images, and this level of detail is not convenient nor required in clothing development. Figure 10.4 displays the final, tripartite FE human body model.

The deformable human body model is meshed with `ELFORM 10`. `ELFORM 10` is a 1-point tetrahedron element. In preliminary examinations, compared to `ELFORM 4` and `ELFORM 13`, this element showed the highest stability for the breast-bra interaction application. `ELFORM 4` indentation test behaviour was too stiff. The element became unstable under larger deformations. `ELFORM 10` and `ELFORM 13` performed similarly in the indentation test, but `ELFORM 13` caused in pretests in an interaction with a bra instabilities (Fig. D.2). Consequently, `ELFORM 10` is chosen.

10.2.2.3 Material Calibration

The material calibration for the deformable breast regions is performed by the calibration method by IdentoPRO and literature values (Sect. 9.2.3.4). A remodelled indentation test performed at the breast can be seen in Fig. 10.5. The indentation test is performed at three points on both breasts. One measurement point is on top of

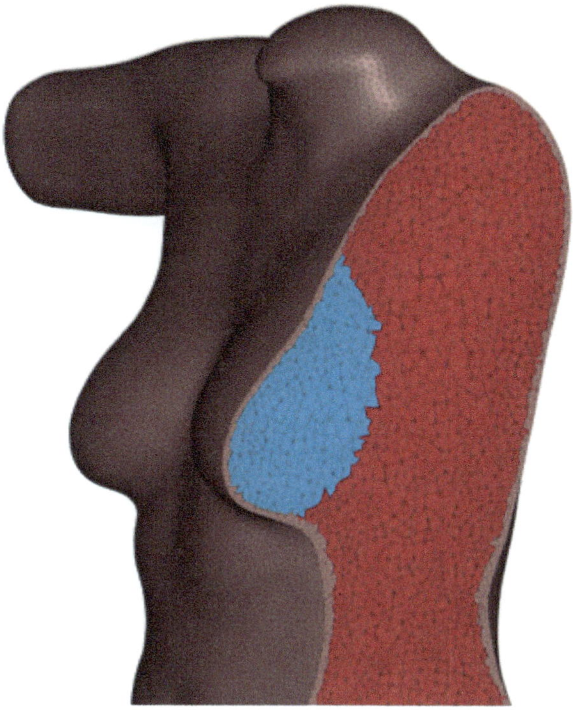

Fig. 10.4 Meshed and part-sectioned human model Finite Element model: red: inner, rigid body, blue: deformable breast, rosa: deformable skin

the breast, and two measurement points are on each side of the breast. The measurement points are marked on the model in Fig. D.3. Five measurements per point are recorded. Tissue indentation depth is 15 mm. The force-displacement curve is measured experimentally to calibrate the FE model. The indentation test is remodelled in the simulation environment. The material model is calibrated by aligning the experimental and numerical force-distance curves. The skin material properties are set to the same values as the breast material properties. Applying the breast material to the skin reduces the parameter set and reduces the calibration effort. Furthermore, the breast plus breast skin is homogenised complying with the indentation test resulting in a sufficient and efficient material model approximation.

As an calibration method, the "Linearisation for closed set of experimental data" 9.3.4.3 is used. The linear-elastic material model `*MAT_ELASTIC` is applied. The linear-elastic material model has low computational cost due to its linear nature compared to non-linear material modules and a small set of three input parameters representing the low complexity of the model. The density, elastic modulus and Poisson's ratio must be set in the material model. The density and Poisson's ratio are based on literature values set to 1×10^{-9} tm^{-3} (Chen et al. 2013) and 0.49 (Cai et al. 2016; Zhang et al. 2022), respectively. The elastic modulus (e in Listing 10.2) is adjusted to align indentation test data and modelled force-distance curve. The stamp is modelled as rigid using the steel material parameter (Listing D.1).

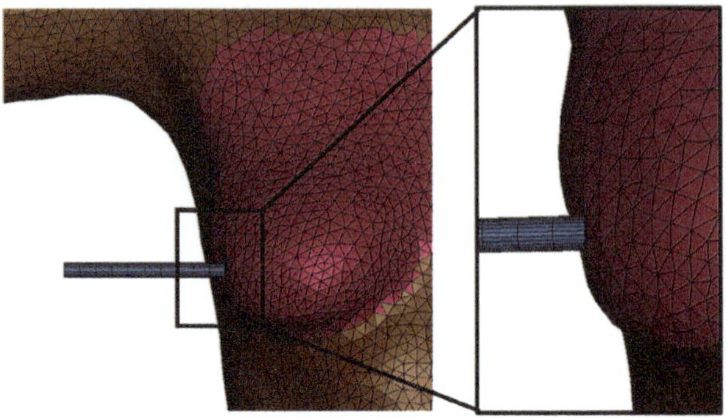

Fig. 10.5 Remodelled indentation test on the breast for calibrating the deformable breast Finite Element model

The method leads to material-specific calibration results based on the test subjects tissue properties. The method is accurate, flexible and due to the simplicity (single parameter) fast.

Listing 10.2 Material model of the breast (Ls-PrePost k-file)

```
*MAT_ELASTIC_TITLE
Breast material
$#      mid          ro          e          pr          da
        db  not used
         11.000000E-9       xxxxx        0.49        0.0
        0.0         0.0
```

10.2.3 *Development of Bra Finite Element Models*

This section presents the development of four bra models and the translation to bra FE models. The section is structured as follows. Outlining geometry development (Sect. 10.2.3.1), meshing (Sect. 10.2.3.2) and finally the material calibration (Sect. 10.2.3.3).

10.2.3.1 Geometry

Four different bras are modelled. Two types of bras in two sizes each are modelled. The bras are described in detail in Sect. 5.3.3. The same nomenclature is applied: bra-2-S, bra-2-M, bra-3-S, and bra-3-M resulting from bra types 2 and 3 in sizes S and M. Bra of type 3 is a commercially available product. The geometry of bra-3 is reconstructed by the method of reconstructing based on cut patterns (Sect. 9.3.1.1), as the original cut pattern is not available (Fig. 10.6). Bra-3-S and bra-3-M seams are opened, separating the bra into its cut pattern parts. The bra is comprised of three main bra parts: bra-band, front part and back part. The parts are digitised using the clothing software CLO3D which generates automatically by digital stitching a 3D geometry

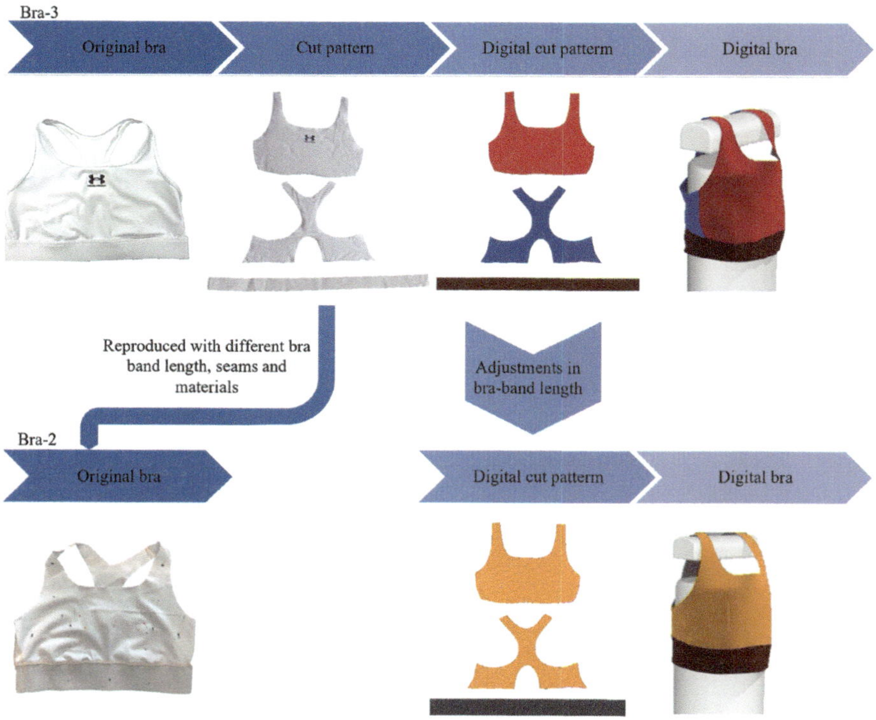

Fig. 10.6 Workflow for the geometry development of bra-2 and bra-3

of the clothing item. The bra is unwrinkled by cylinder and bar objects to avoid pre-strains and bra deformations of the initial geometry. The absence of strains is ensured by measurements of the parts being the same in the 2D and 3D. 3D geometries are exported as *.obj-files. Figure 10.6 displays the geometry development workflow.

Bra-2 is an in-house reproduction based on the cut pattern of bra-3. The bra cut pattern of the front and back parts are the same. The bra-band dimensions, seams and amount of material layers differ. Bra-2 has a 20 mm shorter and 15 mm wider bra-band compared to bra-3. Bra-2 is single-layered and has no seamed edges. Section 5.3.3.2 presents the mechanical properties. Experimental results of the bra fit and dynamic behaviour (Chaps. 5 and 6) show slippage of the bra-band of bra-3. The bra-band moves on top of the breast during the dressing process. The bra-band is, therefore, shortened to stay in place and prevent slippage. After this bra-band redesign, the 3D geometry is generated in the clothing software CLO3D, and the 3D geometry is exported as *.obj-file.

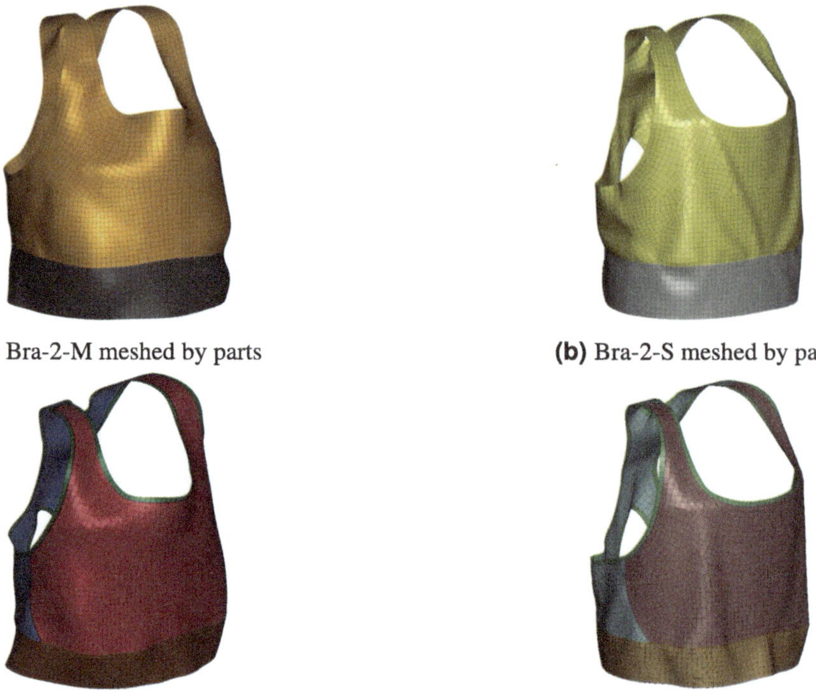

(a) Bra-2-M meshed by parts (b) Bra-2-S meshed by parts

(c) Bra-3-M meshed by parts (d) Bra-3-S meshed by parts

Fig. 10.7 Four modelled bras meshed by parts (each colour indicates a different material behaviour)

10.2.3.2 Meshing

The four bra FE models are imported separately into LS-PrePost and remeshed. Shell elements are chosen as they are identified to be more stable than membrane elements. A mixed quadratic-triangular mesh is used to reflect the bra's complex shape. A 5mm mesh size is found to be optimal, providing stable contact behaviour and realistic deformation while keeping the computational cost reasonable. A bra section before and after meshing is shown in Fig. D.4 in the appendix. The `ELFOM 16` is used as element formulation. `ELFOM 16` is a fully integrated shell element.

Bra-2 is sectioned into two parts. The bra-band and the front/back part (Fig. 10.7a and b). The front and back parts consist of the same material. Bra-3 is sectioned into four parts: bra-band, front part, back part, and seams (Fig. 10.7c and d) with different material properties each. Compared to bra-2, bra-3 has seamed part edges which leads to a stiffer material behaviour. The seams connecting the bra-band and the front/back part are elastic and the material property changes are neglected. The thickness of each part is presented in Table 5.2.

10.2.3.3 Material Calibration

The material calibration is performed using the clothing part testing method (Sect. 9.3.3). Since the original textiles prior to production for bra-3 are unavailable, this method must be used to determine the material parameters for bra-3. Bra-2 is also tested using the clothing part testing method for comparable testing results. Bra-2 could otherwise be tested by standard textile testing as the bra-2 is self-produced and the original textiles are available. The clothing part testing method has the advantage of measuring the post-production material properties.

A cyclic tensile test is performed to characterise the material properties of the bras. A cyclic tensile test reflects the dressing and undressing of a bra. Each cycle simulates the process of putting the bra on and taking the bra off once. The bras are tightly fitted, causing tensile strength to predominate the material behaviour. The cyclic tensile test is performed according to DIN EN ISO 13934-2. Five cycles are performed per bra and the strain limit is set to 50%. The strain limit determination is part of the calibration process (Sect. 9.3.4). The strain of the underbust is calculated by dividing the difference of the initial bra-band length and the underbust circumference of test-subject-A (770 mm) by the initial bra-band length. A strain smaller than 30% for all bra variations is determined. The strain limit is set to 50% to calibrate the material behaviour accordingly to the expected strain values.

The bra-bands are tested in the direction of thorax circumference. The initial clamp width is set to 100 mm with a pre-force of 0.25 N. The textile parts are tested in two directions: vertically (dir1) and horizontally (dir2). Figure 5.13 shows the clamp area for the testing. The clamp width is 60 mm and a pre-force of 0.25 N is set. The material behaviour of the fifth cycle is chosen to represent a bra that was worn multiple times before. The calibration workflow presented in Fig. 9.20 is performed to calibrate the FE model. Results of the cyclic tensile tests are presented in Sect. 5.3.3. A numerical model of the tensile test is remodelled in LS-PrePost. The remodelled tensile test is depicted in Fig. 10.8. Table D.1 shows the numerical tensile test model parameters taken from the samples of the physical experiment. The numerical model does not take into account the surrounding textile material as the physical experiment does. The numerical model is simplified to a strip-tensile test because the surrounding textile causes element instabilities due to the high deformation at the clamp's edges (Fig. D.5). The more stable, simplified model is chosen because the large deformations exceed the deformation of the clothing-dressing process, rendering the deformations of surrounding textiles out of the scope of the calibration process.

Linearisation for a closed set of experimental data (Sect. 9.3.4) is chosen as the material calibration method. Material calibration by linearisation for a closed set of experimental data is fast and straightforward. The method fits the material model within the range of the investigated problem linearly. The chosen material model is `*MAT_ELASTIC` (listing 10.2). Material density is set to the parameters in Table D.1. The Poisson ratio is set to 0.3. The elastic modulus (E-module) is adjusted iteratively until the experimental and numerical force-distance curves are similar. The loading phase

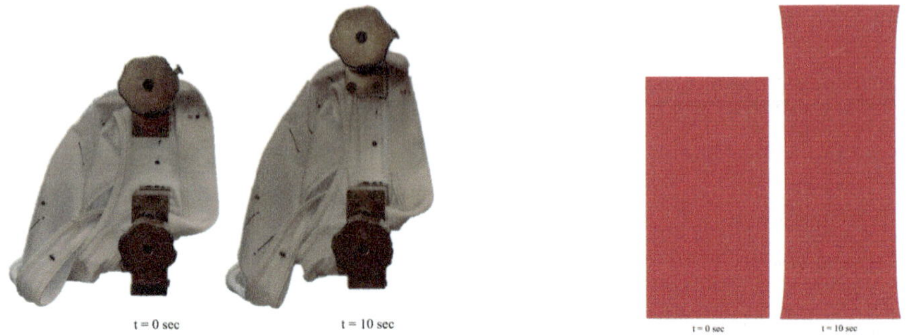

(a) Tensile test - experimental **(b)** Tensile test - numerical model

Fig. 10.8 Tensile test of bra parts: experimental test setup and remodelled Finite Element model

of the cyclic tensile test is selected for model calibration. For validation (Sect. 10.2.5), 4D scan data captured during inhalation—mirroring the loading curve of the tensile test—is used to ensure comparability with the calibration discussed in this section. Each part is fitted with homogeneous material properties. In the case of double-layered textile parts, the layers are experimentally evaluated in combination and modelled as a single homogenised layer to lower the required contact conditions of the numerical model, increasing model stability and reducing the setup time because of fewer required calibration measurements. The properties of the seams are not tested as the complexity is out of the scope of this work. Preliminary tests show that seam properties can be approximated by bra-band properties, because bra-bands are stiffer compared to the front and back parts.

10.2.4 Development of a Combined Deformable Breast Bra Finite Element Model

The developed ID-HB-FEM with deformable breasts and the bra models are combined to model the breast-bra interaction. For each bra model, a combined breast-bra model is set up. Firstly, the dressing process of the ID-HB-FEM with a bra is described (Sect. 10.2.4.1). The dressing process is required for tightly fitted clothing items such as bras to resolve intersections of the human body model and the bra model and exert deformations (Fig. 9.26). Secondly, the boundary conditions of the dressing process described in Sect. 10.2.4.2 are applied. Thirdly, the validation methods for the breast-bra model are presented (Sect. 10.2.5).

Table 10.1 Parameter to model the dressing method by thermal shrinkage

	Bra-2-M	Bra-2-S	Bra-3-M	bra-3-S
Scale factor	1.8	2.2	2.0	2.1
Thermal expansion coefficient α	0.1	0.2	0.2	0.05
Shrinkage time sec	5.88	3.94	3.46	14.84

10.2.4.1 Dressing Method

The ID-HB-FEM with deformable breasts and deformable skin as well as the bra FE models are loaded into LS-PrePost. The dressing method by thermal shrinkage is applied (detailed description in Sect. 9.4.3). Firstly, the bra is expanded by a manually determined scaling factor to resolve intersections of the bra model surface and the human body model surface. The expansion is performed geometrically. Secondly, the bra is shrunk to its original size while the contact between the bra and the body is activated. The shrinkage time is calculated by the algorithm developed in Sect. 9.4.3 and implemented in Python (Listing D.2). The implementation of the thermal shrinkage method in LS-Prepost is shown in Listing D.5. Table 10.1 shows the scale factor and calculated shrinkage time. The scaling factors and shrinkage times differ between bra sizes and geometries. It would be possible to use the same scaling factor for each bra to reduce manual work. Nevertheless, the minimum scaling factor for each bra is picked to decrease the computational cost.

10.2.4.2 Boundary Conditions, Contact and Process Parameter

During the dressing process (Sect. 10.2.4.1), two boundary conditions are set. The first boundary condition is the fixation of the body in space. The translational and rotational degrees of freedom are constrained using the keyword `*BOUNDARY_SPC_SET`. The second boundary condition keeps the strap seams located on the shoulders in place by restricting forward and backward movement of the straps (Fig. D.6). The contact between the bra and the ID-HB-FEM is realised using `*CONTACT_NODES_TO_SURFACE`. The whole bra is defined as `*NODE_SET`, and the body is defined as `*SEGMENT_SET`. The SOFT - parameter in the contact keyword is set to `SOFT = 1`. This parameter is determined in preliminary investigations to increase the contact stability compared to `SOFT = 0` and `SOFT = 2`. The friction between the body and the bra is set to 0.5. Vilhena et al. (2023) determined the friction coefficient for the textile-skin friction. Process parameters are defined by control keywords in listing D.3. Gravity is acting on the breasts throughout the simulation, realised by applying a negative vertical body load (Listing D.4). Gravity not acting on the bra itself does not affect the bra's fit, as the bra is tightly fitted to the body. Turning off gravity for the bra leads to a more stable and faster computation. The termination time is set to the determined shrinkage time (Table 10.1).

10.2.5 Validation of the Combined Deformable Breast Bra Finite Element Model

Three methods are used to validate the accuracy and usability of the developed breast-bra FE model. Firstly, comparisons by bra fit are performed (Sect. 10.2.5.1). Secondly, validation by cross-sections is conducted (Sect. 10.2.5.1). Thirdly, validation by pressure is carried out (Sect. 10.2.5.3).

Validation of bra fit by cross-section leverages 4D scan data. Therefore, in Sect. 5.5.1 determined comparable frames are used. Selected T-pose frames are T-poses in comparable breathing states with underbust circumferences close to 770mm. Having comparable frames is essential for FEM validation. The compared breathing state is taken from the inhalation section of the breathing cycle to ensure comparability to the material calibration process. The frame selection is executed utilising the 4D scan data analysis framework BodyTex4D presented in Chap. 4. Breast-bra FE models are declared valid if the effects of the bras on the real body are mimicked by the numerical model. Identical results are not expected nor targeted as the experimentally measured material properties or pressure sensor data is capturing a snapshot of reality. Humans cannot perform body states consistently multiple times. Therefore, the focus lies on similar clothing effects.

10.2.5.1 Fit Analysis

Overall visual alignment of experimental captured scan data and numerical dressing results are compared by the fit analysis. The four bras are compared to 4D scan data of test-subject-A wearing the bras. Between numerical and experimental breast-bra models, the following criteria are compared:

- bra-band position
- textile part positions
- folds

10.2.5.2 Cross-Section Analysis

The breast deformation is compared using the approach of vertical cross-section comparison (Sect. 9.5.2). A cross-section cutting vertically to the bust point is made

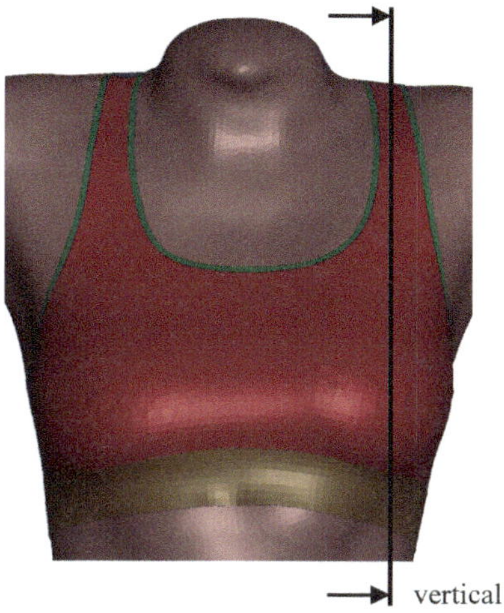

Fig. 10.9 Cross-section planes for the breast deformation analyses

(Fig. 10.9). The effect of bra sizes and the effects of different bra models are focused on.

10.2.5.3 Pressure Analysis

The pressure analysis involves measuring the pressure between the clothing and the body both experimentally and numerically. The four bras are measured and compared. The pressure texsens-g sensor by novel GmbH (OpenRadioss 2024) is used for the experimental pressure measurement. The sensor measurement area is 10 mm. The sensor is designed to measure small pressure values in the range of 1–10 kPa. The pressure measuring rate is set to 50 Hz. While the test subject wears the bras, the pressure between the garment and the skin is measured. The pressure measuring points are shown in Fig. 10.10. The sensor is calibrated beforehand. The measuring points are further labelled based on the body zones and the abbreviated as shown in Fig. 10.10.

Each measurement is performed for at least 45 s. The mean value of the last 30 sec is used to calculate the pressure for the measurement point. The mean values are compared. The numerical pressure is compared to the experimental data at similar measurement points. The target is to verify similar magnitudes of pressure and similar pressure distributions between the different measurement points when comparing experimental and numerical data. The aim is not to replicate exact pressure values, as these vary depending on the state of the body. The study aims to investigate how two different bra models and various bra sizes affect pressure distribution using a numerical model.

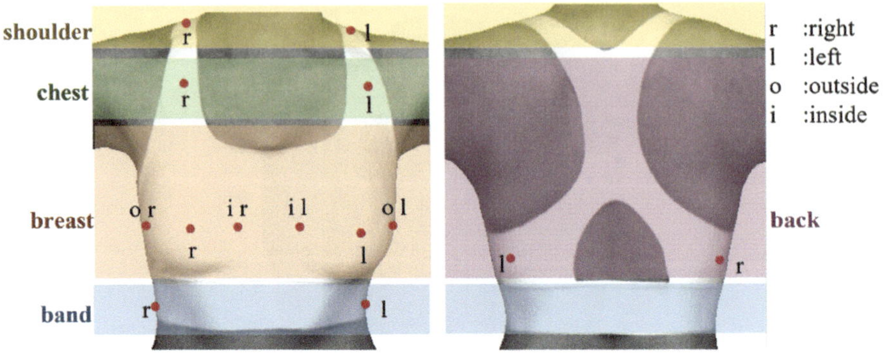

Fig. 10.10 Pressure measuring points and labelling

10.3 Results

The developed FE breast-bra model is presented following the application of the proposed methods. Firstly, the results of the material calibration of deformable breasts (Sect. 10.3.1) and the bra parts (Sect. 10.3.2) are presented. Secondly, the breast-bra FE models are shown and validated after the dressing process (Sect. 10.3.4).

10.3.1 Material Calibration of the Deformable Human Body Finite Element Model

Figure 10.11 shows the experimental and remodelled indentation test results of the right breast of test-subject-A. The force-distance curves of the experimental (EXP, orange) and numerical (SIM, green) measurements are displayed. Three measuring points—top, left side, right side—are measured. For each measuring point, the curve shows good alignment.

The bio-mechanical behaviour of the breast exhibits a slightly non-linear material behaviour for the top measurement point (Fig. 10.11a). The elastic modulus is examined to be 0.05 MPa of the material model `MAT_ELASTIC`. The results show that the linear-elastic material model models breast behaviour well, therefore, the simplified material model is valid to be used for further calculations. Elastic modulus presented in the literature of comparable breast models meshed by solid elements and modelled by linear-elastic material models is 0.0245MPa by Hong et al. (2011) and 0.015 MPa by Zhang et al. (2007). The elastic moduli are within a similar range of magnitude. The discrepancy shows the importance of individual material model calibration for investigations based on the soft-tissue properties of breasts.

10.3.2 Material Calibration of the Bra Finite Element Models

Experimentally determined material property curves of the cyclic tensile test (Sect. 10.3.2.1) are used to calibrate the bra models (Sect. 10.3.2.2).

10.3.2.1 Experimental Analysis of Cyclic Tensile Tests

In Fig. 10.12, the experimental results of the 5th cycle of the cyclic tensile test are displayed. The solid line indicates the tensile test results in the vertical direction (dir1) and the dashed line in the horizontal direction (dir2) (description of dir1 and dir2 in Fig. 5.13). An isotropic material behaviour until a strain of 50% for the back and front of bra-2 and bra-3 is observed (Fig. 10.12a–c).

Loading and unloading show slight differences by forming hysteresis loops. The material behaviour of both bras' back and front parts is slightly diverging from linear material behaviour. The material behaviour of the bra-bands is displayed in Fig. 10.12d. Bra-3-bra-band shows a slightly non-linear material behaviour, similar to the front and back parts of the bras. Bra-2-bra-band shows a highly non-linear material behaviour, having a steep increase of measured force until 5% strain followed

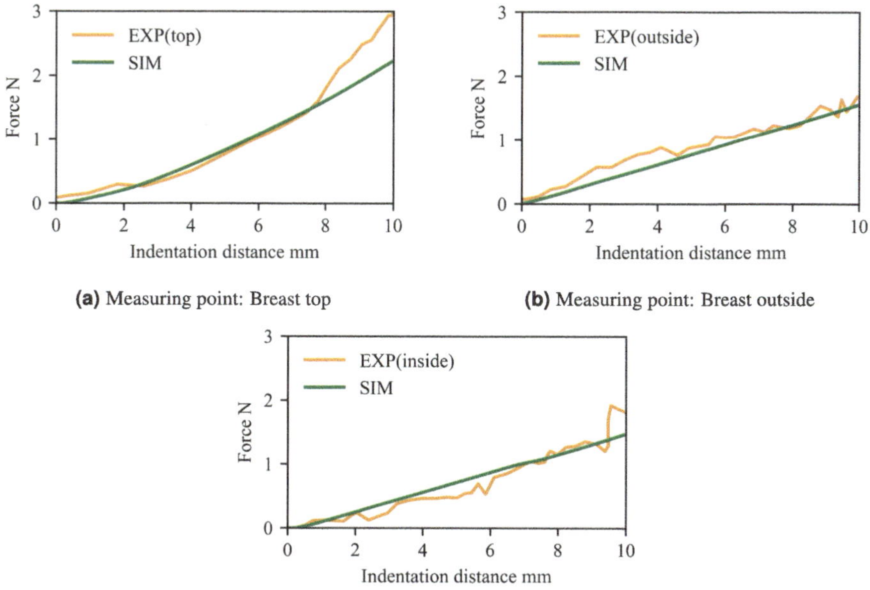

(a) Measuring point: Breast top

(b) Measuring point: Breast outside

(c) Measuring point: Breast inside

Fig. 10.11 Material calibration of the deformable breast area at three different breast measuring points: Comparison of the experimental (EXP) and numerical (SIM) results

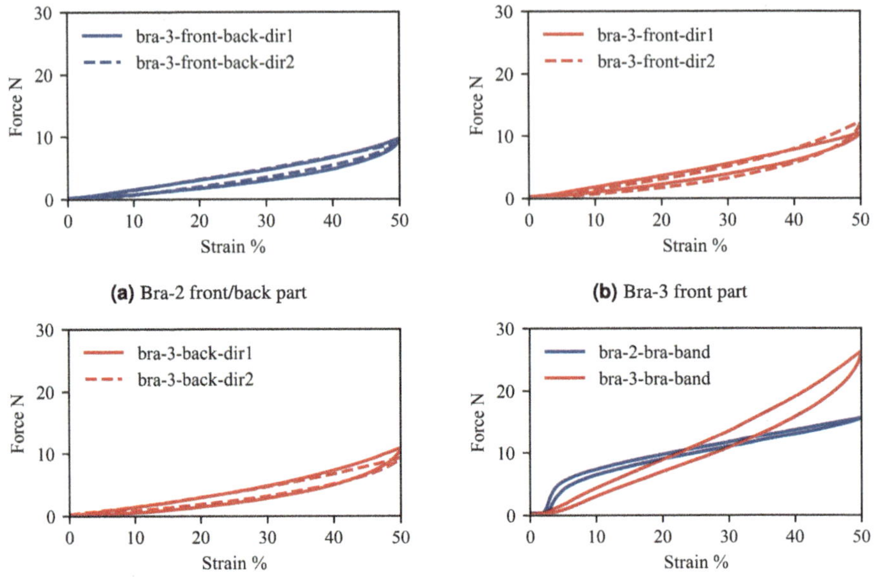

(a) Bra-2 front/back part

(b) Bra-3 front part

(c) Bra-3 back part

(d) Bra-bands of bra-2 and bra-3

Fig. 10.12 Experimental results of the cyclic tensile test of the bra materials (5th cycle). The samples are tested in their two main directions (Fig. 5.13)

by nearly linear material behaviour. The hysteresis loops of bra-2-bra-band and bra-3-bra-band intersect.

Figure 10.13a presents the force-distance curves of the bra bands in the loading part with dots indicating calculated bra-band strains for bra-2-M, bra-2-S, bra-3-M and bra-3-S. Strain calculation is based on the initial bra-band lengths (Table D.2) and the measured reference underbust circumference of 770 mm.

Measured strains are the highest for bra-2-S at 24.2%, resulting in the highest forces with 10.5 N. The force difference between sizes S and M is 1.7 N for bra-2 and 3.5 N for bra-3, whereas, size S produces higher forces consistently. All bra-bands are elongated by 20 mm during the cyclic tensile test. The difference in occurring forces doubles from bra-2 to bra-3. The difference in force is caused by the different bra-band materials. Bra-3-bra-band has a steeper force-distance curve compared to the bra-2-bra-band. The force-distance curves intersect at 24.1%. Bra-2-bra-band experiences higher forces before the intersection and bra-3-bra-band experiences higher forces after the intersection. This analysis highlights how the force exerted by bra-bands onto the chest depends on the wearer's chest and breast dimensions (strain), the pattern (initial bra-band length and width), and the material (force). Size differences impact bra materials inconsistently and intuitively.

Calculated pressures exerted on the body show the dependencies of bra sizes, materials, cut patterns, and wearer's body shapes as well (Fig. 10.13b). The calculation method and parameters used are detailed in Sect. D.1 and Table D.2.

Fig. 10.13 Evaluation of the calculated pressure by different bra-bands in two sizes

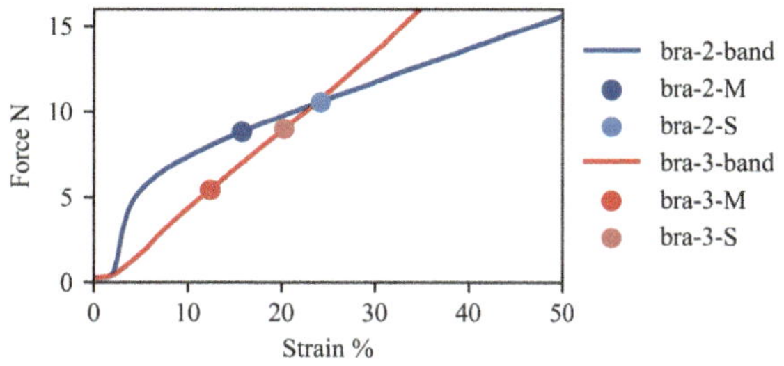

(a) Calculated bra-band strains

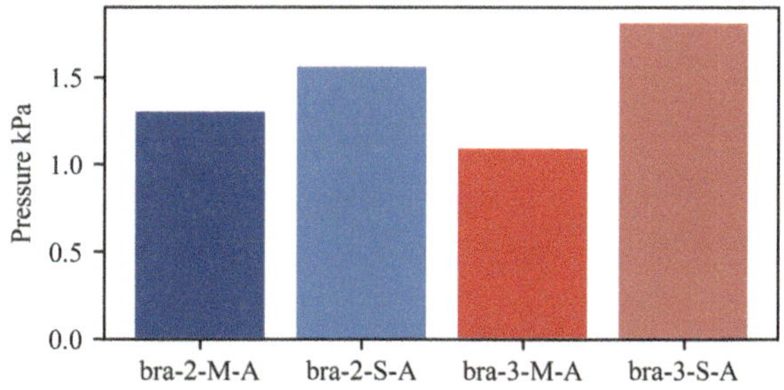

(b) Calculated pressure on body by bra-bands

Even though the absolute force is highest in the bra-2-S-bra-band, the calculated pressure is highest in the bra-3-S-bra-band (1.81 kPa). Pressure depends on the bra-band material as well as the band's contact area with the body. Bra-3-bra-band has 15 mm less width, resulting in higher calculated pressures. The pressures of bra-2-M-bra-band is (1.3 kPa) and bra-2-S-bra-band (1.56 kPa) are higher compared to bra-3-M-bra-band (1.09 kPa). Based on these pressure results, conclusions regarding comfort and fit can be drawn. However, the analyses are not part of this work. They merely illustrate how the fit of a bra is influenced by numerous parameters, highlighting the importance of conducting an individual analysis of bra parameters. The findings are the first explanation for the difficulties of women finding comfortable bras. Current metrics and sizes do not reflect the properties like the force-distance hysteresis of bras.

In conclusion, the materials of the back part and front part show isotropic material behaviour and only a slight divergence from non-linear material behaviour in the region of interest. Consequently, a linear-elastic material model is valid. The loading curve is chosen for consistency as it represents inhaling. Analogous, the linear-elastic material model is appropriate for bra-3-bra-band as a material calibration model. Non-linear material models are recommended for bra-2-bra-band. However, to keep the material models the same for comparability and low computational costs, a linear-elastic material model is applied to the material model of bra-2-bra-band.

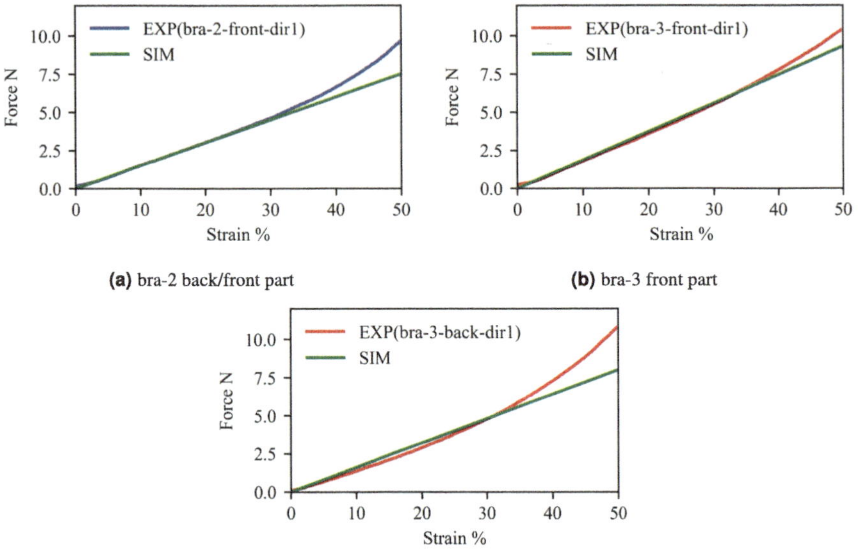

(a) bra-2 back/front part

(b) bra-3 front part

(c) bra-3 back part

Fig. 10.14 Material calibration of the bra front and back part material of bra-2 and bra-3: Comparison between the experimental (green) and numerical (blue and red) results. The experimental curve is the loading curve in direction 1 (Fig. 5.13)

10.3.2.2 Numerical Analysis of Material Calibration of Bra Models

In Fig. 10.14, the tensile test results in the form of force-distance curves of the back part and the front part of bra-2 and bra-3 are shown. The green curves represent numerically measured data. The blue and red curves represent experimentally measured data. It is observed that the force-distance curves are well aligned. In the range of 25–30% the curves become non-linear and diverge. Nevertheless, since the region of interest is assumed to be below 30% strain, the discrepancies in the diverging sections are considered negligible.

The comparison between the numerical (green curve) and experimental (blue/red curves) force-distance curves of the bra-2-bra-band and bra-3-bra-band are presented in Fig. 10.15. The linear-elastic material model is calibrated in the area between the calculated, expected strains of bra-2-S-bra-band (light blue dot) and bra-2-M-bra-band (dark blue dot).

In the case of bra-3-bra-band, the experimental curve is almost linear and is modelled well by a linear material model.

Table 10.2 shows the determined elastic modulus for each bra part. Overall, the linear-elastic material model approach aligns well with the experimental values. The values are used further to model the breast-bra interaction.

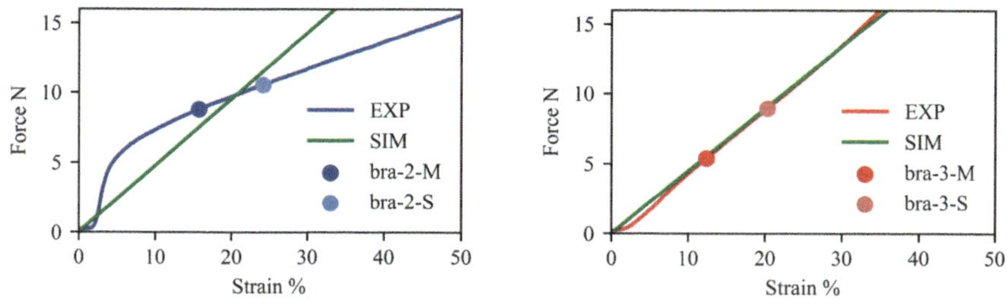

(a) Material calibration results of the bra-band of bra-2 **(b)** Material calibration results of the bra-band of bra-3

Fig. 10.15 Material calibration of the bra-band of bra-2 and bra-3: Comparison between the experimental (blue, red) and numerical (green) results. The points indicate the calculated strain value of the bra-bands at sizes M and S for an underbust circumference of 770 mm

Table 10.2 Elastic modulus of the bra parts of the bra Finite Element models

	Bra variation				
	Bra-2		Bra-3		
	Band	Front/back	Band	Front	Back
Elastic modulus MPa	1.45	0.62	0.67	0.278	0.265

10.3.3 Evaluating the Dressing Process by Thermal Shrinkage

Figure 10.16 shows the bra-dressing process for bra-3-M. The bra shrinks to its original size. The deformable breast and bra interaction starts as soon as the bra contacts the body. The novel dressing method by thermal shrinkage presents its usability. The deformation is stopped by a force equilibrium between the body and the bra. A lower elastic modulus of the bra material results in higher stretch values and less breast deformations, while a stiffer bra material results in more breast deformation and lower bra stretch values. The dressing process worked for every bra model investigated with realistic results, demonstrating the effectiveness and usability of the developed dressing method at hand. The computational cost depends on the scaling factor and scale factor which determines the amount of required shrinkage. The maximum computation time was 2.5 h with an average laptop containing an Intel Core i7 7th generation and 32 GB RAM.

10.3.4 *Validation*

The novel developed numerical dressing method proved to be very proficient. Proficiency is shown through the realistic tight fit of the investigated bras on the body. Interactions between the bra and the human body are modelled. The pressure exerted onto the body by the bra causes realistic deformations showing the validity of the linear-elastic material models. The bra is stretched while the breast is compressed by the interaction. In the following, results of the breast-bra interaction are analysed by optical fit (Sect. 10.3.4.1), cross-sections of the chest (Sect. 10.3.4.2) and by the pressure between breast and bra (Sect. 10.3.4.3).

10.3.4.1 Analysing the Bra Fit

The results of modelling the dressing process and comparing the results to static 4D scans of the real bra-2-M (Fig. 10.17), bra-2-S (Fig. D.8), bra-2-M (Fig. 10.18) and bras-2-S (Fig. D.9) are analysed. Overall, the four modelled bras are realistically, tightly fitted onto the deformable model of the human body, proving the usability of the newly developed dressing method.

The fit of bra-2-M is shown in Fig. 10.17. Optically, the fit is similar between the numerical model and scan data. The bra-band did not slip onto the breast as it is

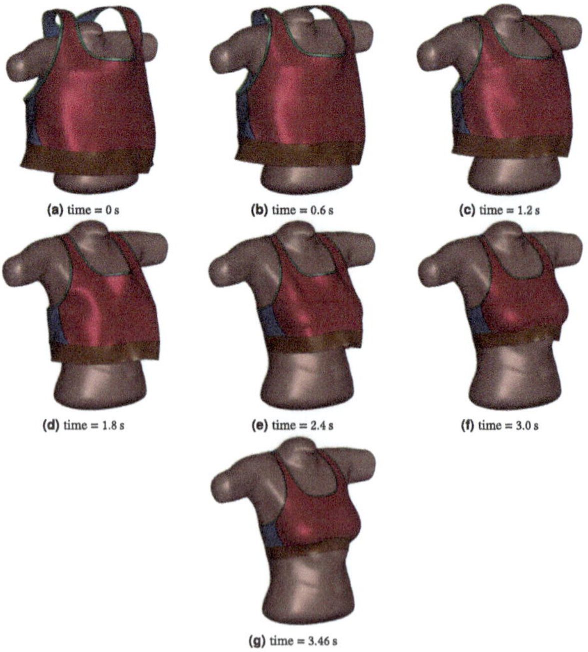

Fig. 10.16 Thermal shrinkage of bra-3-M

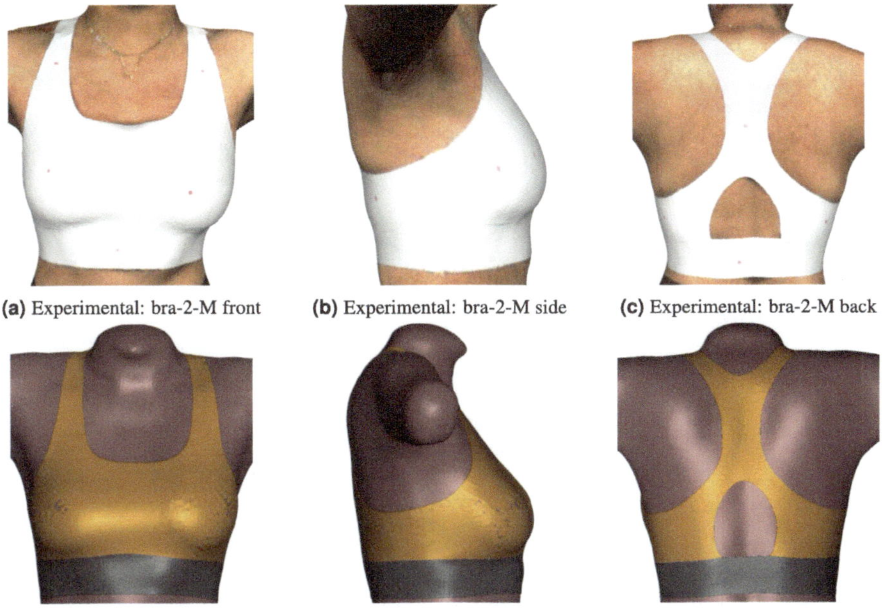

(a) Experimental: bra-2-M front (b) Experimental: bra-2-M side (c) Experimental: bra-2-M back

(d) Numerical: bra-2-M front (e) Numerical: bra-2-M side (f) Numerical: bra-2-M back

Fig. 10.17 Comparison of the fit of bra-2-M **a–c** experimental data and **d–f** Finite Element model

observed for bra-3-S (Fig. 10.17b and e). The front part covers the full breast fully, and the straps are placed on the shoulders next to the neck as in reality (Fig. 10.17a and d). At the back part, the middle part is placed between both shoulder blades. The slightly asymmetrical back part position is shown in the experimental and numerical results as a strong sign of validity of the dressing method (Fig. 10.17c and f). No folds are observed for all bra parts, which indicates a tight fit. In conclusion, the bra has a very good fit alignment comparing numerical model and experimental 4D scan data.

In Fig. 10.18, the fit of bra-3 is displayed. The bra-band of bra-3-M is not entirely under the breast. The bra-band moved slightly upwards on top of the lower breast areas (Fig. 10.18b). Similar slippage is observed in the numerical results also (Fig. 10.18e). In contrast, the bra-band of bra-2 stays under the breast (Fig. 10.17e). The bra-band of bra-3 has a longer initial length, in the case of size M, the strains are lower resulting in lower required forces to reach the same compared to bra-2 leading to the slippage (Fig. 10.15). Observing similar slippage by the numerical models proves the high accuracy of the FE models. Additionally, it is shown that the bra-band is not horizontally orientated. The bra-band is positioned lower on the back side than on the front side. This effect again is observed in both the experimental and numerical results (Fig. 10.18b and e).

The front part of bra-3-M is optically similar comparing the numerical model and experimental results (Fig. 10.18a and d). Compared to bra-2-M, the front part

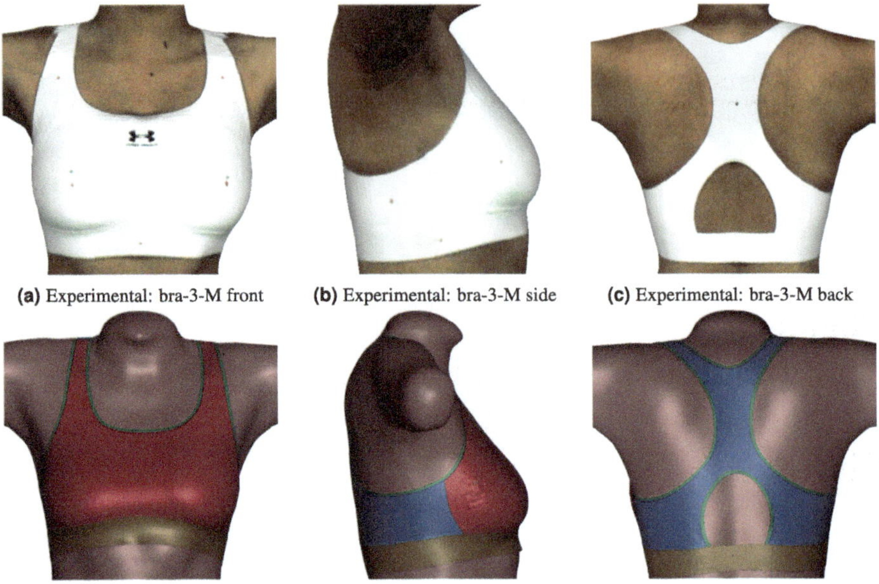

(a) Experimental: bra-3-M front (b) Experimental: bra-3-M side (c) Experimental: bra-3-M back

(d) Numerical: bra-3-M front (e) Numerical: bra-3-M side (f) Numerical: bra-3-M back

Fig. 10.18 Comparison of the fit of bra-3-M **a–c** experimental data and **d–f** Finite Element model

covers more breast area, which is reflected by the numerical model. The cut pattern of the front parts of bra-2 and bra-3 are the same. The difference in fit is assumed to be caused by the material properties and the seam differences. The similarity to the experimental data proves the validity of the material model. The back part of bra-3-M shows similar fit in experimental (Fig. 10.18c) and numerical results (Fig. 10.18f). The cut-out at the bottom of the back part has a similar shape, but in the experimental results, the cut-out is wider at the bottom. The assumed reasons are inaccuracies of the digital cut pattern due to the cut pattern reconstruction process. The cut-out differs in shape in the case of bra-2-M as well. The shape of the back cut-out bra-3-M is slightly wider while bra-2-M is more elongated. Overall, the numerical results of bra-3-M show a very good alignment with the scan data and indicate high usability and accuracy of the numerical model. The fit comparison of bra-2-S and bra-3-S is shown in Figs. D.8 and D.9. Similar effects are observed for S-sized bras as for the M-sized bras, reinforcing the assumption of the validity of the numerical model.

In conclusion, for both the numerical and experimental results, similar differences between bra-2 and bra-3 are observed. The differences result from variations in bra-band lengths and bra material properties. With respect to the fit, the numerical model shows realistic results, proving the calibration and material model selection to be correct. Based on the model, further bra variants could be modelled digitally to investigate the fit in the future. Overall, the following same effects are observed when comparing the numerical and experimental fit investigation:

- **Differences of bra-band fit due to length and material differences are modelled realistically**:
  - Bra-2-bra-band is fully under the breast.
  - Bra-3-bra-band is partially on the breast.
  - Bra-3-bra-band is not horizontally orientated; the back part is located lower than the front part relative to the human body model's back.
- **Differences of front and back part of the bras due to material differences and seams are modelled realistically**:
  - The front part of bra-3 covers a larger breast area compared to bra-2's front part.
  - The back part of bra-2 and bra-3 show differences in shape.
- **Size differences of the same bras are modelled realistically**:
  - Bra-2-S and bra-3-S cover smaller breast areas compared to bra-2-M and bra-3-M.
  - The bra-band of bra-3-S is placed higher on the breast than bra-3-M.
- **The bra is tightly fitted on the body causing deformations of the breasts**.
- **The fit is non-symmetrical due to the non-symmetrical structure of the body**.

10.3.4.2 Analysing Breast Deformation

Figure 10.19 displays the vertical cross-section at the bust point. Results of the breast deformations of experimental captured 4D scan data and FE model are compared. Figure 10.19a–c present the experimental results. The analysis focuses on modelling similar effects of the bra-breast interaction observed during the experiments. The following effects are determined by experimental analysis (Sect. 5.5.2):

- Bras deform breasts
- Smaller bra size leads to higher breast deformation.
- Bra-3 leads to a higher breast deformation compared to bra-2.

Figure 10.19d–f visualises the breast deformations of the numerical model. All named experimental results can also be seen in the numerical results. The breast is compressed due to bra-2 and bra-3. The smaller bra size leads to higher compression of the breasts (light blue line in Fig. 10.19d and light red line in Fig. 10.19e). Figure 10.19f shows higher compression by bra-3-S (red line) compared to bra-2-S. Bra-2 and bra-3 front and back parts are made of similar cut patterns and similar materials resulting in similar deformations. Nevertheless, even the small changes are reflected by the numerical model. The same is true for the two sizes. The absolute breast deformation of the numerically modelled bra is smaller compared to the experimental results. The restricted compression behaviour of the breast may be due to the selected element formulation, element locking, and the chosen boundary conditions. Further research is needed to improve the compressibility of the breast. Nevertheless, any other breast model can be replaced with the current one. Experimental results just reflect a snapshot in time. The experimental results of bra-2 and bra-3 will not have the consistency of boundary conditions (Breathing state, pose, dressing procedure) of the numerical model, experiments with humans are not consistently repeatable.

In conclusion, comparing breast deformation of experimental and numerical results shows the same effects. Influences on breast deformations by bras in general, by different bra sizes and by bra materials are investigated. The numerical model can be used for further digital bra fittings to investigate breast deformation behaviour. Additionally, the numerical model has the huge advantage of providing consistent boundary conditions (breathing state, pose, dressing method), which improves the comparability of bra designs. The numerical model is validated in terms of breast deformation. Future studies should be conducted to improve the compression behaviour of breasts further.

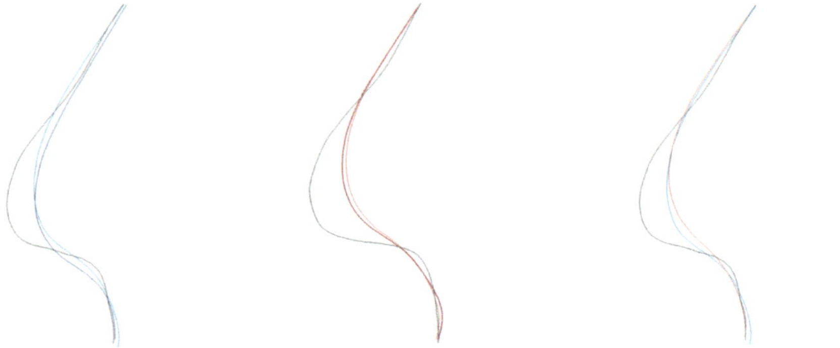

(a) Experimental: bra-2-M (dark blue), bra-2-S (light blue)

(b) Experimental: bra-3-M (dark red), bra-3-S (light red)

(c) Experimental: bra-2-S (light blue), bra-3-S (light red)

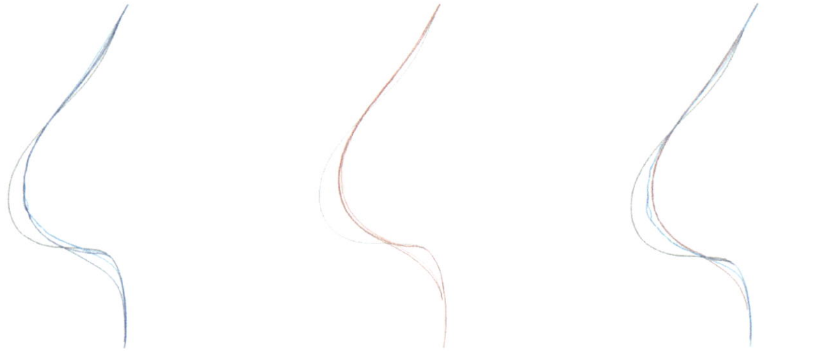

(d) Numerical: bra-2-M (dark blue), bra-2-S (light blue)

(e) Numerical: bra-3-M (dark red), bra-3-S (light red)

(f) Numerical: bra-2-S (light blue), bra-3-S (light red)

Fig. 10.19 Comparison of the breast deformation by bra-2 (blue) and bra-3 (red): **a**–**c** experimental 4D scan data and **d**–**f** Finite Element model

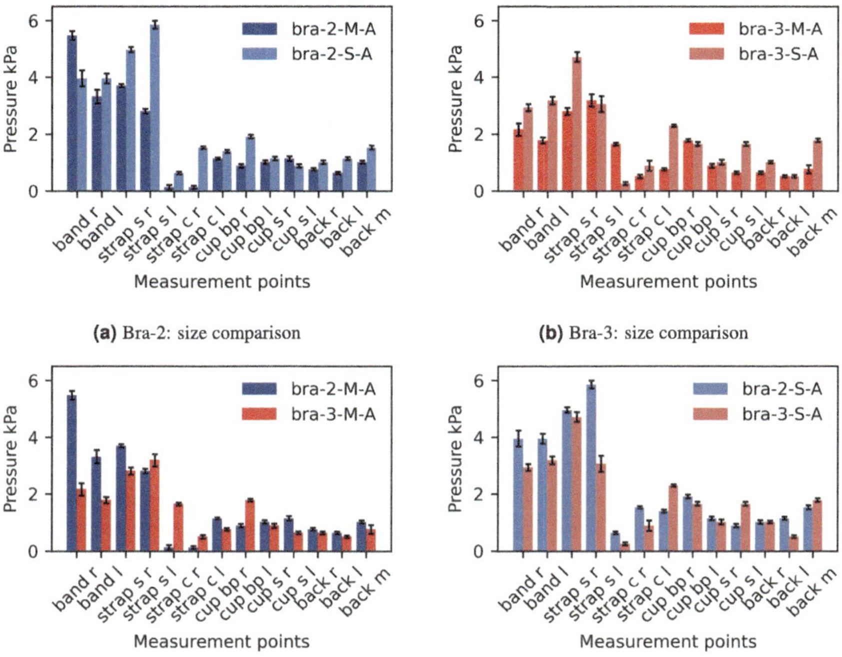

(a) Bra-2: size comparison

(b) Bra-3: size comparison

(c) Bra-2-M and bra-3-M: material comparison

(d) Bra-2-S and bra-3-S: material comparison

Fig. 10.20 Pressure measurement of test-subject-A wearing bra-2 and bra-3 in two sizes

10.3.4.3 Analysing Pressure Distribution

Figure 10.20 presents the results of the experimentally measured pressures occurring between breast and bra. Points of measurement are displayed in Fig. 10.10. Figure 10.20a and b compare the pressures exerted by different bra sizes. Figures 10.20d and c compare the influence of bra material properties of same-sized bras.

The highest pressures are measured in the region of the bra-bands and the shoulder straps consistently as shown in Fig. 10.20. The pressure ranges for bra-bands and straps between 2 and 6kPa. Pressures are lower at bra cups and back areas, mainly under 2kPa. The results are expected because the bra-band material is stiffer than the front and back part materials. Furthermore, the bra-band strain is assumed to be higher. The higher pressure in the shoulder is explained by the smaller strap width while the weight of the breasts is pulling on the straps.

It is assumed that the smaller bra size shows higher pressures compared to bigger bra sizes of the same type. Considering bra-2, this assumption holds true for 13 out of 15 measuring points. Pressures at the left cup side are slightly higher for size M (1.1 kPa) compared to size S with (0.9 kPa). This difference can be an inaccuracy due to the sensor placement or the bra fit. The bra-2-M-bra-band-r is 1.5 kPa higher bra-

2-M-bra-band-l. This pressure difference on the two bra-band sides is unexpected as a homogeneous pressure distribution is assumed. This pressure distribution may be caused by sensor placement errors.

For bra-3, the same tendency of having higher pressures at a smaller bra size can be seen. 10 out of 15 measurement points of bra-3-S are higher than those of bra-3-M. The measuring points bra-2-strap-s-r, bra-2-cup-bp-r and bra-2-back-l have similar pressures, laying in the standard deviation of the measuring points. Bra-2-M-strap-l is 1.4 kPa higher than bra-2-S-strap-l. This pressure difference may be caused by the difference in the arm or body position.

Comparing the measured pressure of the bra-bands to the calculated pressure (Fig. 10.13b), a higher measured pressure can be seen. The reason can be the simplified pressure calculation assuming a circular shape and not an elliptical shape as it is in reality. Nevertheless, similar findings of having a higher bra-2-M-bra-band and bra-2-S-bra-band pressure than bra-3-M-bra-band can be determined. The measured bra-3-S-bra-band pressure is in contrast to the calculated pressure, not the highest, which can be caused due to the up-sliding bra-band fit of bra-3-S (Sect. 10.3.4.1).

In summary, the following key findings of the experimental measurements are determined:

- Pressure measurement inaccuracy occurs due to not repeatable measuring conditions: body poses, breathing behaviour, and sensor placements.
- The region with the highest pressure is at the bra-band and the shoulder straps.
- The pressure range at the bra-band and shoulder area is between 2 and 6kPa.
- The pressure at the measurement point in the cup and back area are mainly below 2 kPa.
- A smaller bra size tends to lead to higher pressure values.

However, Overall, high inaccuracies are inevitable when measuring pressures on a bra-wearing test subject standing in a T-pose. The following influences decrease the accuracy and reliability of the pressure measurement results:

- **Breathing**: Breathing causes the pressure values to oscillate.
- **Arm position**: Arm positioning influences the pressure values due to lifting of the breasts, especially for the shoulder and breast measurement points. Test subjects are incapable of holding their arms perfectly still. Test subjects are incapable of repeating the same position over multiple measurements. To reduce the influence of the arm position problem, build structures to rest the arms on.
- **Feet position and weight shifting**: test subjects do not stand still because the weight is shifted from one foot to another, resulting in a shoulder movement that directly influences the pressure measurements.
- **Placement of pressure sensor**: Placing a sensor between the skin and a tight-fitted garment is challenging. The clothing has to be stretched to manually place the sensor between the textile and the skin. Placing the sensor changes the fit of the bra. The location of the sensor relative to the width and the seams of the strap has an influence on the measurement results.

- **Pressure sensor**: The pressure sensor shows inaccuracies for long measurement duration.

Preliminary tests using a manikin showed similar levels of inaccuracy. Nevertheless, pressure is a major factor for comfort. To compare the pressure values with numerical results, the order of magnitude of the pressure values is used as a validation metric. The differing magnitudes at the various measurement points enable their comparison. The numerical will show similar but more consistent results proving the advantage of performing pressure evaluations numerically.

Figure 10.21 displays the numerically calculated pressure distribution of the four modelled bras. The pressure level is indicated by a colour gradient from blue indicating 0 kPa to red indicating 8kPa. It is generally difficult to compare single pressure values between the numerical and experimental results. Experimental results are obtained by a sensor covering an area of 78.54 mm^2. Obtaining a continuous pressure distribution for the whole breast area as it is produced by the numerical model requires impractical effort if it is impossible at all. Therefore, in the following, the pressure values are not individually compared but tendencies are discussed. Future studies must be conducted to evaluate and develop technologies to measure pressures for the whole breast area precisely. The development of precise pressure measurement technologies is out of the scope of this work.

Comparing the magnitudes of pressure of the numerical results to the experimental results, it is shown that the ranges of magnitude are the same. Numerical and experimental results fall in a range of 0–8 kPa. Pressure peaks occur consistently at the shoulders (Fig. 10.21c, f, i and l) and for bra-3 at the bra-band at the back side (Fig. 10.21h and k). Having higher pressures at the straps and bra-band could be also seen in the experimental pressure analysis.

Looking at the bra-band, the difference in the bra variants is shown. A smaller bra size leads for both bras to larger pressure areas and higher pressures. The bra-band of bra-2 has a larger pressure area on the front side than bra-3. This is due to the bra fit. As in Sect. 10.3.4.1 investigated, the bra-band of bra-2 is placed under the breast. The bra-band of bra-3 is placed slightly on top of the breast. Pressure peaks at the back area compensate for this lower bra-band pressure in the front area of bra-3. This is not seen for bra-2. The pressure distribution is equal as the bra-band is placed under the breast.

Comparing the pressure at the breast area, the highest pressure is shown at bra-3-S (Fig. 10.21j). This analysis fits the deformation results in Sect. 10.3.4.2, where bra-3-S leads to the highest breast deformations. The pressure on the breast area is between 0.5 and 4.5 kPa. The experimental results measure the pressure in the breast area around 2kPa. This fits into the range of the numerical results. However, it is shown that the smaller bra size leads to higher breast pressure (darker green) for bra-2-S and bra-3-S, which is also the trend in the experimental results.

Similar effects are shown on the backside of the numerical model. On the backside, higher pressure distributions of the back part of the bras due to bra-3 are shown. The stiffer seams and material are a reason for that. The influence and higher pressure

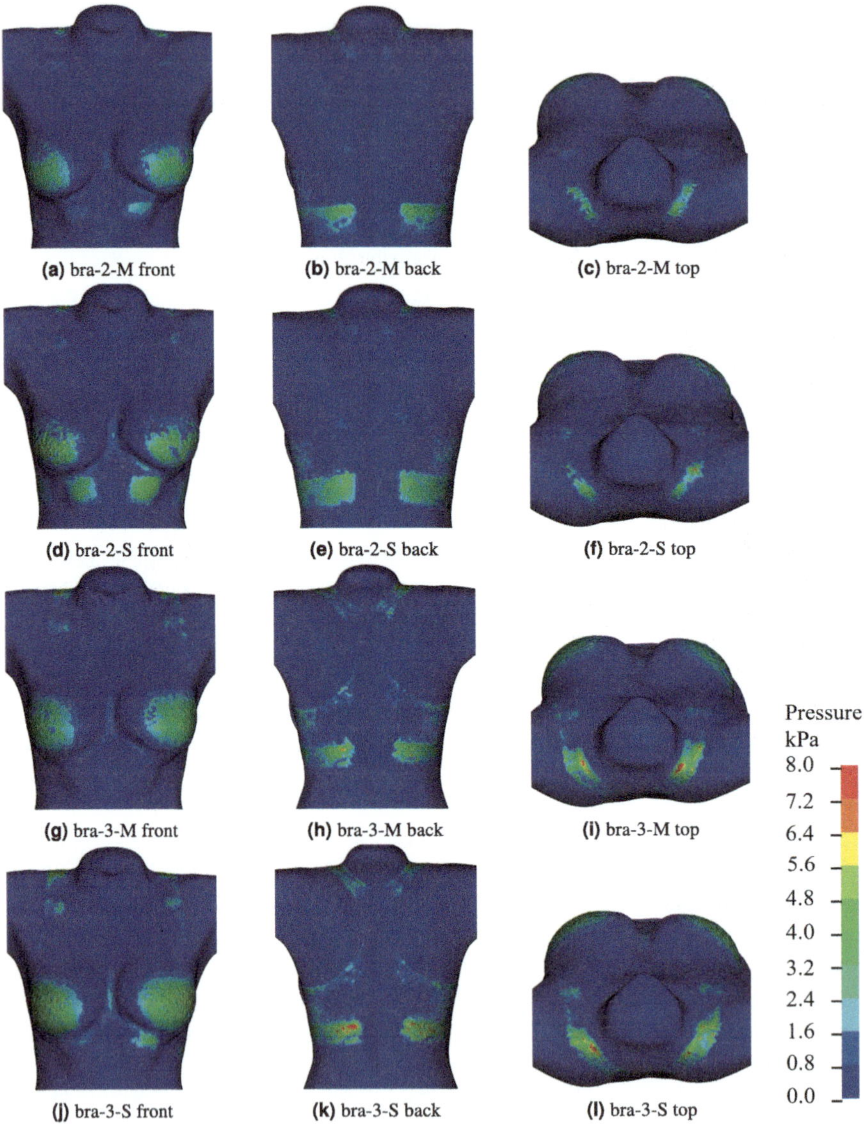

Fig. 10.21 Pressure distribution of the four modelled bras

through the seams of bra-3 can also be seen. A higher pressure at the bra edges occurred at bra-3 compared to bra-2.

Overall, the determined pressure distribution aligns with the deformation and fit analysis. Even though the modelled bras are all very similar in cut pattern and material behaviour, plausible results could be determined. The numerical data greatly improved the accuracy and usability of the pressure distribution results. In the numerical model, the same initial body model is used for all bras, and the pressure distribution on the whole body is determined. In the experimental result, only single measuring points indicate the pressure, which must not represent the region. In conclusion, the following key findings could be investigated:

- **Pressure changes due to size changes are modelled**.
  - A smaller bra size leads to higher pressure.
- **Pressure changes due to bra-band, material differences and seams are modelled**.
  - The bra-band of bra-2 shows a more homogenous pressure distribution than the bra-band of bra-3.
  - Due to the seams of bra-3, the bra is stiffer, leading to higher pressure on the body.
- **Regions with pressure peaks are determined**.
  - At the shoulder pressure, peaks are determined.
  - For bra-3, pressure peaks are also at the back of the bra-band.
- **Numerical modelling gives more comparable and usable pressure information**.
  - The body poses are the same for each bra.
  - The pressure of every body part is investigated.
  - No inaccuracy due to the measuring device or person occurs.

10.4 Conclusion and Discussion

This chapter develops an FE model for determining the breast-bra interaction. Methods developed and described in Chap. 9 are used and their functionality is validated by applying them to the breast-bra interaction use case.

An individualised, deformable human body FE model (ID-HB-FEM) with deformable-modelled breast regions is built up. The ID-HB-FEM is derived from a female test subject standing in a T-pose statically. Data to build up the ID-HB-FEM are single frames captured of the test subject by the Move4D scanner. The model is meshed by volume elements and sectioned into the breast, skin layer, and internal body. The internal body is rigid, and the breasts and the skin layer are deformable. The material calibration of the deformable breasts is performed using literature data and the IndentoPRO (Sect. 9.2.3.4).

The research involves modelling four different bras (CLOTH-FEM), with two types of bras each in two different sizes. The two bra types have the same cut pattern in their front and back parts. Differences are in their bra-band, the materials, and the seam stiffness. The influences of size and material are investigated. The bras are digitally reconstructed using their cut patterns (Sect. 9.3.1.1). Bra-2 represents a redesign developed within the scope of this work, based on the original cut pattern of Bra-3. The 3D shape of the bras is reconstructed in CLO3D and imported into LS-PrePost for further analysis. The bra models are meshed by using shell elements. The material calibration is based on cyclic tensile tests performed on the produced bra (method in Sect. 9.3.3). A linear elastic material model is used for all materials as it fits the experimental data in the region of interest. Linear elastic material models are low in complexity enabling fast model setup and are low in computational cost.

Combining the deformable human body model and the bra model has the challenge of the bra being undersized, meaning, smaller than the body, therefore, the models intersect which prevents analyses. A dressing process needs to be performed to place the bra models realistically on the deformable human body model. The bra is extended until intersections are resolved, subsequently, shrunk by thermal exposure until a force equilibrium between bra and body stops the shrinkage (Method Sect. 9.4.3). The shrinkage process time is calculated and terminated after the calculated duration.

The developed ID-HB-CLOTH-FEM is validated by a fit, deformation and pressure analyses Sect. 10.3.4. Experimental 4D scan data of the test subject wearing the bra variations are captured and used for the validation. Comparable 4D scan data is obtained by applying methods of Chap. 5. Pressure measurements while wearing the bras are performed experimentally and are compared with the numerical results.

By developing the breast-bra FEM the methods to develop an ID-HB-CLOTH-FEM could be validated in their usability and performance. The following bullet points present and discuss the main findings:

- ID-HB-FEM: The method IndentoPRO and literature values show high usability for breast material calibration
- CLOTH-FEM: The modelled bras based on the original cut pattern using clothing software lead to fast generations of 3D clothing item models and decrease the effort for modification of clothing items immensely
- ID-HB-CLOTH-FEM:
  - The dressing method by thermal shrinkage results in a tightly, realistic fit of the bra on the body for all considered bra models
  - Implementation effort and computational cost are low for the dressing method thermal shrinkage
  - 4D scan data and the methods to select comparable frames (Chap. 5) using BodyTex4D are valid and improve the model validation

- The analysis of the pressure distribution resulting from the breast-bra interaction indicates a higher usability of numerical analysis compared to experimental analysis.
- Various effects of the bra interacting with the breasts are modelled:
  - Influences of bras caused by differing materials and seams are modelled
  - Influences of bra sizes are modelled
  - Bra fit is modelled
  - Breast deformation is modelled
  - Pressure distribution between the breast and bra is modelled

The ID-HB-FEM is based on real human body data in a static T-pose captured by the Move4D scanner. The Move4D scanner has the advantage of processing the raw data to homologous meshes, short exposure time, and accurate representations of the human surface. Test-subject-A is scanned, transformed into a numerical model and dressed with four bra models to investigate the effects of the bras on human bodies. Part sectioning of breast regions has to be performed manually. Manuel sectioning takes approximately 2min. If a huge number of bodies should be meshed in the future, a completely automated volume meshing and part sectioning method should be developed. One approach could be predesigning a fully meshed avatar with predefined part-sectioned areas as a template and only having to perform minor adaption based on the scan data. The finished meshed avatar is iteratively shrunk/extended till the surface areas have a minimal distance error. Problems could occur in the element quality after the fitting process. Another approach using **A**rtificial **I**ntelligence (AI) for example by using **m**achine **l**earning (ML) should be investigated as AIs are built for and trained on pattern recognition.

Defects in the area of the fingers are now common for meshes generated by the software of Move4D. Currently, the mesh defects are solved by cropping the torso, neglecting the fingers. However, stabilising the homologous mesh generation in the finger area would render the cropping step obsolete, simplifying the model generation further.

It is challenging to capture the whole breast surface, which is necessary to build up deformable breast models. The larger the breasts, the larger the inframammary fold (underbust crease). If the breasts are sufficiently large, the inframammary fold results in the breast skin contacting the chest skin. Contacting skin is fused by the 3D imaging technologies used within this work, falsifying the model. Leaning forward of the test subject overcomes skin fusion in the inframammary fold, separating the breast's skin and volume from the thorax, resulting in a realistic model of the breast. The upper body is rotated to be upright during the modelling process. The inframammary fold forms realistically, proving the validity of the lean-forward method. A process flow suggestion and the first results of the FE model of a test subject with larger breasts (test-subject-B is described in Sect. 5.3.2) is shown in Fig. 10.22.

Calibrating both the deformable breast and the bra materials, the simplified material model `*MAT_ELASTIC` showed a very good alignment of numerical and experimental data for the material stress and strain ranges of interest. The linear-elastic

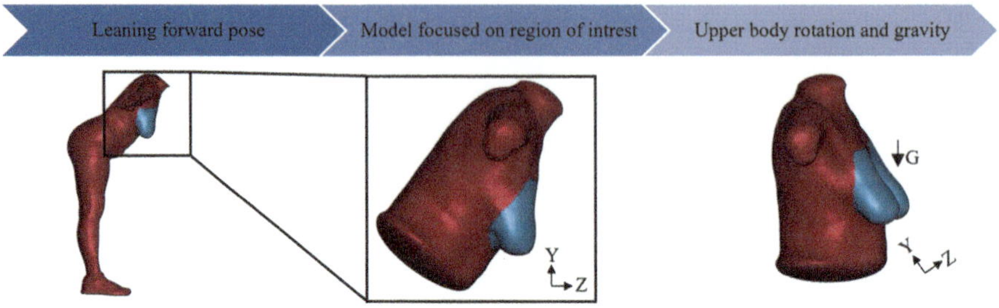

Fig. 10.22 Workflow for the Finite Element model generation for test subject with larger breasts

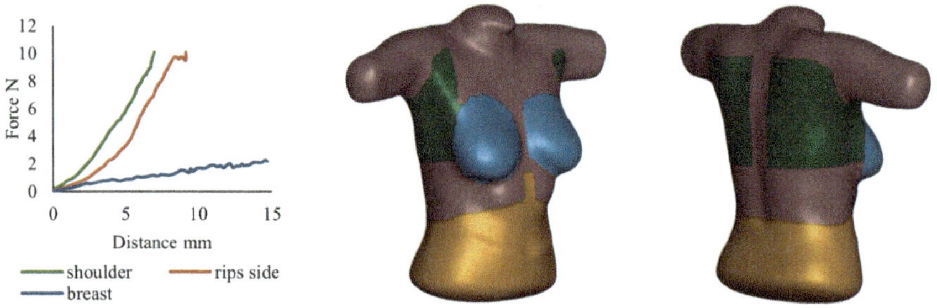

(a) Force-distance curves of the IndentoPRO measured at the shoulder, back and side **(b)** Possible sectioned parts at the front side **(c)** Possible sectioned parts at the back side

Fig. 10.23 Advanced body part sectioning to increase the accuracy of the individualised, deformable human body Finite Element model

material model is low in complexity due to the low number of parameters (density ρ, elastic modulus e, Poisson's ratio υ), which simplifies the calibration process. The outcome of the calibration of the breast material model is applied to the skin material model. Approximating skin by breast properties keeps the model setup time low and leads to isotropic material properties in the breast region while sufficiently realistic model behaviour is observed. Nevertheless, setting the skin material properties to model stiffer material simulates skin more realistically. This results in a model that is closer to reality. However, balancing the complexity and accuracy of simulation models is application specific. In the context of this work, less complexity is preferred to increase the attractiveness of the developed methodology for industrial clothing design processes.

A proposal to generate a more realistic model is to extend the measurements by IndentoPRO to more upper body parts and, consequently, separate the upper body into a higher number of subparts with different material models. The increases in the level of detail would be needed to justify the increased model setup time. The first tests to increase the level of detail of the upper body model are shown in Fig. 10.23. Figure 10.23a displays the curves measured by the IndentoPRO at the shoulder back and side part. Figure 10.23b and c show possible sectioning of the upper body.

Modelling of the bra variations evolves around the material calibration. Four bras of two types in sizes S and M sectioned in front part, back part and bra-band each are modelled. The materials are modelled by linear-elastic material models as they approximate the measured material behaviour in the region of interest with acceptable accuracy. The bra-2-bra-band material becomes non-linear first with respect to a strain till 50% which is the limit of evaluated strains in this chapter. To broaden the field of applications of the developed models non-linear material models need to be tested, calibrated and applied in the future.

The manual work of placing the up-scaled bra without intersections with the human body model is minor. Bra positioning and exchange of bra models could be automated in the future. Currently, knowledge of how to operate the modelling software is required. Additionally, computational times are reduced by increasing computational resources.

For setting up the ID-HB-CLOTH-FEM, the novel method for the dressing process by thermal shrinkage is determined to be low in complexity, requires low manual effort, and is low in computational cost, while being generically applicable. The generic property allows the transfer of the method to other tightly fitted clothing items like compression socks, jeans or sports shirts easily.

In reality, many woman readjust their breast position relative to the bra as part of their dressing process. Readjustments are made to increase comfort. It is assumed that the readjustments lead to a more evenly pressure distribution. Readjustment is not remodelled in this work and the test subjects are told to not readjust their breast during the experiments. An improved extension of the digital dressing process is to model readjustments of the breasts. It can be modelled by applying force vector fields to the breast as shown in Fig. 10.24. The first test to model readjustments of the breasts shows realistic effects and changes the fit of the bra. Nevertheless, comparing various bras without breast readjustment proved to be possible and improves already clothing development, if the presented dressing method is applied. To generate the same dressing method for ID-HB-CLOTH-FEM is a huge advantage of the FE model compared to experimental comparisons, where the dressing process cannot be consistently performed.

The main finding of the validation process of the models is that the experimental reference results are challenging to compare. The experimental results are a snapshot of a short moment in time. Influencing factors like the body pose, breathing state, time of day, room temperature, day of the menstruation cycle, measuring devices, etc., influence the accuracy and comparability of measured data in the case of breast-bra interaction. Using the 4D scan data and the methods developed and integrated into BodyTex4D to select comparable frames showed the usability and necessity of frameworks like BodyTex4D for the validation process. The numerical model shows its advantage of generating comparable results. The body pose, breathing state and dressing process are controllable by the numerical model to be consistent. Additionally, the pressure distribution for every investigated body part is measured simultaneously when using a numerical model. In experiments with real test subjects, the pressure can only be measured at a few measuring points selected by the researcher simultaneously. A single measuring cycle takes 4 min. Measuring ten measuring

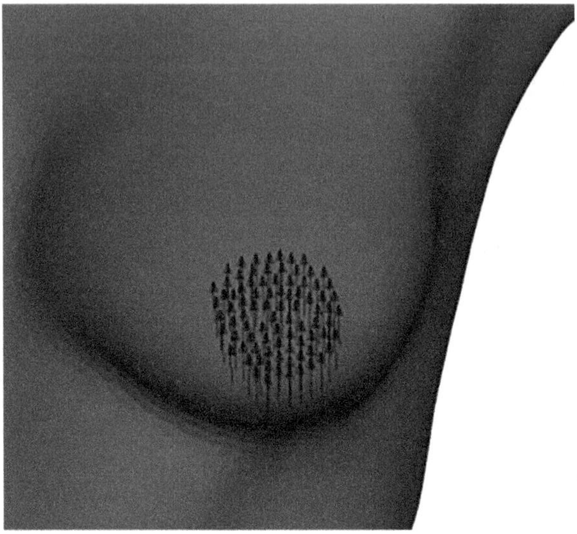

Fig. 10.24 Modelling the manual replacement of a breast during a dressing process of a bra

points for 4 min for four bras results in 160 min. Considering the amount of different breasts, bras, and their possible combinations, experimentally generating a database, while being inaccurate, becomes unfeasible. Numerical models are more accurate and faster in pressure comparisons.

An improvement of the ID-HB-CLOTH-FEM is shown in the size of the breast compression due to the bras. The numerical model could model the same effects of different bras as measured by experiments, but the absolute breast compression is lower. The difference could be due to is chosen element formulation or element locking. Further research has to be conducted regarding the compression behaviour.

With a validated model, more bras can be designed and tested digitally making it feasible for manufacturers to individualise bras more accurately. The ID-HB-FEM or the CLOTH-FEM is replaced fast in the simulation environment. The influence of seams, cut patterns, or materials is explored and can be used in the bra development process. The same bras can be dressed on different ID-HB-FEMs to identify their usability for various breast types and applications.

References

Cai Y (2016) Nonlinear dynamic analysis of bra fitting using finite element models. PhD thesis. The Hong Kong Polytechnic University (2016)

Cai Y et al (2018) A piecewise mass-spring-damper model of the human breast. J Biomech 67:137–143. https://doi.org/10.1016/j.jbiomech.2017.11.027

Cai Y, Yu W, Chen L (2016) Finite element modeling of bra fitting. In: Advances in women's intimate apparel technology. Elsevier, pp 147–168. ISBN: 9781782423690. https://doi.org/10.1016/B978-1-78242-369-0.00009-8

Chen L-H et al (2013) A study of breast motion using non-linear dynamic FE analysis. Ergonomics 56(5):868–878. ISSN: 0014-0139. https://doi.org/10.1080/00140139.2013.777798

Cignoni P et al (2008) MeshLab: an open-source mesh processing tool. In: Scarano V, De Chiara R, Erra U (eds) Eurographics Italian chapter conference. The Eurographics Association (2008). ISBN: 978-3-905673-68-5. https://doi.org/10.2312/LocalChapterEvents/ItalChap/ItalianChapConf2008/129-136

DYNAmore GmbH, an Ansys Company: LS-PrePost. www.dynamore.de. Accessed 26 Aug 2023

Greenbaum AR et al (2003) An investigation of the suitability of bra fit in women referred for reduction mammaplasty. Br J Plast Surg 56(3):230–236. ISSN: 0007-1226. https://doi.org/10.1016/s0007-1226(03)00122-x

Hong L et al (2011) A study of the relationship between clothing pressure and garment bust strain, and Young's modulus of fabric, based on a finite element model. Text Res J 81(13):1307–1319. https://doi.org/10.1177/0040517510399961

OpenRadioss: OpenRadiossTM, June 2, 2024. https://openradioss.org

Parrilla E et al (2019) MOVE 4D: accurate high-speed 3D body models in motion. In: D'Apuzzo N (eds) Proceedings of 3DBODY.TECH 2019–10th international conference and exhibition on 3D body scanning and processing technologies, Lugano, Switzerland, 22–23 Oct 2019. Hometrica Consulting—Dr. Nicola D'Apuzzo, Ascona, Switzerland, pp 30–32. ISBN: 9783033075283. https://doi.org/10.15221/19.030

Ruixin L et al (2020) Computational modelling methods for sports bra-body interactions. Int J Clothing Sci Technol 32(6):921–934. ISSN: 0955-6222. https://doi.org/10.1108/IJCST-09-2019-0143

texsens: mobile load measurement in the textile and garment industry. https://novel.de/products/texsens/. (visited on 12/01/2023)

Vilhena L, Ramalho A (2016) Friction of human skin against different fabrics for medical use. Lubricants 4(1):6. https://doi.org/10.3390/lubricants4010006

Zhang J et al (2022) Non-linear finite element model established on pectoralis major muscle to investigate large breast motions of senior women for bra design. Text Res J 92(19–20):3511–3521. ISSN: 0040-5175. https://doi.org/10.1177/00405175221075049.

Zhang Y et al (2007) 3D finite element modeling of nonrigid breast deformation for feature registration in -ray and MR images. In: 2007 IEEE workshop on applications of computer vision (WACV'07), pp 38–38. https://doi.org/10.1109/WACV.2007.2

Chapter 11
Analysing the Breast-Bra Interaction in Motions Using Finite Element Models, 4D Scan Data and BodyTex4D

Abstract This chapter extends the digital analysis of breast-bra interactions to dynamic movements. Using the methods developed in Chapter 9 for creating individualised, deformable finite element human models, as well as 4D scan data and BodyTex4D, dynamic models are created. The same test subject from Chapter 10 is used, and two bra models, along with breast deformation without a bra, are examined during movement. Material calibration for the breast and skin is performed using 4D scan data and BodyTex4D, proving to be a highly promising approach. Finally, the dynamic human-clothing models are compared, revealing the effects of different bras and their properties. Once again, 4D scan data and BodyTex4D contribute significantly to model validation, enhancing the accuracy of digital simulations. The digital models offer great potential for clothing development, as they enable rapid and resource-efficient testing of various garment designs before physical production.

Keywords FEM · Soft avatar · Individualised · Digital clothing development · Human-body-model (HBM) · Clothing · 4D scanning · Breast-bra

11.1 Introduction and Objectives

This chapter investigates how the static individualised human body-clothing FE model (ID-HB-CLOTH-FEM) can be extended to analyse breast-bra interactions in motion.

Breasts consist of soft tissue. Relative movements between breasts and body occur during activities. Activities include, for example, running, cycling, climbing stairs, or walking normally. In a study by Lorentzen and Lawson (1987), half of the participants experienced breast pain and discomfort due to exercise. The dominant reason for discomfort is high breast movements. An individual bra development is needed, which includes dynamic motions to evaluate the bra's functionality and comfort in motion.

Virtual bra development opens up the possibility of individualisation and functional analyses of clothing items, which improves bra development (Chap. 10). Investigations of motions and dynamic breast-bra interaction behaviour in the virtual bra

A.-M. Schmidt, *Novel Methodologies for Digital Investigations of Clothing-Body-Interactions*, Lecture Notes in Bioengineering,
https://doi.org/10.1007/978-3-031-91794-3_11

development process are necessary to achieve individualisation. Breasts in motion experience large deformations. Spacial deformations (3D) over time (+1D) point to the requirement of a 4D measurement technology, 4D scanning. 4D scan data can capture the breast deformation during motion (Chap. 6). 4D scan data provides experimental data of bras interacting with breasts in the context of the whole body being in motion. 4D scan data has the decisive advantage of being useful for calibrating and validating numerical, dynamic ID-HB-CLOTH-FEMs. Currently, most FE breast-bra models capture breast and body movement using single acceleration sensors placed in a region of interest. Data from the single acceleration sensor has very limited informative value compared to capturing the full body in motion. Until now, to the author's knowledge no literature has used 4D scan data for FE model calibrations or validations.

This chapter introduces methods based on 4D scan data to develop individualised, deformable, human body (ID-HB) -bra (CLOTH) FE models (FEM). The development of ID-HB-CLOTH-FEM explores the usability of dynamic 4D scan data.

11.2 Methods

The following describes the methods to set up ID-HB-CLOTH-FEMs. Figure 11.1 provides an overview of the required steps that make up the method developed by the work at hand. The steps and underlying methods are derived from the general methodology development presented in Chap. 9.

Methods differing compared to the static numerical breast-bra model (Chap. 10) are material properties calibration, boundary conditions of the ID-HB-FEM and validation of the ID-HB-CLOTH-FEM. The bra variations are the same as presented in Chap. 10. Analogue steps of static and dynamic breast-bra FEMs are the geometry development and meshing of the ID-HB-FEM, the development of the CLOTH-FEM and the dressing process to get the ID-HB-CLOTH-FEM. Analogue steps are described in less detail, but references are provided. Newly introduced and further developed methods are described in detail. The finite element environment is the same as in Chap. 10 in Sect. 10.2.1 described.

11.2.1 Scan Plan and 4D Scan Data Analysis

Jumping motions are chosen to analyse the dynamics of breast-bra interactions. Jumping is a common movement, causing breasts to oscillate. Jumping does not introduce additional rotational components as running motions do. The test subject performs the jumps while the upper body resembles the A-pose. The switch from T-pose to A-pose is due to the stability of the A-pose during the jumping movement. T-pose causes significantly more flapping of the arms resulting in more breast movements. Flapping arms is declared as a compounding factor as the movement is unnatural

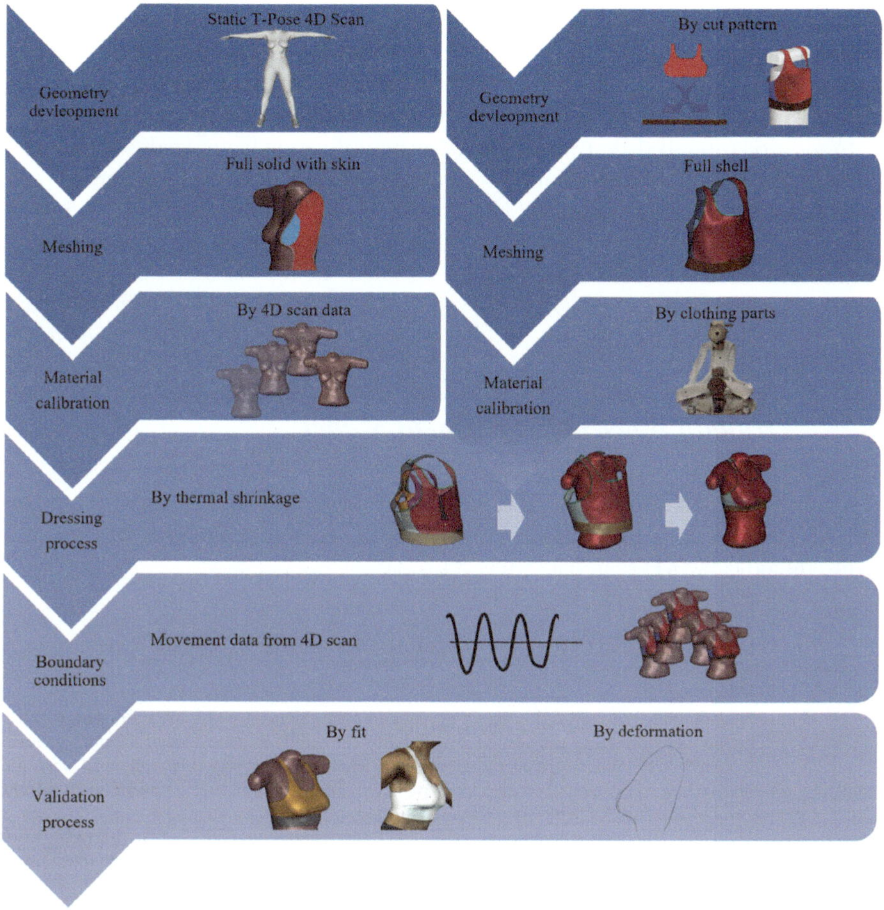

Fig. 11.1 Overview of Chap. 11

in humans. 4D scan data is captured by the Move4D Scanner of IBV (Parrilla et al. 2019). The experiments are performed on test-subject-A presented in Sect. 5.3.2. The following bra types are used:

1. Without-bra
2. Bra-2-M.
3. Bra-3-M.

Detailed descriptions of the bras are given in Sect. 5.3.3. The frame rate is set to 30fps. Each frame is meshed by a homologous mesh and exported as *.obj-file.

For each dynamic movement, the body movement is exported using the movement analysis presented in Sect. 4.7.1. The body movement is analysed based on the key-vertex-ID 13905 (Fig. 6.7a). The vertical component of the motion data is used due to jumps being vertical motions. Selecting a single component reduces compounding factors further, and reduces computational cost resulting in less analysis effort. Three body movement curves in the vertical direction are investigated. The curves describe the body movement of the bra variations without bra, bra-2-M-A and bra-3-M-A.

In addition to analyse body movements, an amplification factor analysis (Sect. 4.3) is used to classify the bra support. The breast movement is calculated by vertex-ID extracted by identification method based on mesh and texture data (Sect. 4.5.3 and Fig. 6.7a). Section 6.4.4 provides a more detailed description of the implementation of the method.

11.2.2 Development of an Individualised, Deformable Human Body Finite Element Model

The development of the ID-HB-FEM is divided into geometrical model development and meshing process (Sect. 11.2.2.1). A novel soft-tissue material calibration method is introduced using 4D scan data (Sect. 11.2.2.2).

11.2.2.1 Geometry Development and Meshing

The geometrical development and meshing of the ID-HB-FEM utilise the methods presented in Chap. 10. The ID-HB-FEM developed in Chap. 10 is used here to demonstrate its extensibility. The basis of the geometrical model is the static 4D scan of the female test-subject-A (Sect. 5.3.2) in T-pose without upper clothing. The meshing method is performed by the full-solid method (Sect. 9.2.2.2). The element formulation is `ELFORM 10`. The model is sectioned in a rigid body, deformable breasts, and deformable skin parts. The meshed human body model is shown in Fig. 10.4.

11.2.2.2 Material Calibration

The breast material properties are calibrated based on 4D scan data and literature values (Sect. 9.2.3.5). The breast material properties are calibrated based on the jumping motion (Sect. 11.2.1).

As a boundary condition for ID-HB-FEM the body movement curve representing test-subject-A jumping without a bra is used in LS-PrePost. The `*BOUNDARY_PRESCRIBED_MOTION_SET_ID` keyword of LS-PrePost applies the boundary condition to the model. The body movement curve is inserted as a time-displacement curve. The frame rate of the 4D scan defines the time sampling. The body is restricted in movement by only allowing movements in a vertical direction without rotational degrees of freedom by the `*BOUNDARY_SPC_SET_ID` keyword of LS-PrePost. The k-file snippet of setting the boundary conditions is shown in Listing 11.1.

Listing 11.1 Material calibration by 4D scanned movement: boundary restrictions

```
*BOUNDARY_PRESCRIBED_MOTION_SET_ID
$#       id
                 heading
         0 Body movement
$#     nsid         dof          vad         lcid          sf
          vid        death        birth
          1           2            2          101          1.0
            01.00000E28            0.0
*BOUNDARY_SPC_SET_ID
$#       id
                 heading
         0 Body movement restriction
$#     nsid         cid         dofx         dofy         dofz
        dofrx        dofry        dofrz
          1           0            1            0            1
            1           1            1
*DEFINE_CURVE_TITLE
vertical displacement curve by 4D scanned data of
   test subject A jumping
$#     lcid        sidr          sfa          sfo         offa
         offo       dattyp        lcint
        101           0          1.0          1.0          0.0
          0.0           0            0
$#                   a1                        o1
                    0.0                       0.0
                    0.1                    20.973
                  0.133                    44.053
                  0.167                    61.563
                  .....                      ....
```

The body part is modelled by rigid material (Listing D.1). Skin and breast parts are modelled by linear-elastic material, *MAT_ELASTIC. The linear-elastic material model is stable and low in complexity, which is beneficial, especially for analysing dynamic movements due to the high volume of data. The linear-elastic material model parameters of density 1×10^{-9}t mm^{-3} and Poisson ratio 0.49 are derived in Sect. 10.2.2.3. These values are based on literature data (Sect. 2.4.2.3). The material model calibration process consists of adjusting the skin's elastic modulus, the breast's elastic modulus and the global damping constant to align the breast's deformation behaviour between numerical and experimental results. Characteristic motion states are selected to compare the breast deformation between datasets (Fig. 11.2). The motion states are minimum and maximum, the null-intersection of the mean-removed oscillation curve between minimum and maximum, and the null-intersection following the maximum. Subsequently, the motion states describe a single motion cycle of upwards and downwards motion. Each of the motion states corresponds to one frame of a dynamic movement.

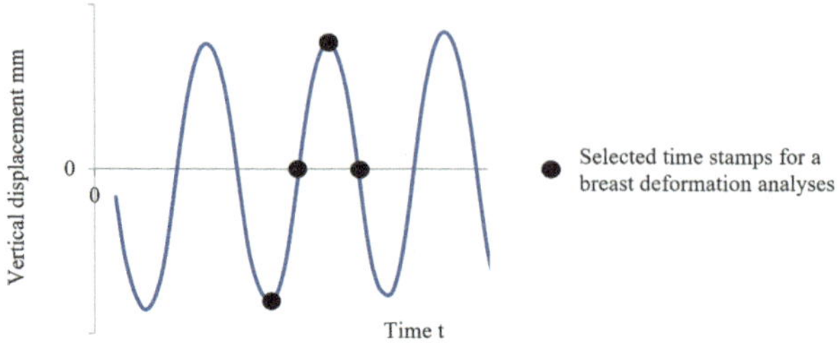

Fig. 11.2 Characteristic frames for the breast deformation comparison between 4D scan data and ID-HB-FEM at a body movement curves: minimum, intersection point moving upwards, maximum and intersection point moving downwards

Breast material elastic modulus is varied in a range of 0.001 MPa to 0.005 MPa, skin material elastic modulus is varied in a range of 0.005–0.5MPa and the damping factor is varied in a range of 0–50.

The material parameter combination that leads to the most similar breast deformation compared to the dynamic 4D scan data is chosen for the following breast-bra interaction analyses.

11.2.3 Development of Bra Finite Element Models

Bra-2-M and bra-3-M are modelled to analyse the dynamic breast-bra interaction by FE models (Fig. 11.3). Generation of the geometries of bra-2-M and bra-3-M for FE models is performed in Sect. 10.2.3.1. The material properties determined in Sect. 10.3.2 are applied. The bra materials are modelled by `*MAT_ELASTIC` and parameterised by measured and determined material properties presented in Table 11.1.

Fig. 11.3 Bra variations used in the dynamic FE analysis

(a) bra-2-M **(b)** bra-3-M

Table 11.1 Material and shell properties of the modelled bras

Bra parts		MAT_Elastic			Shell properties	
		Density kg^3 m^{-1}	Elastic modulus MPa	Possion ratio-	Thickness mm	Element formulation
Bra-2	Bra-band	6.288	1.45	0.3	0.79	16
	Front/back	5.162	0.62	0.3	0.482	16
Bra-3	Bra-band	3.758	0.67	0.3	2.0	16
	Front	3.35	0.278	0.3	1.31	16
	Back	3.178	0.265	0.3	1.13	16
	Seams	3.758	0.67	0.3	2.0	16

11.2.4 Dressing Process

The thermal shrinkage method (Sect. 9.4.3) is applied to dress the body model with the bra models. The boundary settings, bra scaling factor and shrinkage time, are analogue to Sect. 10.2.4.1. The contact and process parameters are analogue to Sect. 10.2.4.2.

11.2.5 Validation of the Dynamic Breast-Bra Finite Element Models

The final temperature of the dressing process is kept constant for the duration of the simulation to prevent further temperature influences. The boundary condition of fixed bra straps is resolved after the dressing process enabling movement of the bra relative to the body.

To analyse the dynamic breast-bra interaction, the rigid body of bra-2-M is moved by the movement curve extracted from the 4D scan of test-subject-A jumping while wearing the physical bra-2-M. Bra-3-M is analysed analogously (Sect. 11.2.1). The bra FE models are validated by comparing the breast deformation and bra fit at the selected motion states shown in Fig. 11.2. The model is validated if similar effects of the breast-bra interaction are observed in the model and the 4D scan occurs due to their coupling via the movement curves.

Table 11.2 Movement curve implementations captured from 4D scan data and their functionalities in the model development process

Bra variations		Movement curves when wearing			Functionality
		Without-bra	Bra-2-M	Bra-3M	
1.	Without-bra	x			Material calibration
2.	Bra-2-M		x		Model validation
3.	Bra-3-M			x	Model validation
4.	Without-bra		x		Bra comparison (2., 4., 5.)
5.	Bra-3-M		x		Bra comparison (2., 4., 5.)

11.2.6 Comparison of the Dynamic Breast-Bra Interactions When Performing the Same Motion

The calibrated and validated FE models with without-bra, bra-2-M and bra-3-M are moved by the movement curve resulting from test-subject-A jumping with bra-2-M. The effects of breast-bra interaction of without-bra, bra-2-M and bra-3-M are made comparable by actuating the simulated movement by the same experimentally measured movement curve each. The breast deformation is analysed at the comparison motion states (Fig. 11.2). The bra fit is evaluated at the comparison motion states based on breast deformation comparing vertical and horizontal cross-sections cutting the centre of the bust point. Table 11.2 provides a summary of presented FE models and their respective purposes.

11.3 Results

The following presents the results of the dynamic breast-bra interaction FE models. Firstly, the jumping movement is analysed experimentally (Sect. 11.3.1). Secondly, results of the dynamic FE model are shown (Sects. 11.3.2–11.3.4).

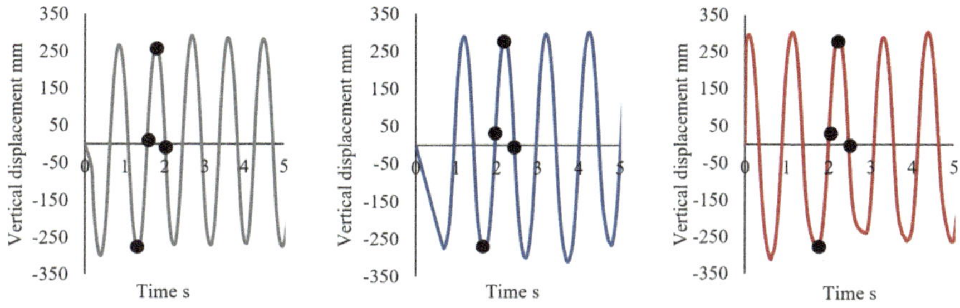

(a) without-bra (b) bra-2-M (c) bra-3-M

Fig. 11.4 Movement curves implemented into the dynamic FE model. The black dots indicate the selected time stamps for the human body model calibration (without-bra) and the breast-bra model validation (bra-2-M and bra-3-M)

Table 11.3 Selection of the time stamps from the 4D scan at the defined movement states

Bra variation	Time stamps s			
	Minimum	Intersection point moving upwards	Maximum	Intersection point moving downwards
Without-bra	1.35	1.55	1.75	2.0
Bra-2-M	1.7	1.95	2.25	2.4
Bra-3-M	1.7	2.0	2.3	2.5

11.3.1 Experimental Analysis of Jumping Motions

The experimental 4D scan data analysis of the jumping movement is presented in the following. Firstly, the extracted body movement curves are discussed. Secondly, the results of the amplification factor analysis are discussed.

The extracted body movement curves of the jumping movement without a bra, bra-2-M and bra-3-M are shown in Fig. 11.4. For each body movement curve, four time stamps are selected based on the motion state criteria displayed in Fig. 11.2. Time stamps corresponding to the motion states for each body movement curve are shown in Table 11.3.

Figure 11.5 presents the results of the amplification factor analysis of test-subject-A's jumping motions. First observation, the movements with a bra result in lower amplification factors compared to the movement without a bra. The observation indicates that bras support the breast. Second observation, bra-2-M has slightly lower support compared to bra-3-M. Less support of bra-2-M fits the results of amplification factor analyses performed in Sect. 6.5.3. In summary, bra-2-M and bra-3-M provide similar breast support.

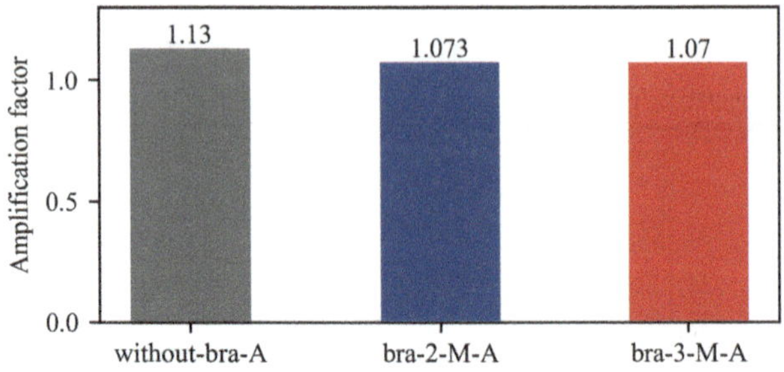

Fig. 11.5 Amplification factor analyses of test-subject-A in a jumping movement

11.3.2 Material Calibration of the Individualised, Deformable Human Body Finite Element Model

The breast deformation is compared between the 4D scan data and the numerical model to calibrate the material of the breast in the context of jumping motions. The body positions of the time stamps representing the motion states (Table 11.3.1) are shown in Fig. 11.6. As each jump is different, the aim of the comparison is not to have the exact same breast deformation between the experiment and the numerical model. The aim is to generate the same effects by the numerical results as observed in the experiment. The main effects observed by jumping motions captured in the 4D scanner on breast deformation are determined to be the following:

- Breasts move asymmetrically to each other (Fig. 11.6c).
- Breast form ovals during the upwards movement (Fig. 11.6f)
- Breasts are slightly lifted relative to the chest during the upwards movement (Fig. 11.6f)
- Breasts are lifted at the apex of the jump (Fig. 11.6c and g)
- Breasts take a round shape at the apex of the jump (Fig. 11.6c)
- When the body moves downwards, the breasts' downward movements lag behind due to the inertia of the breasts. The breast is still slightly lifted between the maximum and minimum (Fig. 11.6h)
- The lowest position of the breast is at the lowest jumping position (1.35 s), the breast has an oval shape and is not lifted (Fig. 11.6a and e)

The numerical model is calibrated considering the main effects of the breast behaviour while jumping. Influences on the calibration of the dynamic numerical model are the skin material model and breast material model, the global damping factor, and the formulation of solid elements.

The element formulation `ELFORM 10` is determined to be the most stable during the movement. `ELFORM 4, 13` resulted in element instabilities due to body accelerations and deformable materials.

Adjusting the elastic modulus of the breast and skin, along with the damping factor, significantly enhanced the flexibility in calibrating breast materials. The evaluated parameter combination and their evaluation are shown in Table 11.4.

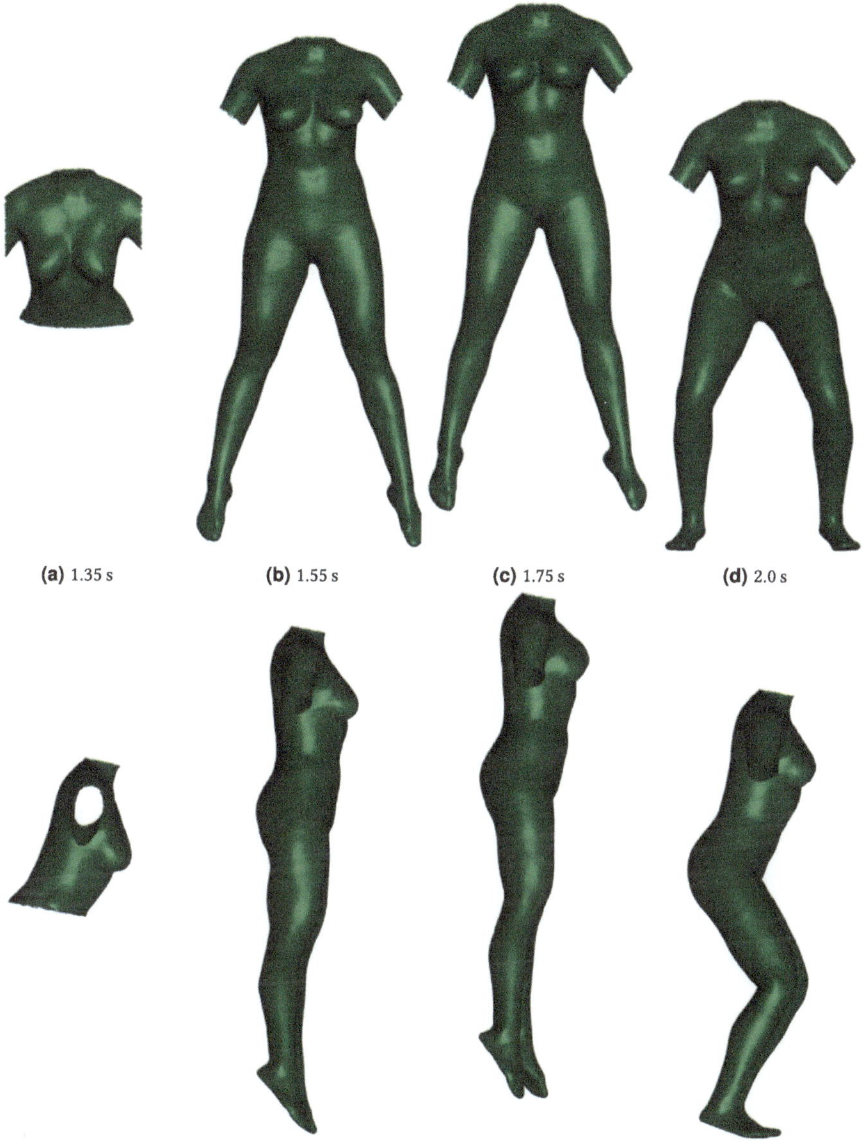

(a) 1.35 s (b) 1.55 s (c) 1.75 s (d) 2.0 s

(e) 1.35 s (f) 1.55 s (g) 1.75 s (h) 2.0 s

Fig. 11.6 Breast deformation while jumping captured by a 4D scanner

Table 11.4 Results and evaluation of the parametric analysis of the breast and skin elastic modulus and damping factor for the deformable breast Finite Element model

Model parameter			Evaluation	
Elastic modulus MPa		Damping	Stability	Breast deformation
Breast	Skin			
0.005	0.005	0	Unstable elements	Large, unrealistic
	0.1		Stable	Realistic, with oscillations
	0.25		Stable	Realistic, with oscillations
	0.5		Stable	Realistic, with oscillations
0.001	0.5		Unstable breast elements	
0.005	0.005	5	Unstable elements	Large, unrealistic
	0.01		Stable	Large
	0.05		Stable	Realistic deformations
	0.1		Stable	Small-medium
	0.25		Stable	Small
	0.5		Stable	Small
0.005	0.005	50	Stable	Large, unrealistic, high inertia
	0.1		Stable	Large, unrealistic, high inertia
	0.25		Stable	Medium, phase shifted, high inertia
	0.5		Stable	Small deformations
	1.0		Stable	No deformation

The static ID-HB-FEM calibration by IndentoPRO for static breast-bra analysis in Chap. 10 results in breast and skin elastic modulus of 0.005MPa, additionally, damping is not required. Applying the same material parameter of the static ID-HB-FEM to the dynamic ID-HB-FEM results in an unstable model with unrealistic breast deformations (Fig. 11.7). The results show the importance of calibrating FE models on an application-specific basis.

Generally, ID-HB-FEM with skin elastic modulus of 0.005MPa are unstable or lead to large, unrealistic breast deformations. The higher the elastic modulus of the

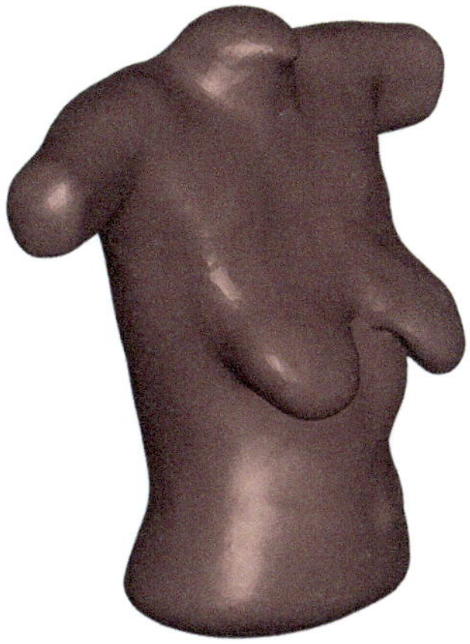

Fig. 11.7 Large, unrealistic breast deformations

skin, the stiffer the breasts behave, independent of the damping behaviour. The elastic modulus of the skin proved to be necessary for dynamic ID-HB-FEM calibration.

Similarly, increasing the damping factor stiffens the breast's behaviour. Damping factors at 50 starts to increase the breast's inertia additionally. Increased inertia leads to elongation of the breasts while the body is moving upwards. The higher the inertia, the longer the elongation. Realistic elongations of the breast are captured by 4D scan data and presented in Fig. 11.6f. Subsequently, the inertia needs to stay within realistic values, limiting the damping factor to 5. In the case of a damping factor equalling zero, the breast moves and deforms unrealistically caused by two superimposed oscillations. A lower frequent oscillation is caused by the jumping movement and a higher frequent oscillation is caused by the inability of energy to dissipate.

In summary, the following effects are determined by the parametric model calibration:

- Elastic modulus of the skin and the global damping factor are the main parameters to calibrate the breast model
- Material parameters of the static ID-HB-FEM are not transferable to dynamic ID-HB-FEM
- An application-specific calibration needs to be performed always
- Damping is necessary in the dynamic breast model to enable energy dissipation
- Too high damping factors lead to too high inertia causing unrealistic phase-shifted breast movement and unrealistic breast deformations
- Damping factor of 5 and elastic modulus of the skin set to 0.05MPa lead to the most realistic results in the case of test-subject-A

The results of the breast calibration process are shown in Fig. 11.8. Similar breast deformations of the numerical model compared to the experimental 4D scan data while jumping are accomplished. Breast deform asymmetrical (Fig. 11.8d). Breasts move up while the body moves up. Breasts hit their apex at the apex of the body

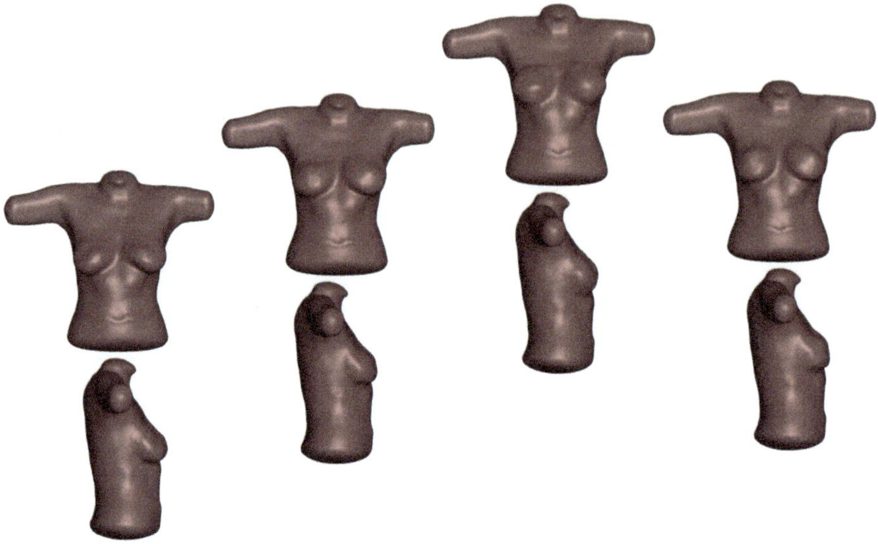

Fig. 11.8 Breast deformation while jumping modelled by the individual, deformable human body Finite Element model with deformable breasts

movement (Fig. 11.8c). Breasts have a round shape when falling down (Fig. 11.8d). The model is used to compare the breast-bra interaction in jumping motions.

11.3.3 Validation of the Dynamic Breast-Bra Finite Element Models

The validation results of the dynamic breast-bra interactions are presented in Figs. 11.9 and 11.10. The numerical and experimental breast deformation are compared by motion state derived from the oscillation curve of jumping motions.

Bra-2-M and bra-3-M exhibit similar breast deformation in numerical and experimental analysis. During landing, the breast moves downward, during the jumping up phase, it moves upward, reaching its maximum deformation at the apex of the jump. The breast remains deformed as it moves downward again. The effect is clearly visible in moving pictures. To represent the effects in static figures is challenging. The results for bra-2-M and bra-3-M are discussed separately, starting with bra-2-M.

The experimental bra-2-M forms a fold on the right side of the bra (Fig. 11.9a–d). Analogously, the fold forms in the dynamic FE model (Fig. 11.9e–h). The folding effect indicates a very good similarity between reality and the numerical model. Additionally, a loose edge at the front part of the bra is observed. The loose edge is due to missing seams at the edges. The same effect is observed in the experimental

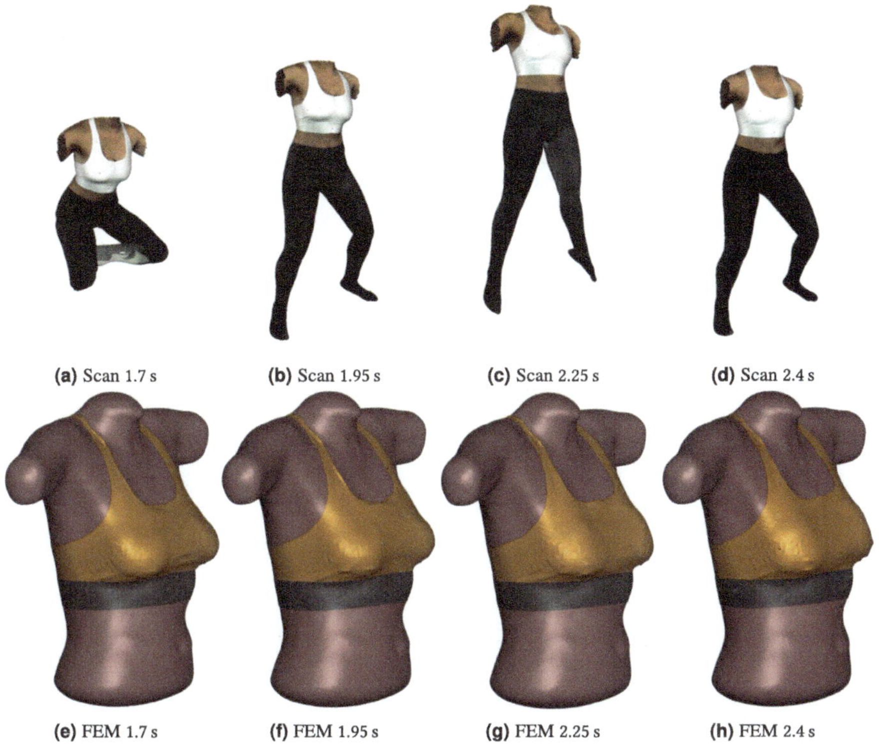

Fig. 11.9 Experimental and numerical comparison of the breast deformation while wearing bra-2-M

measurement. Due to the wider and shorter underbust band of bra-2-M compared to bra-3-M, the bra-band stayed during the whole jumping movement under the breast. Differences are observed in breast compression. The breast compression in the numerical model is smaller compared to the experiments. The difference in absolute breast compression could be due to the breast elements' compression behaviour or caused by the differences in the dressing procedures. Similar differences between the experiment and the numerical model are observed in the static FE analyses (Chap. 10). The similarity of the static and dynamic FE analyses is due to both being based on the same initial ID-HB-CLOTH-FEM.

In conclusion, the bra showed very similar effects in the experiment and in the numerical model, validating the developed model for breast-bra interaction in jumping motions for bra-2-M.

Comparing the numerical and experimental results of bra-3-M similar bra-band positions are observed. The bra-band is sliding up the breast during the jumping motion. Due to the seamed edges of bra-3-M compared to bra-2-M, the front part is stiffer and forms no folds (Fig. 11.10). Neither in the experimental nor in the numerical results folds are formed at any point of the jumping motion. The breast is less compressed

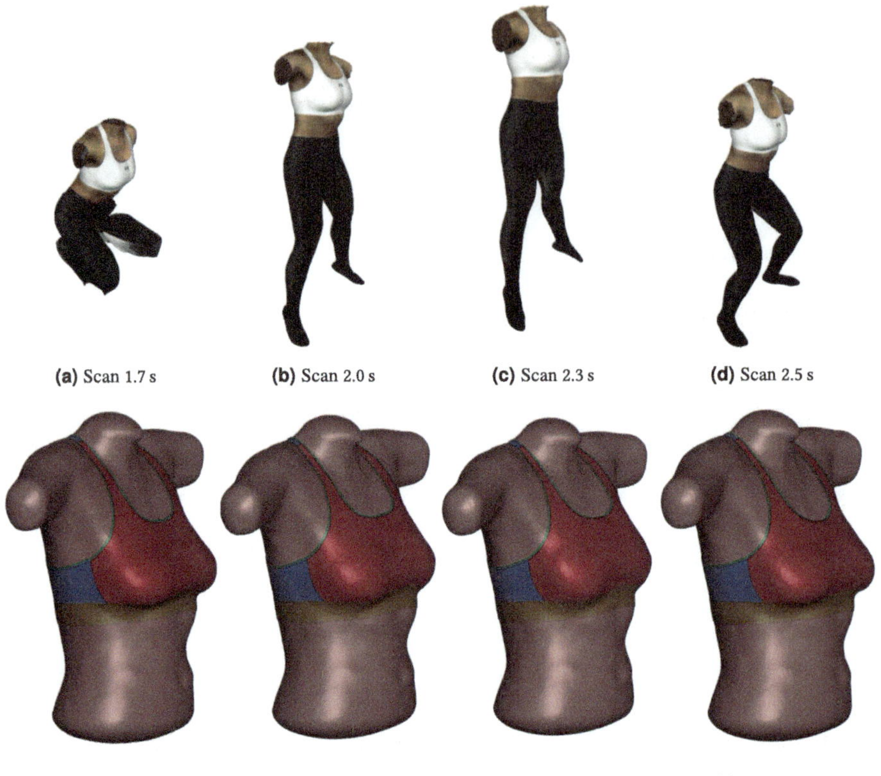

(a) Scan 1.7 s (b) Scan 2.0 s (c) Scan 2.3 s (d) Scan 2.5 s

(e) FEM 1.7 s (f) FEM 2.0 s (g) FEM 2.3 s (h) FEM 2.5 s

Fig. 11.10 Experimental and numerical comparison of the breast deformation while wearing bra-3-M

than the experimental result, analogue to the results of bra-2-M. Nevertheless, bra-3-M exhibits similar effects in jumping motion in the numerical and experimental data. Consequently, the numerical model of bra-3-M is valid.

11.3.4 Comparison of Dynamic Breast-Bra Interactions when Performing the Same Motion

The FE models without-bra, bra-2-M and bra-3-M are excited by the same movement curve. The cross-section results are displayed in Fig. 11.11 for four motion states (Fig. 11.2). Being able to excite different bra models by the same motion curve is a big advantage of numerical models in contrast to an experimental setup, the effect of various bra types is made comparable for the first time by this work. The results are:

- Bra-3-M compresses the breasts the most (Fig. 11.11a–c)
- Breast deformations are the largest when wearing no bra

Results of the fitting analysis are shown in Fig. 11.12. Two jumping states of the three bra variations are compared. Differences in breast deformation and bra folds are observed. Bra-2-M forms more folds, which indicates a less tight fit compared to bra-3-M. The bra-band position of bra-2-M is under the breast while it is on top of the breast for bra-3-M. Bra-2-M and bra-3-M front and back parts have the same cut pattern. Therefore, the difference in the fit is due to the material properties and the difference in the amount and placement of seams. The material properties of the front and back parts of bra-2-M and bra-3-M (Fig. 5.15) indicate similar tensile strength overall. However, a difference in tensile strength is observed in the vertical direction of the front part. Additionally, the seams contribute to a higher tensile strength in bra-3-M. The severity of each factor influencing the dynamic behaviour and breast deformation is part of future parametric analyses.

The pressure distributions are directly comparable between bra-2-M and bra-3-M by the numerical model compared to experimental trials, where comparability is nearly impossible to achieve. Pressure distribution exerted by wearing bra-2-M and bra-3-M at motion state 2.25 s (apex) is displayed in Fig. 11.13. The overall pressures for bra-3-M are higher caused by higher compression of the breast. Pressure peaks up to 9kPa are observed at the back, shoulders and breasts. The pressure distribution is visualised in Fig. 11.13b and d. Furthermore, increased pressures are observed beneath the seams in the region of the back. Pressure peaks lead to discomfort. Even though the bra-band from bra-2-M is shorter, it leads to lower pressure peaks. The superior design of the bra-2-bra-band compared to the bra-3-bra-band causes an improved fit by keeping the bra-band under the breast during the jumping motion and being wider. The results show that an improved fit causes lower pressure peaks, consequently, the comfort is improved. Future motion analyses need to investigate the influence of cut pattern, material properties, seam placements, etc. further based on the developed numerical model to reduce pressure peaks, subsequently, improving bra comfort generally.

11.4 Conclusion and Discussion

A Finite Element Model (FE model) is presented to model the dynamic interactions between breasts and bras. Real-life motion curves of jumping motions are extracted from 4D scan data. The motion curves are used to excite motions in the FE models. A novel method for breast material calibration (Sect. 9.2.3.5) is developed and applied. Two bra types are modelled. The FE models are validated based on comparing breast-bra interactions with 4D scan data of experimental trials using 4D scan data and BodyTex4D. The individualised (on test-subject-A) breast-bra FE models are validated based on breast deformation behaviour and fit of the bra in motion.

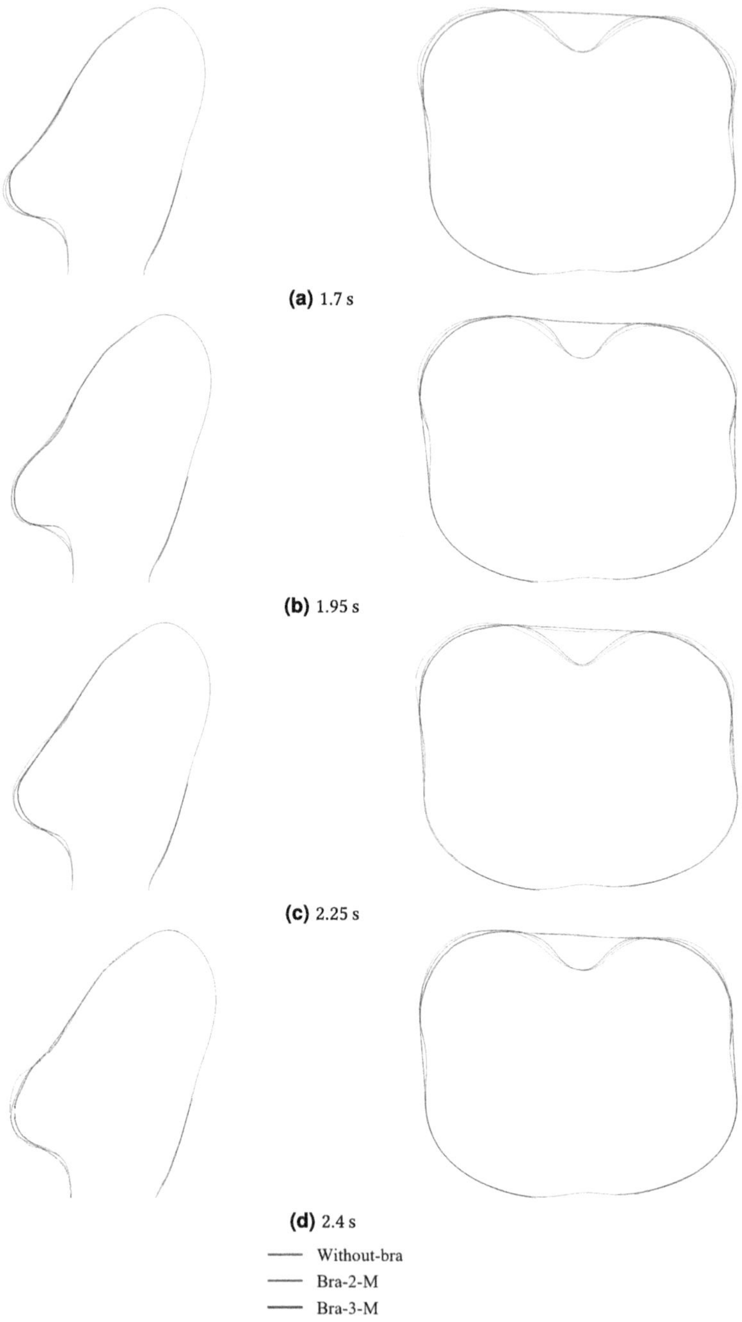

Fig. 11.11 Cross-section comparison of three bra variations while jumping with the same movement curves

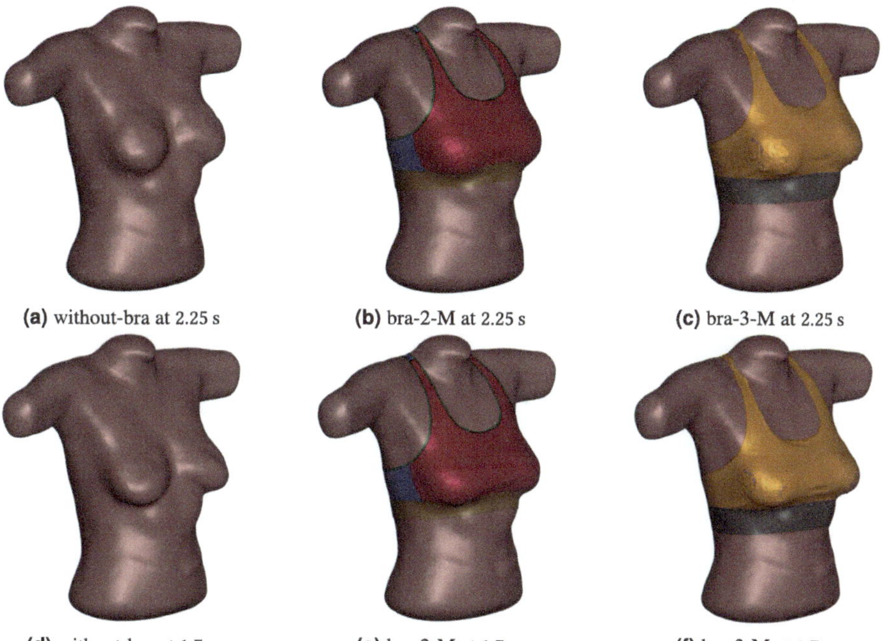

(a) without-bra at 2.25 s (b) bra-2-M at 2.25 s (c) bra-3-M at 2.25 s

(d) without-bra at 1.7 s (e) bra-2-M at 1.7 s (f) bra-3-M at 1.7 s

Fig. 11.12 Comparison of the bra fit in two jumping states: minimum at 1.7 s and maximum at 2.25 s

Finally, the FE models are moved by the same motion curve to compare the effect of the bras on the breasts.

The following main findings are made:

1. **A dynamic ID-HB-FEM (Individualised, deformable human body Finite Element Model) needs different material properties than a static ID-HB-FEM**: Material parameters of the breast measured with the IndentoPRO device to calibrate the static ID-HB-FEM lead to unstable dynamic ID-HB-FEM and unrealistic breast deformations. An application-related material calibration of soft tissue is required.
2. **The breast material calibration using the novel method by dynamic 4D scan data is an application-specific method which provides usability by simplicity**: Breast deformations by motion are large. Deformations are dependent on the material parameters of the breast and skin. Calibrating the numerical model on 4D scan data leads to a well-calibrated model for analysing breast behaviour in motion. Future research can leverage the insight of utilising motion curves to investigate breast-bra interaction. It is advised to include the motion of the bust point as a reference point in the calibration process. Motion curves of the bust point can be exported using BodyTex4D (Chap. 4) developed by this work. The first attempts to compare motion curves of the bust point experimentally

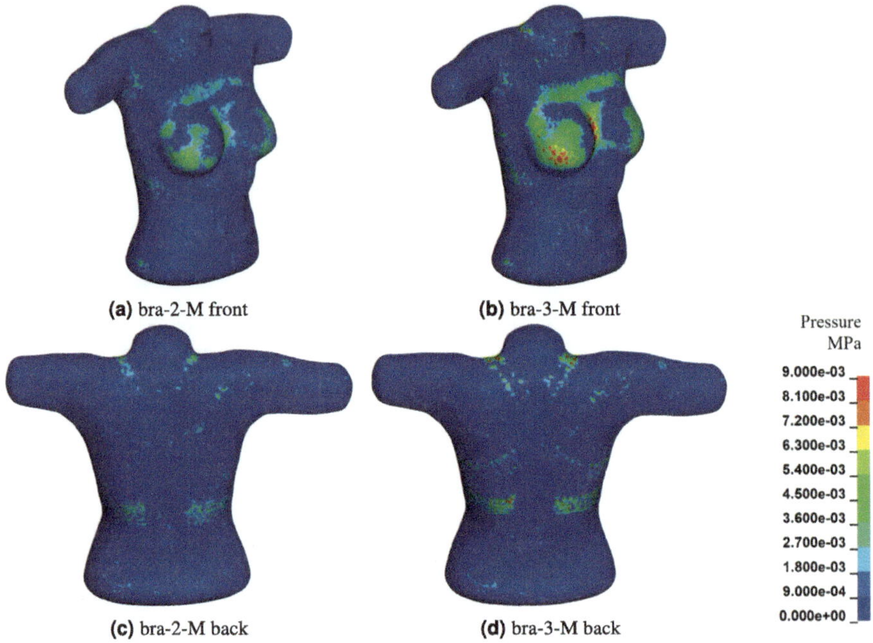

Fig. 11.13 Comparison of the pressure distribution of bra-2-M and bra-3-M at the jumping state: maxmimum at 2.25 s

and numerically do not yield sufficient results due to the increased stiffness of the numerical model. The experimental breast displacement is larger than the numerical breast displacement. The mesh of the breast is connected to the rigidly modelled body, which restricts its motion unrealistically. An option to model the breast motion more realistically is to use the solid-shell ID-HB-FEM resented in Fig. 9.5c. Furthermore, restrictions on breast movement would be lifted by applying sliding surfaces between the breast and the body. Proposed improvements impose more complex contact conditions increasing model instabilities. Further research on aligning numerical and experimental motion curves together is necessary.

3. **Elastic modulus of the skin and the global damping factor show the strongest effect on breast deformations in motions, which offers a simple, directed and fast method to calibrate breast material individualised for given test subjects**: By changing the elastic modulus of the skin material model and the damping factor, individualised breast calibration becomes feasible. The calibration process is still an iterative, manual process which future research needs to automate to enable individual bra development on an industrial scale.
4. **Deformation behaviour of the breast and the fit of a bra in motion can be validated by 4D scan data**: The ID-HB-CLOTH-FEM (combination of ID-HB-FEM and the FEM of a bra, CLOTH-FEM) is excited by the original body

motion curve of test-subject-A performing a jumping motion wearing bra-2-M or bra-3-M. Resulting breast-bra interactions show similar effects for bra-2-M and bra-3-M when compared to 4D scan data of experimental trials. Bra-2-M shows folds in the same areas in the numerical model and in experimental trials. The bra-2-bra-band stays underneath the breast for the whole motion compared to the up-slipping of the bra-3-bra-band which replicates the real-world observations. Bra-3-M forms no folds in the numerical model and the experimental trials. The formation of folds is prevented by seams at the edges of the front part, causing stiffer edges. In summary, breast compression is similar to the bra. Further analyses regarding the element formulation and material parameters need to be conducted. However, due to the very good replication of the real breast-bra interactions, the FE models used are regarded as successfully validated. Calibrating the dynamic breast-bra interaction using 4D scan data demonstrates significant potential, as it allows for comprehensive experimental and numerical deformation and fit analyses.

5. **Using the validated ID-HB-CLOTH-FEM, the breast-bra interactions can be analysed in the same motion, making comparisons possible**. This is a significant advantage of the numerical model compared to experiments. The same motion can be compared. In this comparison, differences in the bra variations could be seen. Differences in the breast deformation, bra fit and pressure distribution can be investigated and compared. The cut pattern or material parameter can be changed in the numerical model, and the bra comfort can be improved due to reduced pressure peaks. Using this model, digital bra development can be improved. Nevertheless, as this chapter is focused on the FE model development rather than the bra optimisation, these analyses are made in the following research.

References

Lorentzen D, Lawson L (1987) Selected sports bras: a biomechanical analysis of breast motion while jogging. Phys Sports Med 15(5):128–139. ISSN: 0091-3847. https://doi.org/10.1080/00913847.1987.11709355

Parrilla E et al (2019) MOVE 4D: accurate high-speed 3D body models in motion. In: D'Apuzzo N (ed) Proceedings of 3DBODY.TECH 2019-10th international conference and exhibition on 3D body scanning and processing technologies, Lugano, Switzerland, 22–23 Oct 2019. Hometrica Consulting-Dr. Nicola D'Apuzzo, Ascona, Switzerland, 2019, pp 30–32. ISBN: 9783033075283. https://doi.org/10.15221/19.030

Chapter 12
Part III: Conclusion, Discussion and Outlook

Abstract This chapter summarizes and discusses the findings and methods developed in Chaps. 9–11 on the digital analysis of clothing interaction with a deformable human model and realistic clothing models. The proposed methods are compared to the current state of the art, highlighting the research gaps that have been addressed. Additionally, the economic feasibility of digital modelling approaches is explored. The investigation revealed that the developed generic methods for constructing an individualized, deformable human body model with a realistic clothing cut pattern demonstrate high usability in the design of tightly fitted clothing. Incorporating 4D scan data and BodyTex4D further enhances the clothing development process.

Keywords FEM · Soft avatar · Individualised · Digital clothing development · Human-body-model (HBM) · Clothing · 4D scanning · Breast-bra

12.1 Conclusion

The development of an individual, deformable human body-clothing Finite Element model (ID-HB-CLOTH-FEM) to improve the tightly fitted clothing development is presented in part III. Firstly, generalised methods are derived for the model setup (Chap. 9). Secondly, derived methods are applied to model the breast-bra interaction for static poses (Chap. 10). The static model is the extent to model the dynamic breast-bra interaction (Chap. 11). Experimental 4D scan datasets are used consistently to improve the accuracy of the models. Motion curves and comparable 4D scan data points for model generation, model calibration and model validation are obtained by scan data handling methods and algorithms developed in part II.

Chapter 9 develops and evaluates generalised methods to set up an ID-HB-CLOTH-FEM. The evaluation is focused on proving the usability of the methods for clothing development. The methods are visualised in Fig. 9.33 and summarised by the list below in the form of modelling steps and the developed modelling approaches followed:

A.-M. Schmidt, *Novel Methodologies for Digital Investigations of Clothing-Body-Interactions*, Lecture Notes in Bioengineering,
https://doi.org/10.1007/978-3-031-91794-3_12

1. Geometry development:
   (a) ID-HB-FEM: 4D scan data
   (b) CLOTH-FEM: cut pattern
2. Meshing:
   (a) ID-HB-FEM: Full solid
   (b) CLOTH-FEM: Full shell
3. Material calibration:
   (a) ID-HB-FEM:
      i. Static analysis: IndentoPRO
      ii. Dynamic analysis: 4D scan data
   (b) CLOTH-FEM: Tensile testing
4. Dressing Process: by thermal shrinkage
5. Validation:
   (a) Deformation analysis based on 4D scan data made comparable by novel criteria
   (b) Fit analysis based on comparable 4D scan data

Chapter 10 applies the model setup, model validation and analyses methods developed in Chap. 9 to breast-bra interaction for static T-pose scenarios. Four different bras are modelled and evaluated. The evaluation in Chap. 10 compares static frames of similar breathing states determined in Chap. 5. The models show very good alignment of numerical and experimental deformation behaviour, bra fit and pressure values. The numerical model is subsequently used to investigate the effects of different bra sizes, materials and seam structures on fit, support and comfort. Differences between the numerical model and 4D scan data of the absolute level of breast compression are observed. The numerically modelled breast compression is generally lower. To model pressure distributions covering the whole body is an outstanding advantage of the numerical models compared to point-wise measurements in experimental trials. Causes of measurement errors and outliers in the data of experimental trials are made explainable by the numerical model. This chapter demonstrates that the generalised methods developed in Chap. 9 are applicable and transferable to static breast-bra interaction analyses.

Chapter 11 analyses dynamic breast-bra interaction by applying methods for dynamic numerical model analyses developed in Chap. 9. The analyses show that breast and skin material parameters are required to differ for modelling static or dynamic breast-bra interactions, respectively. Material calibration of the breast models and validating the breast-bra Finite Element (FE) model is based on experimental 4D scan data. Two bra types are modelled. Utilising 4D scan data for model generation and validation proves to be highly effective, as it allows for the simultaneous investigation of the deformation behaviour of the entire body. This approach makes 4D scan data-based development both accurate and time-efficient. The dynamic ID-HB-CLOTH-FEMs show very good results in the deformation behaviour of the breast, but further research in breast movement comparison is required. The effects of the modelled bras interacting with the modelled breasts are similar to the experimentally captured breast-bra interaction regarding folds, bra-band position and overall fit.

The significant advantages of the developed methods to build up a static or dynamic ID-HB-CLOTH-FEM are:

1. easy exchange of human body model (ID-HB-FEM)
2. easy exchange of bra model (CLOTH-FEM)
3. stable dressing method
4. improved model validation by selecting comparable poses and motions using 4D scan data and BodyTex4D
5. improved model validation by performing deformation and fit analyses with 4D scan data
6. pressure peaks exerted on the body by bra strains are directly comparable, leading to hints for design adjustments of the bra to fit the individual's needs

The developed and generalised methods are valuable, highly usable and accurate tools for developing tight-fitting clothing proven by the example application of breast-bra interaction in static poses and dynamic movements. The method steps are consistent for all static and dynamic ID-HB-CLOTH-FEMs, enabling further automation of methods.

12.2 Discussion

The discussion is divided into comparing the developed methods with existing methods 12.2.1, the transferability of the developed methods 12.2.2 and the economic feasibility 12.2.3.

12.2.1 Addressed Research Gaps

The following concludes and discusses the main research gaps addressed in this part III.

1. **Development of generalised methods to generate ID-HB-CLOTH-FEMs**: ID-HB-CLOTH-FEMs are useful in clothing development due to enabling analyses of individual fit, body deformations and pressure distributions in pre-production digitally. Currently, available ID-HB-CLOTH-FEMs are specifically designed for single-use cases of single bra types on a single test subject. Current methods are not designed with usability for clothing development in mind. Chapter 9 generalises methods to develop ID-HB-CLOTH-FEMs, which contain (partially) automated

development steps, exhibit stable FE calculations and are easily transferable to multiple use cases of clothing development.

2. **Introducing 4D scan data to ID-HB-CLOTH-FEMs**: Combining 4D scan data with ID-HB-CLOTH-FEMs is performed for the first time by this work and offers great advantages. The advantages permeate the entire ID-HB-CLOTH-FEM-based development process. Model generation requires static body scans. 4D scanners exposure times are very short compared to 3D scanners increasing the accuracy and decreasing the duration of full-body scans. 4D scan data-based calibration of material models for example for dynamic breast analyses is accurate due to accurate full-body scans in motion. Soft tissue's material behaviour can be accurately and application-related calibrated by deformation comparison. Damping factor-based breast material calibration is performed by studies in the literature. The breast is lifted by hand and released. The breast movements in running motions are measured by acceleration sensors. Subsequently, the current state of the art requires additional sensors, which influence the movement and are limited in the number of measuring points. 4D scan data enables deformation and movement analyses of the entire body simultaneously. In summary, leveraging 4D scan technology enhances the usability of ID-HB-CLOTH-FEMs in the clothing industry by increasing the accuracy of the digital development process and decreasing its complexity.
   Model validation benefits greatly from 4D scan data. The validation is based on the deformation behaviour of the soft tissue is novel as 4D scanning enables full-body deformation analysis with high accuracy for the first time. A comparison between the numerical model and the experiment is crucial for the usability of FE models. Model validation is challenging due to the non-repeatability of human body states. Humans cannot repeat poses or movements exactly. Comparison metrics are required. The metrics increase in informative value with more precise and higher amounts of data. Comparability metrics for 4D scan data frames are developed and automated by BodyTex4D (Chap. 4). Increased automation and the potential of further automation increase the efficiency of model validation processes. The higher the quality of experimental data, the more accurate the validation becomes.
3. **Development of novel methods for the material calibration of ID-HB-FEMs**: The most critical factor for the usability of ID-HB-FEMs is the individual material parameters of different body regions. Deformable material parameters differ depending on age, body composition and genetics among many others. Material data based on literature values, MRI or not-portable devices for application-specific indentation tests are presented in the literature. Using literature data is not individual, using MRI images can increase the model composition accuracy, but is expensive. Furthermore, the body position in an MRI is very different from a body in motion wearing tightly fitted clothing. Application-specific indentation test devices such as those used for amputees are also expensive and not flexible. Two novel sources of data for model calibration are used in this work. Firstly, the portable indentation test device IndentoPRO is used to calibrate the material model of the FE models in regard to soft-tissue material. It is implemented into

the FE model calibration process and proves to be very usable. IndentoPRO is less expensive than other devices (500€), portable, easy to use, and measures different material behaviours by force-distance curves with ease.
Secondly, a material calibration method by the deformation behaviour of the body in motion based on 4D scan data is developed. Calibration by 4D scan data of dynamic movements produces good results as it is an application-related calibration method. The IndentoPRO and 4D scan data methods are performed iteratively by manually adjusting the breast and skin material model. An optimisation algorithm to automatically find the best fit between numerical and experimental results should be developed in the future.

4. **Using original clothing cut patterns in ID-HB-CLOTH-FEMs**: Currently, geometries of tightly fitted clothing items are mostly generated by an offset of the body's surface. Offsetting the body's surface is unrealistic and insufficient to develop clothing items as clothing-body interactions are ignored. Basing the generation of clothing FE models on original cut pattern-based is essential for enhancing the clothing development to efficiently test new cut pattern designs digitally. Clothing software such as CLO3D, VStitcher, or Style3D offers clothing item geometries based on cut patterns. Access to the original cut pattern enables the integration of listed software tools into the FE model. This work presents automation approaches to integration. The clothing models are exported to file formats supported by FE software. Clothing companies which already use the software tools can adapt presented approaches immediately to improve their digital cut pattern design process by adding the Finite Element Modelling step.
5. **Combining the ID-HB-FEM and CLOTH-FEM by the digitalisation of the dressing process**: Combining the ID-HB-FEM and CLOTH-FEM is challenging for tightly fitted clothing items, as the clothing is undersized resulting in intersections of the model geometries. The intersections need to be resolved to enabling interaction modelling. Resolving the intersections realistically is done by a novel developed, digital dressing process. The new dressing method is based on thermal shrinkage. Therefore, sensitivity to the coefficient of thermal conductivity is added to the material model. The method consists of scaling the clothing item up until the clothing item model is fully outside of the ID-HB-FEM, resolving intersections. Subsequently, the model is shrunk by reducing temperature for a certain period of time. The time is calculated to reflect the duration it would take the clothing item to shrink to its original size. The scaling process and calculation of the shrinkage time are implemented in Python, which increases the degree of automation and re-usability. The dressing method by thermal shrinkage is compared to other dressing methods and determined to be the most stable, realistic, and fastest digital dressing method. Compared to existing approaches by shrinking and extending the ID-HB-FEM, the shrinkage method of the bra is faster as it does not include as many elements, is more stable and resembles the real dressing process.
6. **Improvement of the accuracy of static breast-bra FE models**: Improvements in static breast-bra FE models are achieved by enhancing ID-HB-FEM, bra models, calibration processes and validation processes. To the best of the author's

knowledge, nearly all of the breast-bra models found in the literature focus on body modelling rather than bra modelling. The developed static breast-bra FE model is focused on its usability in clothing development, focusing on a fast and easy model setup, an accurate representation of the bra and high quality in the validation. The IndentoPRO is unknown to the literature as a material calibration method; most sources used literature values or calibrations based on breast behaviour. Therefore, the perspective of the work at hand is especially useful for the clothing industry. The bras are modelled by original cut patterns. Influences of seams are modelled. Modelling seams are unrepresented, as well as for existing clothing software such as CLO3D or VStitcher. Seams show a strong influence on clothing fit and clothing-body interactions. Compared to the existing one, another improvement of the static breast-bra FE model is the comprehensible validation via pressure and deformation analyses. The pressure is critically discussed, and the deformation analyses are made after selecting comparable 3D models with the method presented in Sect. 5.4.1. The developed breast-bra models are validated in regards to the bra fit, breast deformation and pressure distribution and, consequently, usable for future tightly fitted clothing development. The bra FE model or ID-HB-FEM is easily replaced, improving digital clothing development further.

7. **Improvement of dynamic breast-bra FE models by using 4D scan data and accurate CLOTH-FEM**: Analysing the effect of bras on the breast-bra interaction in motion is rare in the literature. Most dynamic models focus on the breast movement behaviour in motion without influences of clothing. 4D-scan-data-based calibration of breast material models with respect to dynamic deformation behaviour is unrepresented. Additionally, no dynamic FE models with an accurate representation of the bra model are documented in the literature. Finally, exciting an ID-HB-FEM dressed with different bra models by the same motion curve extracted from 4D scan data leads to comparable pressure distributions on the body and comparable breast deformation behaviour caused by the bra models. The bras are evaluated by fit, breast deformation and functionality.

12.2.2 Transferability

The transferability of the ID-HB-CLOTH-FEM generation is essential for the integration in clothing development. Generalised, application-independent methods to enhance clothing development are presented in Chap. 9. Each presented modelling step is transferable to other test subjects, body parts, or clothing items.

During the geometry development process of the ID-HB-FEM, any body part can be selected, transformed to a geometrical human body model and modelled by deformable material. The surface of the body part of interest is required to be captured completely. For example, in the case of the investigation of the breast, touching skin areas caused by the inframammary fold is required to be resolved to enabling scanning of the full surface. Larger breasts form larger inframammary folds hiding skin area from the sight of scanning devices. Consequently, additional preparation steps must be added to the model development process. How to resolve the inframammary fold is presented in Sect. 10.4 and Fig. 10.22. In the case of scanning legs, collisions between the meshes of the inner tights is advised against. Nevertheless, tactics as exemplary shown for larger breast exists to scan any body part and include the body part in the FE model generation workflow.

The meshing process and material calibration of the ID-HB-FEM is automated by using the LS-PrePost software tool. The body part selection process has to be adapted on a case-by-case basis. Re-usability of the selection process is implemented by a template-base method (Fig. 9.6) for full-shell meshing approaches. Generally, the part to be modelled deformable is separated from the rest of the body. The deformable material calibration can generally be performed by force-distance curved measured with the IndentoPRO indentation test device. IndentoPRO is a portable device, low in cost and is flexibly applicable on any body part.

The CLOTH-FEM is highly flexible as well. Any cut pattern can be designed using common clothing software and 3D models can be generated. Any compressive clothing item can be imported into the FE model. An experienced researcher needs approximately 10 min to set up the model, excluding material calibration. A possible application is Jeans development. Legs, especially persons suffering from obesity, are highly individually shaped. Individualised clothing is necessary to provide sufficient comfort. In the field of compression socks, individualisation is necessary. Smart sleeves are designed for different means of communication in the field of smart textiles. The examples share the requirement of providing a good, individual fit, which is essential for functionality.

The material calibration of the CLOTH-FEM is generalised. Standard testing methods are applicable depending on the produced clothing item's available form (e.g. pre-production cut pattern or post-production clothing item). Tightly fitted clothing material behaviour is generally dominated by tensile strength. Therefore, the developed material calibration method is transferable to any analogue clothing application.

The dressing process to generate the ID-HB-CLOTH-FEM by thermal shrinkage is applicable for many clothing items. The clothing item model is scaled until it can be placed without intersections around the body model and then shrunk until a force equilibrium is reached. Figure 12.1 shows the example clothing item after the thermal shrinkage. The same keywords as for the breast-bra investigation are applied in this simulation, which proves good transferability.

The validation process can be performed using the pressure or deformation method. Whether a motion comparison is needed depends on the use case. For example, motion analyses can be performed when vibrations in soft tissue should be restricted, as is the case for arms or legs while running.

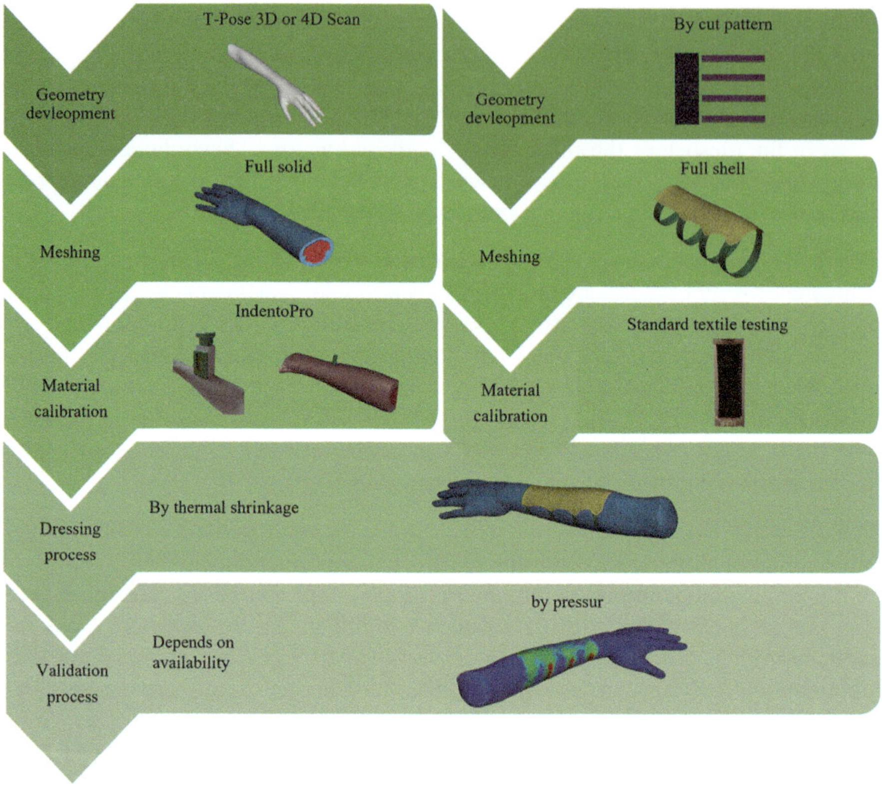

Fig. 12.1 Demonstrating the transferability of the generalised methods to develop an ID-HB-CLOTH-FEM for modelling sleeve-arm interaction

In summary, the developed methods are transferable to:

- **different bodies of test subjects**
- **different body parts**
- **different clothing items**
- **different clothing materials**
- **different clothing parts**
- **different clothing sizes**

Prove of transferability of the presented generalised ID-HB-CLOTH-FEM methods by applying the methods to another use case is presented Fig. 12.1. The model is based on 4D scans, meshed with solid elements and material calibration is done by IndentoPRO. Further possible applications are compression socks, tights, T-Shirts and many more.

Various software tools can be used. Clothing geometry can be generated from cut patterns by software and exported to *.obj-files. *.obj-files are compatible with the developed BodyTex4D analysis framework. CLO3D is used in this work, but VStitcher or Style3D are valid alternatives. For example, Python scripts are implemented to automatically combine boundary conditions with *.obj data and output LS-PrePost compatible k-files to run simulations. However, Ansys has human-readable input file formats as well, for which the Python script could be adopted or any other format. Next to commercial software tools, open-source alternatives are available. For example, OpenRadioss can run LS-Dyna files (Herakovich 2017).

12.2.3 Economic Value

Clothing development is as manifold as available clothing items, rendering a quantitative comparison is feasible. The clothing development process differs depending on

- the company size
- the company resources
- the companies involved
- regulatory boundary conditions
- the demand for the product to be developed
- the number of customers
- the scale of production
- the degree of global distribution of production steps
- and many more

Nevertheless, a comparison of the product development process with and without an ID-HB-CLOTH-FEM for a medical, individualised bra is derived. Table 12.1 presents the bra development using manual, iterative prototyping. Table 12.2 uses the ID-HB-CLOTH-FEM in the product development. Medical, individualised bras are manufactured to treat breast cancer or to reduce back pain. Fit, comfort and fulfilment of functionality are important in the individual bra development. The following calculations serve as an example and a rough estimation of the benefits of digital clothing development.

Traditional product development processes start with the patient having to make an appointment with the orthopaedist. The Orthopaedist measures the patient with measurement tapes commonly. The measurement depends on the orthopaedist's craftsmanship. The breast's material properties are respected based on experience. Especially in cancer treatment, the breast exhibits highly individual material properties due to tissue changes or scars of operations, which influence the bra's fit and functionality. The measurement data is transferred to develop the individualised bra cut pattern. The bra is produced and tested again on the patient, and adjustments are made. This process is repeated iterative (rep.) several times. Time spans between repetitions are of the order of days to weeks due to organisation, appointment arrangements and

Table 12.1 Cost estimation for the product development of a medical bra using iterative prototyping

Bra development steps		Time	Costs	
			Approx. hourly costs	Total costs
1.	Measurement patient	30 min	Orthopaedist: 70€/h	35€
	Transfer data and communication	1d		
2.	Bra cut pattern design	60 min	Garment technican: 70€/h	70€
	Transfer data and communication	1d		
3.	Bra Production	60 min	Bra material: 20 € Tailor: 20€/h	40€
	Transfer data and communication	1d		
4.	Test on patient	30 min	Orthopaedist: 70€/h	35€
5.	Repeat step 2-4	1 rep.: 3d 3h 5 rep.: 15d 10h 50 rep.: 154d 14h	Orthopaedist Designer Bra material Tailor Orthopaedist	1 rep.: 145 € 5 rep.: 725 € 50 rep.: 7250 €
6.	Final production	1h	Tailor: 20€/h	20€
Total		1 rep.: 3d 4.5h 5 rep.: 15d 11.5h 50 rep.: 154d 15.5h		1 rep.: 200€ 5 rep.: 780€ 50 rep.: 7305€

logistics. Table 12.1 displays the approximated costs and time for the bra development. Depending on the number of iterative steps, the repetition will increase. For 5 repetitions to produce a single bra 15.5 d are necessary at least. The costs accumulate to 780€. Assuming that one order of an individualised bra is placed per day, the extrapolated cost over a year (230 working days) results in 3565 d of work and costs of 179.400€. Fix costs such as sewing machines or room costs are not included in this calculation as the costs are similar to the product development process based on ID-HB-CLOTH-FEM.

Bra development process based on ID-HB-CLOTH-FEM (Table 12.2) starts by 3D or 4D scanning the patient's body. IndentoPRO is used to calibrate the breast material properties. Measurement results are consistent and accurate, independent of the craftsmanship of the attending orthopaedic. Measurement costs are equal to the manual process, but the fixed one-time costs to purchase the measurement equipment start from 30.500€for a 3D scanner and an IndentoPRO. The bra cut pattern design can be made by a garment technician in clothing software, additional clothing software costs may occur. The ID-HB-CLOTH-FEM is set up by textile engineers. Several parameter combinations to calibrate or alter the functionality of the CLOTH-

Table 12.2 Cost estimation for the product development of a medical bra using the breast-bra Finite Element model (green: fixed costs, red: continuous costs)

Bra development steps		Time	Costs: Continious – Approx. hourly costs	Costs: Continious – Total costs	Costs: Fixed
1.	3D/ 4D scan of patient Breast material measurement	30 min	Orthopaedist : 70€/h	35 €	3D Scanner/4D scanner: one time: 30.000 -500.000 € IndentoPro: one time: 500€
	Transfer data and communication	1d			
2.	Bra cut pattern design	60 min	Garmment technican: 70€/h	70 €	Clothing software: per year: 3000 €
3.	ID-HB-CLOTH-FEM set up	30 min	Textile engineer: 100€/h	50 €	FE Software: per year: 3000€
4.	ID-HB-CLOTH-FEM running one parameter set (including model modifications)	3h	9 min textil engineer for the result evaluation and parameter changes	15 €	Clothing and FE Software
5.	Repeat step 4	1 rep.: 3h 5 rep: 15h 50 rep: 150h	Textile engineer: 1 rep.: 9 min 5 rep.s: 45 min 50 rep.s: 7.5h	1 rep: 15€ 5 rep: 75€ 50 rep: 750€	Clothing and FE Software
6.	Final production	1h	Tailor: 20€/h	20 €	
Total:		1 rep: 1d 6h 5 rep: 1d 18h 50 rep: 7d 9h		1 rep: 140€ 5 rep: 200€ 50 rep: 875€	per year: 6000€ one time: 30.500 - 500.500€

FEM can be calculated without additional manual work. To test a single parameter combination approximately 3 h of computation time (Intel i7 7th generation, 32 GB, Dell Laptop) and 9min of manual work for evaluation are required. Generating a bigger set of parameter combinations is significantly less expensive and time-consuming than the iterative prototyping process due to the inherent option of paralleling the computation or scaling up computational resources. Additionally, pressure distributions over the whole body can be calculated. Generally, higher amounts of data are available to develop the bra further. The data can be saved as a database persistently to accelerate the development process over time. The digital method allows for more parameter combinations to be tested in shorter time intervals. In summary, considering 5 repetitions are needed for the final bra design, 1.5 d are needed without parallelisation, resulting in labour costs of 200€. Consequently, the digital design process is 10 times faster and three times cheaper in regard to labour costs. Finite Element and clothing software add yearly costs of 6000€. Assuming 5 repetitions have to be performed to develop an individualised bra, 500€can be saved with the digital design process based on FEM. After only 12 patients a year, the software costs would be even. Assuming a company treats 1 patient per day over a year, resulting in the demand for 230 bras. A total amount of 230· 500€- 6000€= 109.000€could be saved.

The 3D or a more accurate 4D scanner can be purchased by this surplus. By purchasing a 4D scanner, the dynamic ID-HB-CLOTH-FEMs can be used to improve bra development applications in motion. Overall, using the ID-HB-CLOTH-FEM results in shorter development times, less waste due to prototyping and has higher potential of better fit and better functionality due to parametric bra changes based on more accurate data. Presented methods enhance the development of individualised, tightly fitted clothing greatly. Due to the faster, cheaper and more accurate development process, more humans can be dressed in individualised clothing items. The higher the level of individualisation, the better the clothing fits, functionality improves and comfort increases. The process is applicable to functional clothing such as compression socks, bras or sports tights or everyday clothing such as jeans, leggings or shapewear.

Online shopping is continuously increasing for everyday clothing. Returns due to false fit cause high amounts of waste and costs. Better-fitting, individualised clothing with higher functionality may be more expensive for a producer at first. However, as returns due to false fit decrease, waste is lowered, and costs are decreased. For the consumer, the clothing item may be more expensive than a standard item, but due to the better fit, fewer clothing items must be ordered.

If a company is in need of high-accuracy clothing items, purchasing a 4D scanner is advised. 4D scanners are expensive, subsequently, the clothing item's application would be in motion. For static ID-HB-CLOTH-FEM, a 3D scanner is feasible. Nevertheless, 4D scanners are more accurate due to shorter exposure times and the option of selecting comparable frames from large dataset. The 4D scanner is highly advantageous in dynamic model calibration and improved model validation. The 4D scanner is especially usable in companies or institutions which perform research and development, requiring highly accurate comparisons, validation of ID-HB-CLOTH-FEMs and detailed analyses of clothing products.

12.3 Outlook

Due to the generalised method, the ID-HB-CLOTH-FEM can be applied to different application fields. The automated model setup can be improved further. Additionally, a User Interface to increase accessibility for people without programming background to the interface between the clothing and FE software can be implemented. Clothing can be developed more accurately based on ID-HB-CLOTH-FEMs. Clothing parameters influencing fit, pressure, and body deformation can be examined using static and dynamic models. Clothing developed can thus performed more individually with increased functionality and less waste due to prototyping. A 4D scanner with fewer cameras can be used for the dynamic FE models to reduce costs. Studies on the number of cameras and quality needed to perform accurate dynamic breast-bra analyses for the dynamic model calibration can be performed.

Reference

Herakovich CT (2017) Thin-walled pressure vessels. In: A concise introduction to elastic solids: an overview of the mechanics of elastic materials and structures. Springer International Publishing, Cham, 2017, pp 77–81. ISBN: 978-3-319-45602-7. https://doi.org/10.1007/978-3-319-45602-7_13

Part IV
Conclusion and Outlook

Chapter 13
Conclusion

Abstract This chapter provides a comprehensive summary of the entire research work. The developed methods for both experimental and digital analysis of body deformations and clothing-body interactions are reviewed. In particular, the combination of 4D scan data, BodyTex4D, and the developed digital human models has proven to be highly beneficial for clothing development. 4D scan data enable the experimental acquisition and evaluation of new insights into body deformations and textile interactions, while digital models allow for pre-production garment testing, significantly expanding the range of clothing parameter testing before manufacturing.

Keywords FEM · Soft avatar · Individualised · Digital clothing development · Clothing · 4D scanning · Breast-bra · Automated scan data analysis

This research work investigates how the development process of tightly fitted, functional clothing can be enhanced through the use of 4D scan data and individualised, deformable human body finite element models (ID-HB-FEM). Methods for experimental and numerical (Finite Element) clothing-body-interaction analyses are developed. The experimental investigation utilises 4D scan data to explore body deformations with and without clothing. In numerical investigations, ID-HB-FEM is developed and combined with clothing Finite Element models (CLOTH-FEM) based on original cut patterns. The resulting individualised, deformable human body-clothing Finite Element model (ID-HB-CLOTH-FEM) is validated by 4D scan data. Based on the validated ID-HB-CLOTH-FEM investigations of the influences of clothing cut pattern, materials, seams and the wearers' physiology on the body deformation, clothing fit and pressure to indicate the functionality and comfort of the clothing item is investigated. The experimental as well as the numerical clothing-body analyses are focused on improving individualisation, automation and usability of clothing development processes. Figure 13.1 presents an overview of the cross-links of the research work.

This conclusion provides an overview of the findings of this research work. More detailed discussions can be found in each part and each chapter. Parts conclude and discuss the chapters, while each chapter contains the most detailed and technical conclusion and discussions.

A.-M. Schmidt, *Novel Methodologies for Digital Investigations of Clothing-Body-Interactions*, Lecture Notes in Bioengineering,
https://doi.org/10.1007/978-3-031-91794-3_13

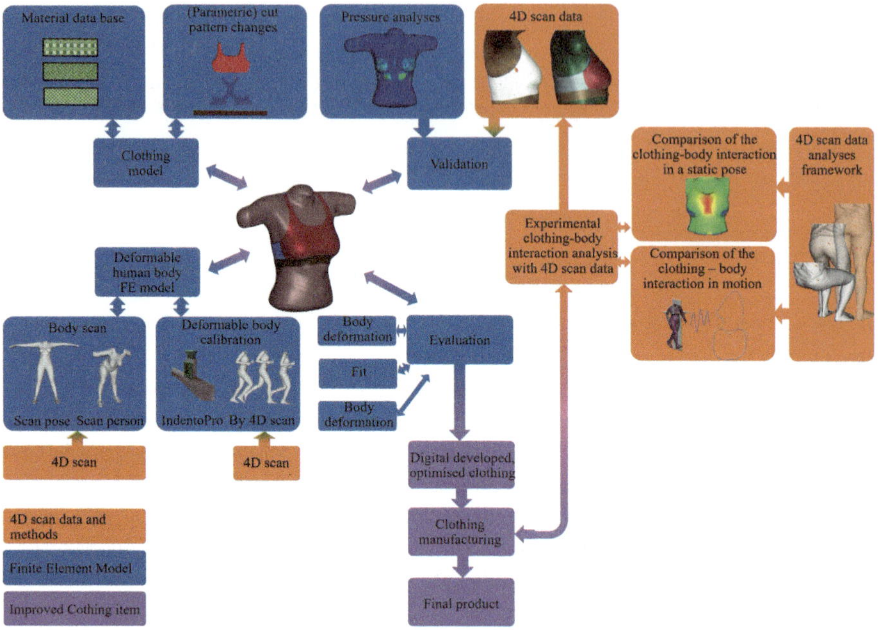

Fig. 13.1 Overview of (novel) methods to improve the development of tightly fitted clothing using 4D scan data and Finite Element Modelling

The use of 4D scan data introduces the challenge of having to handle large amounts of data. Tools for extracting informations of interest for clothing development are rarely found in the literature. Currently, 4D scanned data processing is performed manually or functionally very limited software provided by 4D scanner manufacturers could be used. In the case of Move4D, the software measurement options are predefined locations on the body, adaption is not possible. Additionally, extending the software via APIs (e.g. by Python) or by other means is not intended by the manufacturer, limiting integration into automation of the custom application at hand. "Novel Methodologies for Digital Investigations of Clothing-Body-Interactions" are implemented within a 4D scan data analysis framework, called BodyTex4D, to lift the burden of missing tools. BodyTex4D improves automation of 4D scan data handling in the context of clothing development (Chap. 4). BodyTex4D is implemented in Python, open-source, modular and extendable. The methods implemented into the framework are generalised and applicable to many body parts, clothing applications and test subjects. The advantages of the developed BodyTex4D are listed below:

- BodyTex4D can analyse 4D scan data (semi-)automatically
- BodyTex4D can analyse body or clothing-body interactions by the generalised analysis methods: point movement tracking, length analysis or shape analysis
- BodyTex4D is transferable to any region of interest (ROI) visible in 4D scan data
- BodyTex4D is extendable by application-specific analysis tooling

Example applications to prove the usability of BodyTex4D are breast-bra interactions in static T-poses (Chap. 5) and dynamic running motions (Chap. 6). Two test subjects wearing up to five different bras are analysed. Chapter 5 tracks the state of breathing of test subjects using BodyTex4D to improve the accuracy of breast deformation comparison due to an increased comparability of body states. Chapter 6 presents a novel breast deformation analysis while running. Comparable motion states are selected based on body motion curves acquired through BodyTex4D. Analysing breast deformation in motion is novel due to the ability to capture deformations in motions by 4D scanners. Additionally, a novel method to classify bras by support is developed, tested and implemented into BodyTex4D. Investigating breast-bra interactions lead to the following findings and results:

- BodyTex4D improves the experimental analysis of the breast-bra interaction
- 4D scan data improves the analysis of breast-bra interaction in motion by breast deformation and bra fit analyses
- Methods to select comparable breathing and movement states increase the comparability of bra functionalities
- An amplification factor analysis is developed to classify bras by support with respect to the intended use (e.g. everyday life or sports)
- Breast-bra interaction influencing factors are bra cut patterns, seams, sizes and material and can be displayed using 4D scan data and BodyTex4D

Finite Element Models (FEM) is a powerful tool to analyse clothing-body interaction before production. Therefore, accurately modelling the deformable human body and tightly fitted clothing is crucial. This research work develops a generalised methodology to create ID-HB-CLOTH-FEM (Chap. 9). Novel methods for developing the ID-HB-FEM using 4D scan data are introduced. Material calibration of individual body physiology is achieved by IndentoPRO for static analyses and by 4D scan data for dynamic analyses. CLOTH-FEMs are derived from original cut patterns using software like CLO3D. Thermal shrinkage as a novel digital dressing method for combining ID-HB-FEMs and CLOTH-FEMs is devised, proving to be a stable and transferable dressing method. Validation of the ID-HB-CLOTH-FEM is improved through the selection of comparable body states from 4D scan data. Analysis of interactions between tightly fitted clothing items and individual bodies in terms of fit, deformations, and pressure distribution are based on validated ID-HB-CLOTH-FEM. Digital investigation enhances the development process of tightly fitted clothing by reducing time and costs while increasing individualisation and comfort. The generation of the ID-HB-CLOTH-FEM focuses on maintaining minimal complexity while achieving an unprecedented degree of accuracy usable for future research and industrial clothing design. Lower model complexity results in reduced setup time and lower computational cost, enhancing usability and accessibility for clothing development processes. The novel methods to improve tightly fitted clothing development based on ID-HB-CLOTH-FEM are:

- ID-HB-FEMs geometries are based on 4D scan data: The geometries show high accuracy due to the short exposure times of 4D scanners compared to 3D scanners

- ID-HB-FEMs have deformable soft tissue to model the interaction with tightly fitted clothing realistically
- ID-HB-FEM are calibrated using IndentoPRO, a portable measuring device for soft tissue, or the innovative use of 4D scan data
- CLOTH-FEM are based on original cut patterns, including seams, which is rare in clothing-body Finite Element models
- A digital dressing approach by thermal shrinkage is developed to merge ID-HB-FEMs with CLOTH-FEMs
- ID-HB-CLOTH-FEMs validation process is enhanced through 4D scan data and BodyTex4D
- Transferability of the methods to set up an ID-HB-CLOTH-FEM for different body parts and clothing items is proven

The methods for setting up an ID-HB-CLOTH-FEM are applied to analyse breast-bra interaction in static T-poses and during jumping motions. Developing the breast-bra FE models demonstrates the applicability of the developed methods. The influences of bra sizes, cut patterns, and materials can be modelled in both static and dynamic scenarios. Pressure analysis, owing to consistent boundary parameters in an FE model, shows high usability compared to experimentally performed pressure analysis, which is influenced by many confounding factors (Sect. 10.3.4.3). Using 4D scan data and selecting comparable breathing states improves the validation of the static breast-bra FE model. The dynamic breast-bra FE model is calibrated and validated by the deformation behaviour of the breasts captured by 4D scanners. In conclusion, two approaches to improve the development of tight-fitted clothing items are developed. The first approach is based on 4D scan data to improve knowledge about body deformations and to analyse the clothing-body interaction experimentally. The second approach generates Finite Element Models to enable the analysis of clothing-body interaction digitally. The approaches are focused on the usability of clothing development processes and are highly automated. The approaches can be used separately or combined. 4D scan data supports human body-clothing FE model generation, calibration and validation to enhance accuracy and realism. The influencing factors such as body shape, body composition, clothing cut pattern, material, seams, etc. can be analysed experimentally and digitally, in static poses and motions. Finally, the presented "*Novel Methodologies for Digital Investigations of Clothing-Body Interactions*" is the groundwork for the digitisation of clothing development.

Chapter 14
Outlook

Abstract This chapter provides a brief outlook on the future applications of the developed methodologies and their potential contributions to improving clothing development.

Keywords Digital clothing development · Individualised clothing · Soft avatar · AI

In further research, the presented generalised 4D scan data analysis methods and the development of the individualised, deformable human body-clothing Finite Element model (ID-HB-CLOTH-FEM) can be further improved in the degree of automation. The ID-HB-CLOTH-FEM can be used for digital clothing development by the clothing industry. Different clothing materials should be tested to build up a material property data base. The effect of pressure distribution and deformation should be investigated in more detail. Experiments to confirm and calibrate the pressure distribution calculated by FEM should be conducted. Changes in cut patterns should be investigated further. The bra-band can be widened or shortened, consequently, parameterised cut pattern designs could be created and tested. The developed methods and models allow for fast and accurate digital clothing development. The human body models should be tested on more individuals. The tests should investigate influences of body shapes and soft-tissue properties to increase tight-fitted clothing's fit and comfort. More complex material models could be used to model the non-linearity of clothing-body interactions. Computational time can be lowered by applying more computational resources. At each modelling step, the parameter of the ID-HB-FEM or the clothing item can be changed until the functionalities of the digitally developed clothing are sufficient. The manufactured clothing item can be compared to existing items using the developed 4D scan data analysis methods. The 4D scan data can lead to more accurate FE models due to improved model validation.

It is possible to generate huge datasets of clothing-body FE models with varied clothing parameters and body parameters when having validated clothing-body FE models. Artificial Neuronal Network (ANN) networks can be trained on the generated data base to improve the clothing selection depending on the individual body shape, material and clothing functionality (Fig. 14.1). Ruixin et al. (2020) presented an approach to training an ANN network for an improved clothing selection. The ANN

A.-M. Schmidt, *Novel Methodologies for Digital Investigations of Clothing-Body-Interactions*, Lecture Notes in Bioengineering,
https://doi.org/10.1007/978-3-031-91794-3_14

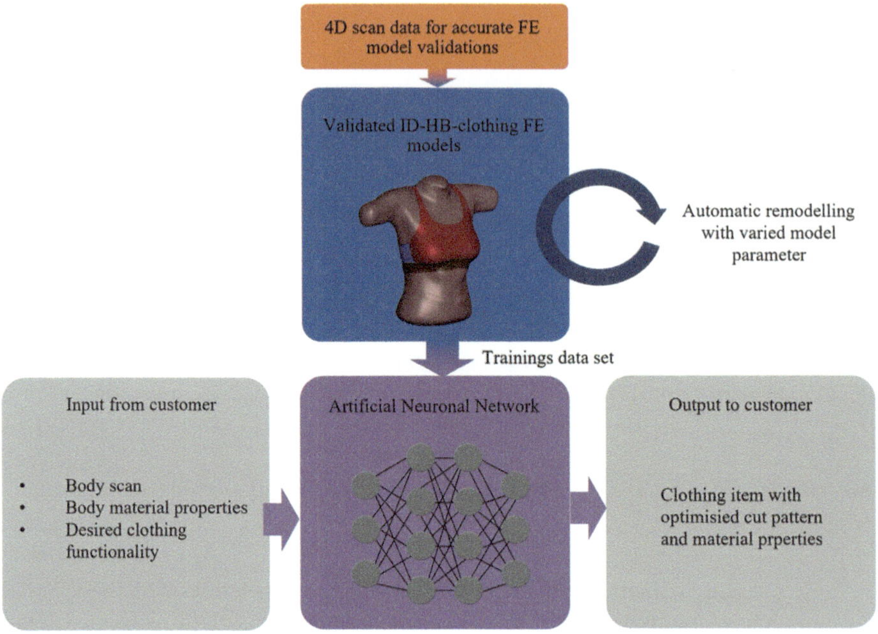

Fig. 14.1 Training an Artificial Neuronal Network with validated clothing-body FE models

network can quickly export clothing material to fulfil a certain functionality. The final stage of the developed methodology would be that physical prototypes are obsolete, saving time and costs. Furthermore, every clothing store, every orthopaedic supply store, etc. would be equipped with 4D scanners to automatically generate individualised clothing items.

Reference

Ruixin L et al (2020) Computational modelling methods for sports bra-body interactions. Int J Cloth Sci Technol 32(6):921–934, ISSN: 0955-6222. https://doi.org/10.1108/IJCST-09-2019-0143

Appendix A
Developing BodyTex4D: A Framework for Automated 4D Data Analysis

Listing A.1 Sorting points by nearest neighbour (sort.py)

```python
import numpy as np

def sort_points_by_nearest_neighbor(original_points: np.ndarray, start_point_index: int = 0) -> np.ndarray:
    sorted_array = np.empty_like(original_points)
    sorted_array[0] = original_points[start_point_index]

    num_points = len(original_points)

    unsorted_indices_mask = np.ones(num_points, dtype=bool)
    unsorted_indices_mask[start_point_index] = False

    for i in range(1, num_points):
        unsorted_points = original_points[unsorted_indices_mask]

        distances_to_unsorted_points = np.linalg.norm(
            unsorted_points - sorted_array[-1], axis=1)

        nearest_neighbor_index = np.argmin(distances_to_unsorted_points)

        sorted_array[i] = unsorted_points[nearest_neighbor_index]

        # transform index from sorted context to unsorted context
        nearest_neighbor_index = np.asarray(unsorted_indices_mask).nonzero()[
            0][nearest_neighbor_index]
        unsorted_indices_mask[nearest_neighbor_index] = False

    return sorted_array
```


A.-M. Schmidt, *Novel Methodologies for Digital Investigations of Clothing-Body-Interactions*, Lecture Notes in Bioengineering,
https://doi.org/10.1007/978-3-031-91794-3

Listing A.2 Circumference calculation after sorting points (circumference.py)

```
import numpy as np
from sort import sort_points_by_nearest_neighbor

def calculate_euclidean_length_of_polyline(
    polyline_points: np.ndarray) -> float:
        """
     Calculates the Euclidean length of a polyline.

     Parameters:
     polyline_points (np.ndarray): An array of shape
      (N, 2) representing the coordinates of the
    polyline points.

     Returns:
     float: The Euclidean length of the polyline.

     The Euclidean length of a polyline is
    calculated by summing the Euclidean distances
    between consecutive points.
     The numpy function np.linalg.norm calculates
    the Euclidean distance between two points.
     The expression polyline_points[1:] -
    polyline_points[:-1] calculates the pairwise
    differences between consecutive points.
     The axis=1 argument in np.linalg.norm
    calculates the norm along the second axis,
    resulting in a 1D array of distances.
     The np.sum function then sums up all the
    distances to obtain the total length of the
    polyline.
     """
     return np.sum(np.linalg.norm(polyline_points
    [1:] - polyline_points[:-1], axis=1))

def calculate_circumference_of_points(points: np.
    ndarray) -> float:
     points_sorted = sort_points_by_nearest_neighbor
    (
         points)

     circmference =
    calculate_euclidean_length_of_polyline(
    points_sorted)
     return circmference
```

Appendix B
Analysing the Breast-Bra Interaction in a T-Pose Using 4D Scan Data and BodyTex4D

Table B.1 Material information of the tested bra variations

	Bra parts	Material
Bra-1	Front	95% Cotton, 5% Elastane
	Back	95% Cotton, 5% Elastane
	Band	No information
Bra-2	Front/back	80% Polyester, 20% Elastane
	Band	57 % Polyethylen, 43% Elastane
Bra-3	Front	88% Polyester,12% Elastane
	Back	Outter layer: same as front
		Inner layer: 92% Polyster, 8% Elastane
	Band	Outter layer:front material, elastic band: no information
Bra-4	Front	Outter: 84% Polyamid, 16% Elastane
		Inner: 100% Polyamid, Lace: 100% Polyamid
	Back	84% Polyamid, 16% Elastane
	Band	Outter layer: front material,elastic band: no information
Bra-5	Front	Shell:79 % Polyester, 21% Elastane
	Back	Mesh Part:82 % Polyester
	Band	No information
Bra-6	Front	81% Polyamide, 19% Elastane
	Back	Mesh: no information
	Band	No information

A.-M. Schmidt, *Novel Methodologies for Digital Investigations of Clothing-Body-Interactions*, Lecture Notes in Bioengineering,
https://doi.org/10.1007/978-3-031-91794-3

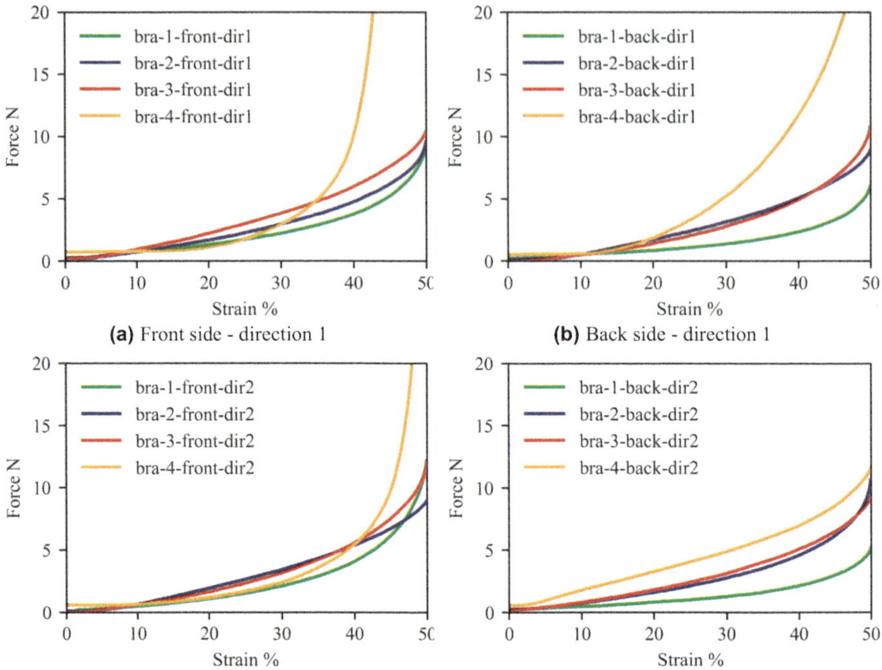

(a) Front side - direction 1 **(b)** Back side - direction 1

(c) Front side - direction 2 **(d)** Back side - direction 2

Fig. B.1 Experimental tensile test: Deloading phase

Listing B.1 Automated selection of frames closest to a target circumference

```

def get_closest_frame_to_target_circumference_
during_inhaling(
frames: list[np.ndarray],
under_bust_circumferences: pd.DataFrame,
target_circumferences: float
) -> tuple[np.ndarray, float]:
mean_under_bust_circumferences =
    under_bust_circumferences.rolling(window=5,
    center=True).mean()

inhaling_frame_indices = np.where(np.diff(
    mean_under_bust_circumferences) > 0)[0]

inhaling_frames = frames[inhaling_frame_indices]

mean_under_bust_circumferences_inhaling =
    mean_under_bust_circumferences[
    inhaling_frame_indices]

```

```
closest_frame = np.argmin(np.abs(
    mean_under_bust_circumferences_inhaling -
    target_circumferences))

closest_frame = inhaling_frames[closest_frame]

clostest_circumference_inhaling = round(
    mean_under_bust_circumferences[closest_frame],
    2)

return closest_frame,
    clostest_circumference_inhaling
```

Appendix C
Methodology Development for Deformable Human Body-Clothing Finite Element Models

Listing C.1 Template-based part selection script: Import OBJ to k-file - faces

```
def extract_faces_from_obj(obj_path: str) -> np.ndarray:
    faces = []
    with open(obj_path, "r") as obj_file:
        for line in obj_file:
            if line.startswith('f '):
                split_line = line.split()
                face = (int(vertex.split("/")[0])
                        for vertex in split_line[1:4])
                faces.append(face)
    return np.array(faces)
```

Listing C.2 Template-based part selection script: Part definition

```
def write_ELEMENT_SHELL_with_breast_pid(
        target_k_file: str,
        obj_file: str,
        pid_body: int,
        pid_right_breast: int,
        pid_left_breast: int,
        template_folder: str,
        element_ids_right_breast: list[int] | str,
        element_ids_left_breast: list[int] | str) -> None:
    if isinstance(element_ids_right_breast, str):
        element_ids_right_breast = np.load(
            template_folder+element_ids_right_breast).tolist()
    if isinstance(element_ids_left_breast, str):
      element_ids_left_breast = np.load(
          template_folder+element_ids_left_breast).tolist()
```


A.-M. Schmidt, *Novel Methodologies for Digital Investigations of Clothing-Body-Interactions*, Lecture Notes in Bioengineering,
https://doi.org/10.1007/978-3-031-91794-3

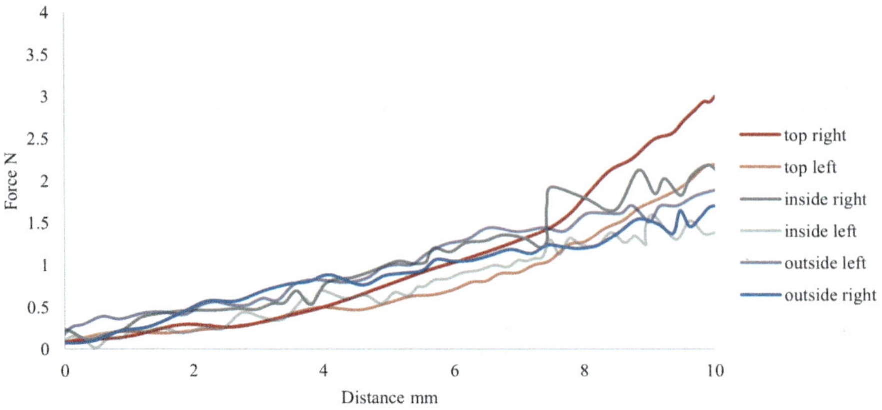

Fig. C.1 Indentation test results of the breast

```
    faces = extract_faces_from_obj(obj_file)
    with open(target_k_file, "a") as k_file:
        k_file.write("*ELEMENT_SHELL\n")
        for line_number, item in enumerate(faces, 1):
            # pid = part id
            if line_number in element_ids_right_breast:
                pid = pid_right_breast
            elif line_number in element_ids_left_breast:
                pid = pid_left_breast
            else:
                pid = pid_body
            k_file.write(
                f"{line_number},{pid},{item[0]},{item[1]},{item[2]},0,0,0,0\n")
```

See Fig. C.1.

Listing C.3 Reading clothing vertex information from a k-file and saving them in an array

```
import numpy as np

def read_point_cloud(file_path: str) -> np.ndarray:
    with open(file_path, 'r') as file:
        lines = file.readlines()

    # Find the line index where *NODE starts
    start_index = next(i for i, line in enumerate(lines) if line.startswith('*NODE'))

    # Find the line index where the next * occurs
    end_index = next(i for i, line in enumerate(lines[start_index + 1:]) if line.startswith('*')) + start_index

    # Extract node IDs and coordinates (assuming comma-separated values)
    data_lines = lines[start_index + 1:end_index+1]
    point_cloud_data = np.array([list(map(float, line.split(',')[0:])) for line in data_lines])

    return point_cloud_data
```

Listing C.4 Writing the coordinates of a scaled point cloud into the boundary keyword

```
import numpy as np
import pyvista as pv

def write_BOUNDARY_PRESCRIBED_FINAL_GEOMETRY(
    output_key_file: str,
    scaled_point_cloud: pv.PolyData,
    LCID: int,
    death_time: float,
) -> None:
    node_ids = scaled_point_cloud["Node IDs"]
    scaled_point_cloud_point = scaled_point_cloud.points

    with open(output_key_file, "a") as file:
        file.write("*BOUNDARY_PRESCRIBED_FINAL_GEOMETRY\n")
        file.write("\n")

        for item, ID in zip(scaled_point_cloud_point, node_ids):
            x_coordinate = round(item[0], 4)
            y_coordinate = round(item[1], 4)
            z_coordinate = round(item[2], 4)
            file.write(
                f"{int(ID)},{x_coordinate},{y_coordinate},{z_coordinate},{LCID},{death_time}\n")
```

Appendix D
Analysing the Breast-Bra Interaction in a T-Pose Using Finite Element Models, 4D Scan Data and BodyTex4D

Listing D.1 Rigid material model to model the non-deformable body part (Ls-PrePost k-file)

```
*MAT_RIGID_TITLE
Rigid - Body
$#     mid          ro          e          pr          n
      couple          m       alias
          29.00000E-9  210000.0       0.0        0.0
          0.0        0.0
$#     cmo         con1        con2
```

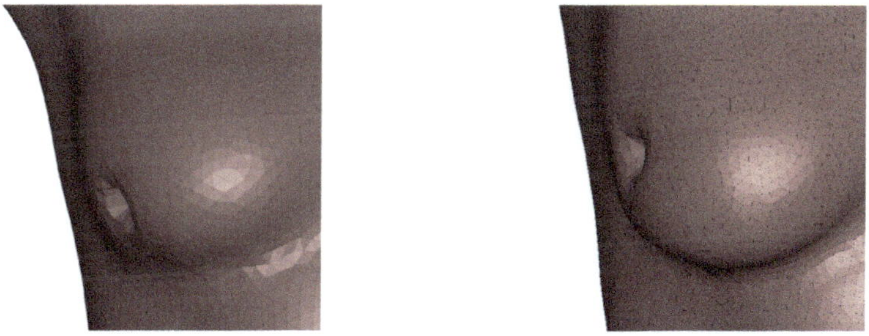

(a) Indentation test modelled with a 7 mm tetrethedron mesh **(b)** Indentation test modelled with a 5 mm tetrethedron mesh

Fig. D.1 Comparison of the deformation behaviour of mesh sizes of the deformable breast Finite Element model

A.-M. Schmidt, *Novel Methodologies for Digital Investigations of Clothing-Body-Interactions*, Lecture Notes in Bioengineering,
https://doi.org/10.1007/978-3-031-91794-3

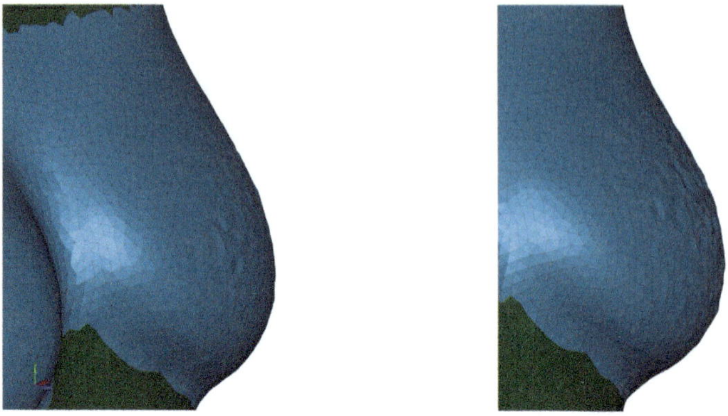

(a) ELFORM 10 after the bra dressing **(b)** ELFORM 13 after the bra dressing

Fig. D.2 Comparison of the deformation behaviour of different human body mesh element formulations

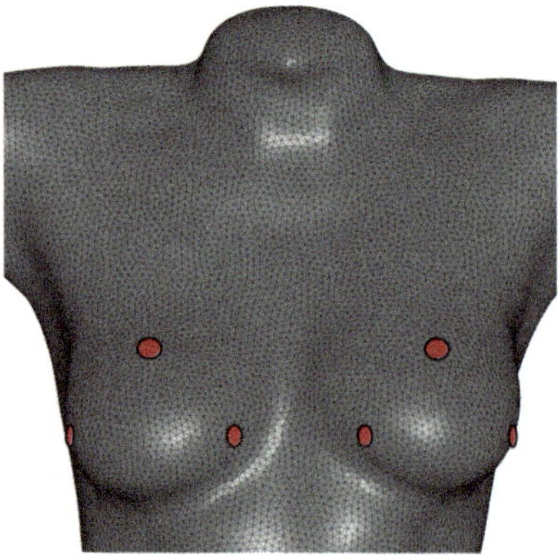

Fig. D.3 Measuring points of the breast material behaviour using the IndentoPRO

Listing D.2 Calculation of time to original size in the thermal shrinkage process

```
def calculate_shrinkage_time(
        length_initial: float,
        scale_value: float,
        alpha: float,
        dt: float = 0.001,
        iterations: int = 5000) -> tuple[float,
 float] | None:
    total_length_shrunk = 0.0
    scaled_length = scale_value * length_initial

    for i in range(1, iterations):
        remaining_length = scaled_length -
 total_length_shrunk
```

```
            next_shrinkage_diff = alpha * dt * remaining_length
            total_length_shrunk += next_shrinkage_diff

            if total_length_shrunk < length_initial:
                return dt * i, dt * (i - 1)

    return None
```

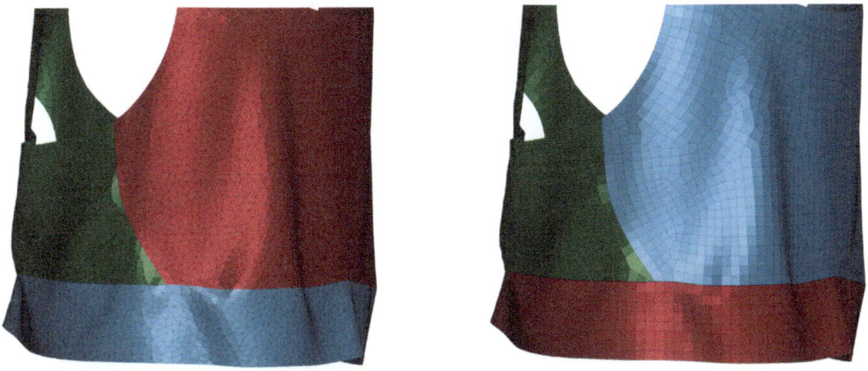

(a) Bra model before remeshing **(b)** Bra model after remeshing

Fig. D.4 Comparison of the same bra Finite Element model before and after remeshing

Table D.1 Input parameter to remodel the tensile test of the bra parts

Numerical model parameter		Unit	Bra-2		Bra-3		
			Band	Textile	Band	Front	Back
Textile	Width	mm	55	30	40	30	30
	Length	mm	100.7	61.7	100.7	62.5	61.5
	Thickness	mm	0.79	0.482	2.0	1.31	1.13
	Density	E-10 t/mm^3	6.288	5.162	3.758	3.35	3.178
Velocity		mm/min	50	50	50	50	50

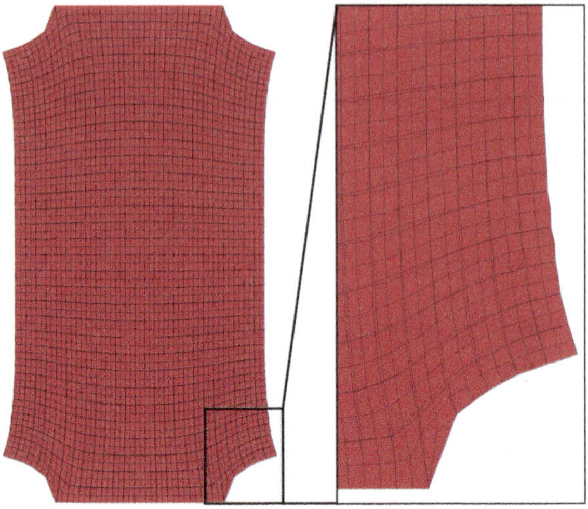

Fig. D.5 Element instabilities due to high deformation at the clamp fixation

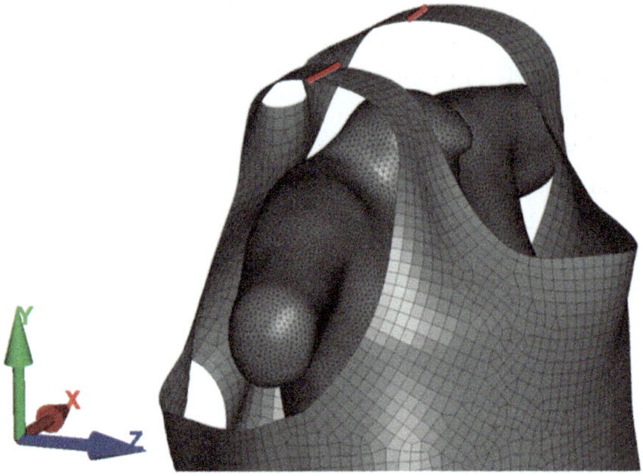

Fig. D.6 Boundary restriction for the bra Finite Element models: constraint translational z-direction

Listing D.3 Control keywords for the static FE breast-bra interaction analysis (Ls-PrePost k-file)

```
*CONTROL_ACCURACY
$#      osu       inn    pidosu      iacc     exacc
          1         4         0         0       0.0
*CONTROL_CONTACT
$#   slsfac    rwpnal    islchk    shlthk    penopt
     thkchg     orien    enmass
        0.1       0.0         1         0         1
          0         1         0
$#   usrstr    usrfrc     nsbcs    interm     xpene
      ssthk      ecdt   tiedprj
          0         0         0         0       4.0
          0         0         0
$#    sfric     dfric       edc       vfc        th
      th_sf    pen_sf
```

```
$#   ignore     frceng    skiprwg     outseg     spotstp
     spotdel     spothin
          0          0          0          0          0
$#     isym     nserod     rwgaps     rwgdth       rwksf
       icov      swradf      ithoff
          0          0          1        0.0        1.0
          0        0.0          0
$#   shledg     pstiff     ithcnt     tdcnof       ftall
     unused     shltrw     igactc
          0          0          0          0          0
                   0.0          0
*CONTROL_DAMPING
$#   nrcyck      drtol     drfctr     drterm      tssfdr
     irelal      edttl     idrflg
        250      0.001      0.995        0.0         0.0
          0       0.04          0
*CONTROL_ENERGY
$#     hgen       rwen     slnten      rylen       irgen
      maten      drlen      disen
          2          1          2          2           2
          1          1          1
*CONTROL_HOURGLASS
$#      ihq         qh
          8        0.1
*CONTROL_OUTPUT
$#    npopt     neecho     nrefup     iaccop       opifs
     ipnint     ikedit     iflush
          0          0          0          0         0.0
          0        100       5000
*CONTROL_SHELL
$#   wrpang      esort      irnxx     istupd      theory
        bwc      miter       proj
       20.0          0         -1          0           2
          2          1          0
*CONTROL_SOLID
$#    esort    fmatrix    niptets     swlocl      psfail
    t10jtol       icoh     tet13k
          0          0          4          1           0
        0.0          0          0
*CONTROL_TERMINATION
$#   endtim     endcyc      dtmin     endeng      endmas
      nosol
        5.0          0        0.0        0.01.000000E8
          0
*CONTROL_TIMESTEP
$#   dtinit     tssfac       isdo     tslimt       dt2ms
       lctm      erode      ms1st
        0.0        0.9          0        0.0         0.0
          0          0          0
*CONTROL_UNITS
$#   length       time       mass       temp
mm         sec        ton        k
```

Listing D.4 Gravity keywords for the FE breast-bra interaction analysis (Ls-PrePost k-file)

```
*LOAD_BODY_Y
$#     lcid          sf     lciddr          xc          yc
         zc         cid
          6         1.0          0         0.0         0.0
        0.0           0
*DEFINE_CURVE_TITLE
Gravity
$#     lcid        sidr        sfa         sfo        offa
       offo      dattyp      lcint
          6           0        1.0         1.0         0.0
        0.0           0          0
$#                   a1                    o1
                    0.0                9806.0
                    1.0                9806.0
                   10.0                9806.0
```

Listing D.5 Ls-PrePost keywords to model thermal shrinkage (Ls-PrePost k-file)

```
*DEFINE_CURVE_TITLE
Thermal
$#     lcid        sidr        sfa         sfo        offa
       offo      dattyp      lcint
          7           0        0.0         1.0         0.0
        0.0           0          0
$#                   a1                    o1
                    0.0                   0.0
                    1.0                  -1.0
                    5.0                  -5.0
*LOAD_THERMAL_LOAD_CURVE
$#     lcid      lciddr
          7           0
*MAT_ADD_THERMAL_EXPANSION_TITLE
Bra-band
$#      pid        lcid       mult        lcid       multy
       lcid       multz
         10           0        0.2           0         1.0
          0         1.0
*MAT_ADD_THERMAL_EXPANSION_TITLE
Bra-front
$#      pid        lcid       mult        lcid       multy
       lcid       multz
         11           0        0.2           0         1.0
          0         1.0
*MAT_ADD_THERMAL_EXPANSION_TITLE
Bra-back
$#      pid        lcid       mult        lcid       multy
       lcid       multz
         12           0        0.2           0         1.0
          0         1.0
*MAT_ADD_THERMAL_EXPANSION_TITLE
Bra-seams
$#      pid        lcid       mult        lcid       multy
       lcid       multz
         13           0        0.2           0         1.0
          0         1.0
```

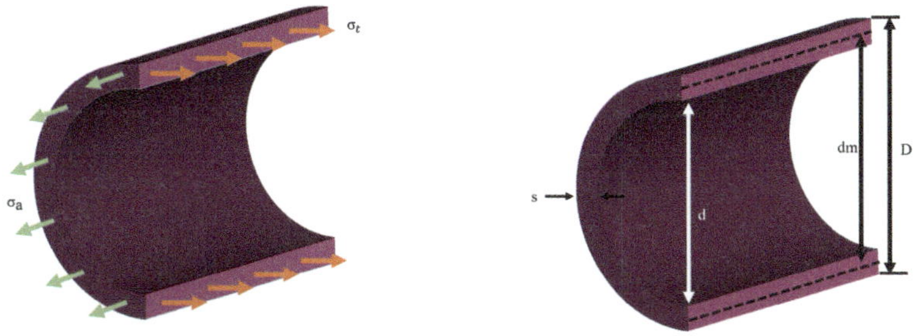

(a) Stresses in a thin-walled pressure vessel **(b)** Parameter used in the think-walled pressure vessel equation

Fig. D.7 Parameter description for the thin-walled pressure vessel equation (figures recreated in reference to Kesselformel 2024)

D.1 Experimental Analysis of Cyclic Tensile Tests

The pressure of the bra-bands on the body is calculated using the equation to calculate the pressure of a thin-walled vessel (Herakovich 2024). Eq. D.1 for tangential stresses (Fig. D.7a) is used.

$$\sigma_t = \frac{p \cdot d_m}{2 \cdot s} = \frac{F_t}{A} \tag{D.1}$$

With:

σ_t: circumference (hoop) stress
p: internal pressure
d_m: mean diameter calculated by $d_m = d + s = \frac{D+d}{2}$ (Fig. D.7b)
s: wall thickness
F_t: force
A: Area

Equation D.1 is converted to calculate the inner pressure (Eq. D.2):

$$p = \frac{2 \cdot F_t \cdot s}{d_m \cdot A} = \frac{2 \cdot F \cdot s}{(d + s) \cdot A} \tag{D.2}$$

Applying Eq. D.2 to the bra-band pressure calculation, the following simplifications are made and parameters are set:

σ_t: circumference (hoop) stress
p: internal pressure
d_m: The mean diameter is calculated by $d_m = d + s$ simplifying the shape of the underbust area to be a circle, resulting in $d_m = \frac{circum_{ref}}{\pi} + s$
s: The wall thickness is the bra-band thickness

Table D.2 Calculating the bra-band pressure based on measurement data

	Bra-2		Bra-3	
	M	S	M	S
Bra-band initial length $l_{bra-band}$ mm	665	620	685	640
Bra-band width $l_{bra-band}$ mm	55	55	40	40
Bra-band thickness mm	0.79	0.79	2.0	2.0
Bra-band strain $\epsilon_{bra-band}$ to reach 770 mm %	15.8	24.2	12.4	20.3
Force for $\epsilon_{bra-band}$ measured in Fig. 10.13a N	8.8	10.53	5.38	8.92
Calculated pressure by Eq. D.3 kPa	1.3	1.58	1.09	1.81

F_t: Force induced into the bra-band by stretching it to the reference circumference ($circum_{ref}$) of 770 mm. The strain is calculated by the initial bra-band width ($l_{bra-band}$) and the reference circumference using $\epsilon = \frac{circum_{ref} - l_{bra-band}}{l_{bra-band}}$. The force can be extracted from the measurement data presented in Fig. 10.13a.
A: The area is calculated by $A = s \cdot b_{bra-band}$ with $b_{bra-band}$ is the bra-band width.

Using these parameters, Eq. D.3 approximated the bra-band pressure:

$$p = \frac{2 \cdot F_t}{(\frac{circum_{ref}}{\pi} + s) \cdot b_{bra-band}} \tag{D.3}$$

With the following measured data, the pressure of the bra-bands of bra-2-M, bra-2-S, bra-3-M and bra-3-S can be calculated (Table D.2). Results can be seen in Fig. 10.13b.

D.2 Validation

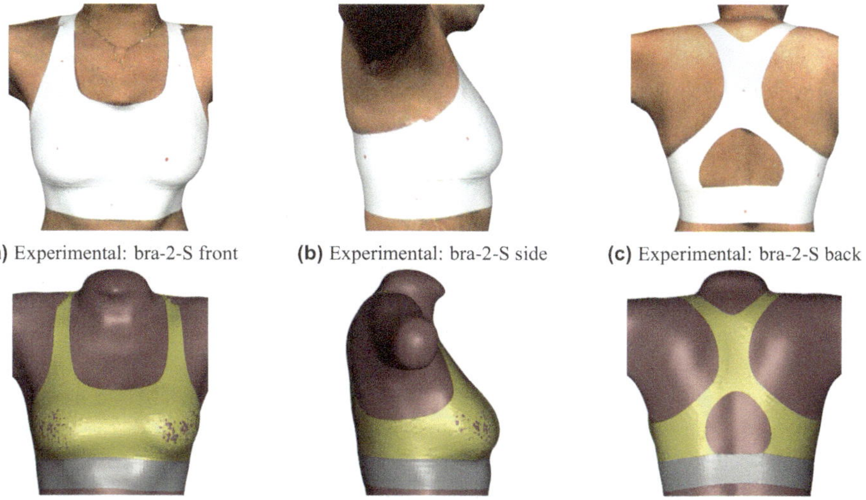

(a) Experimental: bra-2-S front (b) Experimental: bra-2-S side (c) Experimental: bra-2-S back

(d) Numerical: bra-2-S front (e) Numerical: bra-2-S side (f) Numerical: bra-2-S back

Fig. D.8 Comparison of the fit of bra-2-S: **a–c** experimental data and **d–f** Finite element model

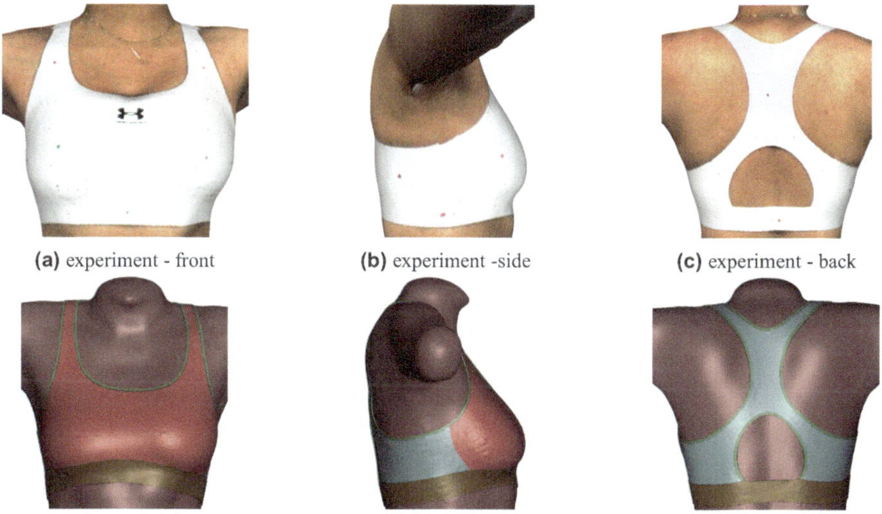

(a) experiment - front (b) experiment -side (c) experiment - back

(d) numerical model - front (e) numerical model -side (f) numerical model - back

Fig. D.9 Comparison of the fit of bra-3-S: **a–c** experimental data and **d–f** Finite element model

Reference

Herakovich CT (2017) Thin-walled pressure vessels. In: A concise introduction to elastic solids: an overview of the mechanics of elastic materials and structures. Springer International Publishing, Cham, 2017, pp 77–81. ISBN: 978-3-319-45602-7. https://doi.org/10.1007/978-3-319-45602-7_13. https://doi.org/10.1007/978-3-319-45602-7_13

Kesselformel (2024) https://de.wikipedia.org/wiki/Kesselformel#cite_note-2 (visited on 07/19/2024)

GPSR Compliance

The European Union's (EU) General Product Safety Regulation (GPSR) is a set of rules that requires consumer products to be safe and our obligations to ensure this.

If you have any concerns about our products, you can contact us on ProductSafety@springernature.com

In case Publisher is established outside the EU, the EU authorized representative is:

Springer Nature Customer Service Center GmbH
Europaplatz 3
69115 Heidelberg, Germany

Batch number: 09125277

Printed by Printforce, the Netherlands